LOTSYA – A BIOLOGICAL MISCELLANY

Edited by Frans Verdoorn — Volume 4

DR J.P. LOTSY

PIONEER
PLANT GEOGRAPHY

To all of my
Colleagues at Kew

Plate 1. — J. D. Hooker at the age of 32. Copy of a print in the portrait collection at Kew. From a sketch by William Tayler, 1849.

PIONEER
PLANT GEOGRAPHY

The Phytogeographical Researches of
Sir Joseph Dalton Hooker

by W. B. TURRILL D.Sc., F.L.S.

Royal Botanic Gardens, Kew, Surrey

SPRINGER-SCIENCE+BUSINESS MEDIA, B.V.
1953

American Distributors:
THE CHRONICA BOTANICA CO.,
WALTHAM, 54, MASS., U.S.A.

CONTRIBUTIONS FROM THE INT. BIOHISTORICAL COMMISSION, No. 1

ISBN 978-94-017-6697-5   ISBN 978-94-017-6758-3 (eBook)
DOI 10.1007/978-94-017-6758-3

PREFACE

When Dr. F. VERDOORN first asked me to write for publication an account of Sir JOSEPH DALTON HOOKER as phytogeographer, with such extracts from his work as would fairly illustrate his great contribution to this branch of botany, it was impossible to comply. World War II was at its height and I was away from Kew engaged in other work, in which, however, as chance would have it, the distribution of plant life played a prominent part. Dr. VERDOORN was patient and neither he nor I abandoned the idea of bringing together in one volume the cream of J. D. HOOKER's investigations in plant geography. It is only right that I should thank Dr. VERDOORN for the initial idea and his stimulus, and for arranging for publication of the result.

Difficulties of presentation have had to be met, as explained in the first chapter. Fairly to indicate HOOKER's work in this field and his standpoint, at a given date, on controversial questions necessitated extensive, yet judiciously selected, quotations from his writings. It is hoped that, on the whole, the selection will meet with approval though it is realized that some subjective element in the choice is unavoidable. Still more difficult was the attempt to appraise HOOKER's methods and results from the modern standpoint. In many, geographically widely separated regions of the world, he blazed the trail. Others followed with the advantages of his experience and results to guide them and often with better equipment. More or less of a century has passed since the voyage of the 'Erebus', 'Terror', the journey to the Himalaya, and the trip to Morocco. Botanists of many nationalities have written on the plant life of these and other areas in books and papers scattered in hundreds of volumes. No doubt many of these have been over-looked and considerations of space have prevented some that were consulted from being quoted. Perhaps enough comments on post-Hookerian researches have been given to form an accurate general account of the present position and sufficient references to enable the enthusiast or specialist to trace details reasonably quickly.

Such then are the general purposes of this book: to lay concisely before the reader the major contributions of J. D. HOOKER to the study of plant distribution and to evaluate these contributions from the subsequent work of many botanists.

To the present writer there is some feeling attached to the fact that this work has been written in his private study at home. Sir JOSEPH DALTON HOOKER lived in this house in the decade, 1855-65, when he was Assistant Director of The Royal Botanic Gardens, Kew.

W. B. TURRILL,
Herbarium House,
Kew, Surrey.

CONTENTS

Chapter I: Introduction

Chapter II: The Distribution of Arctic Plants

Chapter III: Syria and Palestine

Chapter IV: India

Chapter V: Africa

Chapter VI: North America

Chapter VII: The Galapagos Islands

Chapter VIII: Antarctica

Chapter IX: Miscellaneous and General

Chapter X: Summary and Conclusions

* * *

LIST OF PLATES

* * *

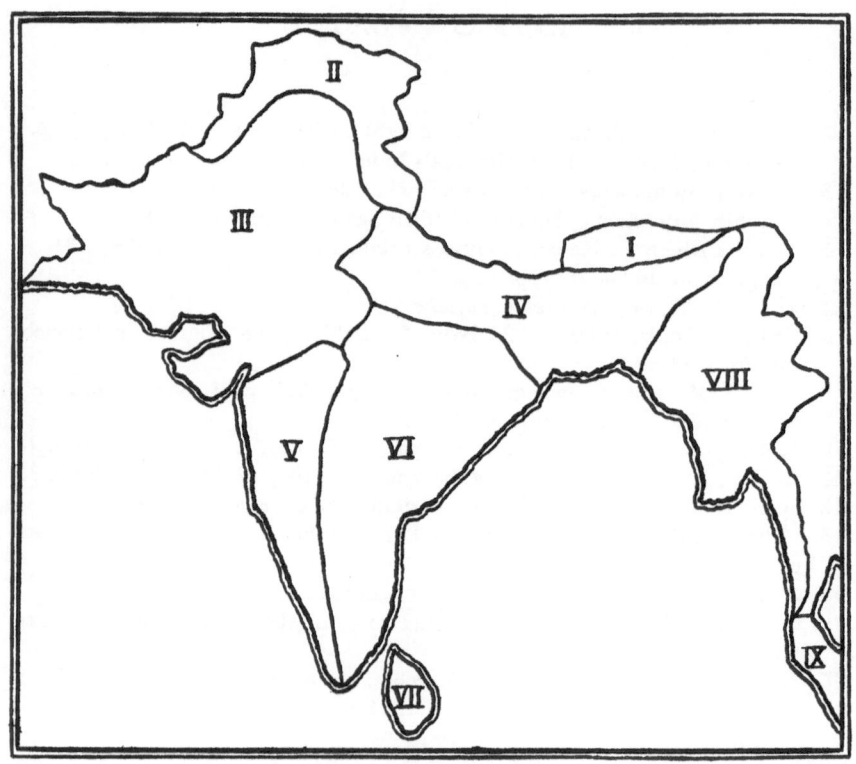

Indian Botanical Provinces, according to J. D. Hooker (1904)
I. The Eastern Himalayan Province; II. The Western Himalayan Province;
III. The Indus-plain Province; IV. The Gangetic Plain Province; V. Malabar
Province; VI. The Deccan Province; VII. The Ceylon Province; VIII. The
Burmese Province; IX. The Malayan Peninsula Province.

Chapter I

INTRODUCTION

The distribution of plants on the surface of the earth and in the waters thereof is one of the widest and most interesting subjects with which the botanist, and especially the taxonomist or ecologist, is concerned. Not only does the modern taxonomist realize the practical importance, as an aid in determination, of knowing exactly whence his specimens are derived, but he has found by experience that there is often a close correlation between range and the best classification he can make of the taxonomic units (taxa) he accepts in his revisions. With wider viewpoints, the constitution and history of the floras of geographical areas can only be unravelled by analysis of the ranges of the constituent species. Floristic analyses lead naturally to a consideration of possible causes and the phytogeographer has then to draw upon the results obtained in almost every other branch of science, and even to accept evidence from other realms of knowledge as well. There are three main stages in phytogeographical research: the collection of facts, the classification, or arrangement, or analysis of the facts, and the interpretation or explanation of the facts leading to generalizations. Though these stages can be traced in quite early published work on plant distribution, as in modern researches, there have been great changes at every stage, or more precisely there have been additions to the kinds of facts utilized, to the methods of analysis, and to the type of explanation. It is, therefore, not without importance to examine the methods used and the results obtained by one outstanding phytogeographer of an earlier period. As will be seen in the sequel, some rather surprising facts emerge which may well act as a stimulus to further research.

JOSEPH DALTON HOOKER was born at Halesworth, Suffolk, England, on 30 June 1817, and was the second son of WILLIAM JACKSON HOOKER and his wife MARIA SARAH. He died at Sunningdale in his ninety-fifth year on 10 December 1911 and was buried in the family vault in the churchyard of St. Anne's on Kew Green, close to the Gardens whose reputation he did so much to enhance. There is no need for our present purpose to outline his long and active life. "The Life and Letters of Sir J. D. HOOKER" by LEONARD HUXLEY (London, John Murray, 1918, 2 vols.) is a full, and very readable, biography.

1

A shorter account was published in the Kew Bulletin 1912: 1–34.

J. D. HOOKER was a man of very great energy and ability and was extremely versatile. Though concerned with only one aspect of his professional life in the preparation of this work, the present writer has been greatly impressed by the width of HOOKER's interests, by his powers of observation, and by his untiring devotion to botanical investigations over a wide range of what are now often regarded as almost separate sciences. Nevertheless, HOOKER was primarily a taxonomist and his largest works are entirely or very largely systematic in the restricted sense. This is true not only for the Flora of British India and for the Genera Plantarum (with G. BENTHAM) but also for the "Antarctic" Floras of Tasmania and New Zealand, to which were attached the renowned "Introductions" with phytogeographical and other data and which are referred to in detail below. HOOKER's phytogeography is essentially "floristic". He wrote mainly before the word "ecology" (or "oecology") was invented and certainly before this branch of biology, with its frequent bearings on phytogeography, had attained its modern contents and aims. The same is true of cytogenetics. DARWIN's "Origin of Species" was published in 1859 and much (not all) of HOOKER's phytogeographical work antedates the general acceptance of the theory of evolution. Indeed, one of the features of special interest of our studies is that the "Hookerian phytogeographical period" strides across the "pre-evolutionary" and the "post-evolutionary" epochs of botany. HOOKER was not the founder of phytogeography. This, or a similar title, has sometimes been given to A. VON HUMBOLDT (1769–1859). It is, however, true to say that HOOKER's researches in plant distribution were of first-class importance because they were largely devoted to botanically hitherto little known, or even quite unknown, areas, were mostly concerned with countries he had himself visited, and were based on his own detailed taxonomic investigations.

It is worth recalling the following statement by J. D. HOOKER himself:

"When still a child, my father used to take me excursions in the Highlands, where I fished a good deal, but also botanised; and well I remember on one occasion, that, after returning home, I built up by a heap of stones a representation of one of the mountains I had ascended, and stuck upon it specimens of the mosses I had collected on it, at heights relative to those at which I had gathered them. This was the dawn of my love for geographical botany" (Quoted in L. HUXLEY, vol. 1, p. 5).

The problem of how best to present J. D. HOOKER's phytogeographical researches had to be given much thought. To have arranged them in chronological order would have had some advantages. Thus it would have facilitated a study of the changes that occurred in his views on "species problems". The main object, however, of this work is a study and criticism of HOOKER's contributions to plant geography. To avoid undue repetitions and to maintain a logical plan it has been decided to arrange the material primarily on a geographical basis. The following summary of the contents of subsequent chapters will serve both as an outline of HOOKER's phytogeographical publications and as a guide to the main body of this book.

— *Arctic* (CHAPTER II) —

Outlines of the distribution of Arctic plants (Trans. Linn. Soc. Lond. 23: 251–348, 1862).

— *Syria and Palestine* (CHAPTER III) —

On the cedars of Lebanon, Taurus, Algeria, and India (Nat. Hist. Rev. 2: 11–8, 1862).

On three oaks of Palestine (Trans. Linn. Soc. Lond. 23: 381–7, 1862).

Botany (of Syria and Palestine), *in* SMITH's Dictionary of the Bible (1863 and later editions).

— *India* (CHAPTER IV) —

Dr. HOOKER's Botanical Mission to India (W. J. HOOKER's London Journal of Botany 6: 604–8, 1847).

Observations made when following the Grand Trunk Road across the hills of Upper Bengal, Parus Nath, in the Soane valley; and on the Kymaon of the Vindhya hills (Journ. Asiatic Soc. Bengal 17: 2, 355–411, 1848).

Extracts from J. D. HOOKER's Diary (*in* Journ. Asiatic Soc. Bengal 18: 419–46, 1849 and Journ. Roy. Hort. Soc. 7: 1–24, 1852).

Himalayan Journals (London, 1854).

Flora Indica (with T. THOMSON, London, 1855).

A sketch of the flora of British India (London, 1904: Oxford, 1907).

— *Africa* (CHAPTER V) —

Journal of a tour of Morocco and the Great Atlas (with J. BALL) (London, 1878).

The ascent of Great Atlas (Proc. Roy. Geogr. Soc. 15: 212–21, 1871 and Rep. Brit. Assoc. 1871: 179–80).

On the vegetation of Clarence Peak, Fernando Po; with descriptions of the plants collected by Mr. GUSTAV MANN on the higher parts of that mountain (Journ. Linn. Soc. Bot. 6: 1–23, 1862).

On the plants of the temperate regions of the Cameroons Mountains and islands in the Bight of Benin (Journ. Linn. Soc. Bot. 7: 171–240, 1864).

On the subalpine vegetation of Kilima Njaro, E. Africa (Journ. Linn. Soc. Bot. 14: 141–6, 1874).

Preface to D. OLIVER, List of the plants collected by Mr. THOMSON F.R.G.S., on the mountains of Eastern Equatorial Africa (Journ. Linn. Soc. Bot. 21: 392–406, 1885).

— *North America* (CHAPTER VI) —

Notes on the botany of the Rocky Mountains (Nature 16: 539–40, 1877 and Amer. Journ. Sci. 14: 505–9, 1877).

Presidential Address to the Royal Society (Proc. Roy. Soc. 26: 444–6, 1878).

The distribution of the North American flora (Proc. Roy. Inst. 8: 568–80, 1879 and Gard. Chron. N.S. 10: 140–2, 216–7, 1878).

The vegetation of the Rocky Mountain Region and a comparison with that of other parts of the world (Bull. U.S. Geol. and Geogr. Surv. of the Territories: 6 1–62, 1880; with AsA GRAY).

— *Galapagos Islands* (CHAPTER VII) —

On the vegetation of the Galapagos Archipelago, as compared with that of some other tropical islands and of the continent of America (Trans. Linn. Soc. Lond. 20: 235–62, 1851).

— *Antarctica* (CHAPTER VIII) —

Botanical contributions *in* A voyage of discovery and research in the Southern and Antarctic Regions, by Captain Sir JAMES CLARK ROSS, London, 1847.

Flora Antarctica, London, 1844–7.

Flora Novae-Zelandiae, London, 1853–5.

Flora Tasmaniae, London, 1855–60.

An account of the botanical collections made in Kerguelen's Land during the Transit of Venus Expedition in the years 1874–5 (Phil. Trans. Roy. Soc. 168: 1–15, 1879).

On the botany of Raoul Island, one of the Kermadec group in the South Pacific Ocean (Journ. Linn. Soc. Bot. 1: 125–9, 1857).

— *Miscellaneous and General* (CHAPTER IX) —

Essay review *in* HOOKER's Journ. Bot. 8: 54–64, 82–8, 112–21, 151–7, 181–91, 214–9, 248–56 (1856).

On Geographical Distribution (Rep. Brit. Assoc. 1881: 727–38).

Lecture on Insular Floras (Gard. Chron. 1867: 6–7, 27, 50–1, 75–6, and as a separate 1896).

— *Summary and Conclusions* (CHAPTER X) —

It is unavoidable that the divisions thus made result in chapters of very unequal length. There appears to be no serious disadvantage in this that cannot be overcome by the use of subheadings, in the longer chapters particularly.

Quotations from HOOKER's works are given in inverted commas when his own words and sentences are used. After a break in any quotation the pagination of the quotation just ended is given in parenthesis. This will enable quotations to be checked or the context of any one to be studied with ease. As far as possible, very short quotations have been avoided. This reduces the risk of misinterpreting the original author by abstracting too drastically.

In the portions of every chapter devoted to discussion and to summaries (often very brief) of more recent work, the published papers and books referred to can be traced in the list of references at the end of this book. These references are arranged under the names of authors in alphabetical sequence. They have been given fully to avoid any justifiable irritation caused by incomplete references. On the other hand, it is not claimed that the list is anything approaching a complete bibliography of the phytogeography of the areas investigated by HOOKER. It is, however, probable that by using these references, and the references they themselves contain, a fairly complete bibliography of the plant geography of the areas could be compiled with a minimum of trouble.

The debt of those who have prepared textbooks of plant geography to J. D. HOOKER is great—if not always acknowledged. Many of his pioneer investigations have now been absorbed in generally accepted knowledge together, of course, with the investigations of other research workers. The following text-books are generally well-known but the list may be useful for students since many of them refer to problems first dealt with by HOOKER:

GRISEBACH, A. (1872): Die Vegetation der Erde, 2 vols. (Leipzig).
ENGLER, A. (1879, 1882): Versuch einer Entwicklungsgeschichte der extratropischen Florengebiete (Leipzig).
DRUDE, O. (1897): Manuel de Géographie Botanique (Paris).
SOLMS-LAUBACH, H. Graf VON (1905): Die leitenden Gesichtspunkte einer allgemeinen Pflanzengeographie (Leipzig).
SCHIMPER, A. F. W. (1903): Plant-Geography upon a physiological basis, Engl. transl. (Oxford).
WARMING, E. (1909): Oecology of Plants, Engl. transl. (Oxford).
IRMSCHER, E. (1922–9): Pflanzenverbreitung und Entwicklung der Kontinente (Hamburg).
MARTONNE, E. DE (1925): Traité de Géographie Physique, 3. Biogéographie (Paris).
HAYEK, A. (1926): Allgemeine Pflanzengeographie (Berlin).
GRAEBNER, P. (1929): Lehrbuch der allgemeinen Pflanzengeographie (Leipzig).
RÜBEL, E. (1930): Pflanzengesellschaften der Erde (Berlin).
WARMING, E. und GRAEBNER, P. (1930–3): Lerhbuch der ökologischen Pflanzengeographie (Berlin).
RAUNKIAER, C. (1934): The Life Forms of Plants and Statistical Plant Geography, Engl. transl. (Oxford).
WULFF, E. V. (1943): An Introduction to Historical Plant Geography, Engl. transl. (Waltham, Mass.).
CAIN, S. A. (1944): Foundations of Plant Geography (New York and London).
WULFF, E. V. (1944): [Historical floras of the countries of the globe] (Moscow and Leningrad; in Russian).
GOOD, R. (1947): The Geography of the Flowering Plants (London).

Chapter II

THE DISTRIBUTION OF ARCTIC PLANTS

J. D. HOOKER never visited the Arctic region but he worked intensively at many collections made in various parts of the Arctic in the first six decades of the 19th century and now at Kew. The results of his studies are summarized in his paper "Outlines of the Distribution of Arctic Plants", published in Trans. Linn. Soc. Lond. 23: 251–348 (1862). Apparently HOOKER defined the "Arctic" very definitely by the Arctic Circle, at least the map accompanying his paper suggests this as does also the sentence "The arctic flora forms a circumpolar belt of 10° to 14° latitude, north of the arctic circle." HOOKER estimates the flora within the Arctic Circle to consist of 925 Cryptogams and 762 Phanerogams (214 Monocots. and 548 Dicots.).

"Regarded as a whole, the arctic flora is decidedly Scandinavian; for Arctic Scandinavia, or Lapland, though a very small tract of land, contains by far the richest arctic flora, amounting to three-fourths of the whole; moreover upwards of three-fifths of the species, and almost all the genera, of Arctic Asia and America are likewise Lapponian, leaving far too small a percentage of other forms to admit of the Arctic Asiatic and American floras being ranked as anything more than subdivisions, which I shall here call districts, of one general arctic flora."

"Proceeding eastwards from Baffin's Bay, there is, first, the Greenland district, whose flora is almost exclusively Lapponian, having an extremely slight admixture of American or Asiatic types: this forms the western boundary of the purely European flora. Secondly, the Arctic European district, extending eastward to the Obi river, beyond the Ural range, including Nova Zembla and Spitzbergen; Greenland would also be included in it, were it not for its large area and geographical position. Thirdly, the transition from the comparatively rich European district to the extremely poor Asiatic one is very gradual; as is that from the Asiatic to the richer fourth or West American district, which extends from Behring's Straits to the Mackenzie River. Fifthly, the transition from the West to the East American district is even less marked; for the lapse of European and West American species is trifling, and the appearance of East American ones is equally so: the transition in vegetation from this district, again, to that of Greenland is, as I have stated above, comparatively very abrupt."

"The general uniformity of the arctic flora, and the special differences between its subdivisions may be thus estimated: the arctic Phaenogamic flora consists of 762 species; of these 616 are Arctic European, many of which prevail throughout the polar area, being distributed in the following proportions through its different longitudes: –

Arctic Europe . . . 616: Scandinavian forms 586; Asiatic and American 30 = 1 : 19.57
Arctic Asia . . . 233: Scandinavian forms 189; Asiatic and American 44 = 1 : 4.2
Arctic W. America 364: Scandinavian forms 254; Asiatic and American 110 = 1 : 2.3
Arctic E. America 379: Scandinavian forms 269; Asiatic and American 110 = 1 : 2.4
Arctic Greenland . 207: Scandinavian forms 195; Asiatic and American 12 = 1 : 16.2"
(pp. 251–2)

HOOKER throughout this paper emphasizes "the exceptional character of the Greenland flora" which, according to his presentation, shows little trace of the botanical features of North America but an "almost absolute identity with those of Europe." His explanation is based on the view that the Scandinavian flora is the flora *par excellence* of the Arctic Region both in number of species and in antiquity and he agrees with DARWIN first "that previous to the glacial epoch it was more uniformly distributed over the polar zone than it is now; secondly, that during the advent of the glacial period this Scandinavian vegetation was driven southward in every longitude, and even across the tropics into the south temperate zone; and that on the succeeding warmth of the present epoch, those species that survived both ascended the mountains of the warmer zones, and also returned northward, accompanied by aborigines of the countries they had invaded during their southern migration" (p. 253).

HOOKER had previously reached the conclusion that "the Scandinavian flora is present in every latitude of the globe and is the only one that is so", and he believed that this conclusion, combined with the views expressed by DARWIN (Origin of Species, ch. 11 *in* ed. 1; ch. 12 *in* ed. 6) and summarized in the quotation given in the last paragraph, not only explained:

"The distribution of arctic plants within and beyond the polar zone, it can also be made, without straining, to account for that distribution and for many anomalies of the Greenland flora, *viz.* (*1*) its identity with the Lapponian; (*2*) its paucity of species; (*3*) the fewness of temperate plants in temperate Greenland, and the still fewer plants that area adds to the entire flora of Greenland; (*4*) the rarity of both Asiatic and American species or types in Greenland; and (*5*) the presence of a few of the rarest Greenland and Scandinavian species in enormously remote alpine localities of West America and the United States" (p. 254).

HOOKER has not much to say in detail regarding the ecological factors limiting ranges. Two paragraphs, however, must be quoted since they give a clear statement of general climatic factors and the importance of their local modifications.

"A glance at the annual and monthly isothermal lines shows that there is little relation between the temperature and vegetation of the area they intersect, beyond the general feature of the scantiness of the Siberian flora being accompanied by a great southern bend of the annual isotherm of 32° in Asia, and the greatest northern bend of the same isotherm occurring in the longitude of west Lapland, which contains the richest flora. On the other hand, the same isotherm bends northwards in passing from Eastern America to Greenland, the vegetation of which is the scantier of the two; and passes to the northward of Iceland, which is much poorer in species than those parts of Lapland to the southward of which it passes." —

"The June isothermals, as indicating the most effective temperatures in the arctic regions (where all vegetation is torpid for nine months, and excessively stimulated during the three others), might have been expected to indicate better the positions of the most luxuriant vegetation: but neither is this the case; for the June isothermal of 41°, which lies within the arctic zone in Asia, where the vegetation is scanty in the extreme, descends to 54° N. lat. in the meridian of Behring's Straits, where the flora is comparatively luxuriant; and the June isothermal of 32°, which traverses Greenland north of Disco, passes to the north both of Spitzbergen and the Parry Islands. In fact, it is neither the mean annual, nor the summer (flowering), nor the autumn (fruiting) temperature that determines the abundance or scarcity of the vegetation in each district, but these combined with the ocean temperature and consequent prevalence of humidity, its geographical position, and its former conditions both climatal and geographical" (p. 255). HOOKER adds that Phanerogams "are probably nowhere found far north of lat. 81°."

In discussing "distribution of Arctic Flowering Plants in various Regions of the Globe" HOOKER notes:

"There is but one distinct genus confined to the arctic regions, the monotypic and local *Pleuropogon Sabini*; and there are but seven other peculiarly arctic species, together with one with which I am wholly unacquainted, *viz. Monolepis Asiatica*. The remaining 762 species are all of them found south of the circle; and of these all but 150 advance south of the parallel of 40° N. lat., either in the Mediterranean basin, Northern India, the United States, Oregon, or California; about 50 are natives of the mountainous regions of the tropics; and just 105 inhabit the south temperate zone" (p. 258).

Again, he emphasizes the importance and wide distribution of the Scandinavian flora and gives (p. 259) the following table of generic distribution:

"Scandinavian Arctic Genera in Europe 280
Found in North United States (approximately). 270
Found in Tropical American Mountains (approximately) 100
Found in Temperate South America (approximately) . . 120
Found in Alps (approximately) 280
Cross Alps (approximately) 260
Found in South Africa (approximately) 110
Found in Himalaya, &c. (approximately) 270
Found in Tropical Asia (approximately) 80
Found in Australia, &c. (approximately) . . . . 100"

A very full and careful analysis is made of Arctic plants by groupings, tabulations, and "observations on the species". The Arctic Region is divided into five districts. The discussions of the floras of these districts are very interesting as they indicate clearly the very wide approach HOOKER made to phytogeographical problems even in 1861 and the tremendous extent of his knowledge of floristic botany. Only an outline of the facts and conclusions can be given here.

1) Arctic Europe.— This is relatively a small area including Lapland, northern Russia to the mouth of the Obi river, Spitzbergen, and Novaya Zemlya. Two floras are represented: the Arctic Norwegian and the Arctic Russian. The latter, from the White Sea eastwards, contains nearly twenty species that are not Lapponian. HOOKER states that both Greenland and Iceland belong "botanically" to the "Arctic Lapland Province" though he (artificially) excludes them.

2) Arctic Asia.— From the Gulf of Obi eastwards to Bering's Straits the climate is marked by excessive mean cold and the district "contains by far the poorest flora of any on the globe". "The low autumn temperature must present an almost insuperable obstacle to the ripening of seeds within this segment of the polar circle" (p. 263). "The rarity of *Gramineae* and especially of *Cyperaceae* in this region is its most exceptional feature; only 21 of the 138 arctic species of these orders having hitherto been detected in it" (p. 265).

3). Arctic West America.— "This extends from Cape Prince of Wales, on the east shore of Behring's Straits, to the estuary of the Mackenzie river, and as a whole it differs from the flora of the province to the eastward of it by its far greater number both of European and Asiatic species, by containing various Altai and Siberian plants which do not reach so high a latitude in more western meridians, and by some temperate plants peculiar to West America" (p. 265).

"The rarity of monocotyledons, and especially of the glumaceous orders, is almost as marked a feature of this as of the Asiatic flora: of the 138 arctic species of *Glumaceae* only 54 are natives of West Arctic America" (p. 267).

4). Arctic East America (exclusive of Greenland).— "This tract of land is analogous to the Arctic Asiatic in many respects of positions and climate, but is very much richer in species. It extends from the estuary of the Mackenzie River to Baffin's Bay, and its flora differs from that of the western part of the continent, both in the characters mentioned in the notice of that province, and in possessing more East American species. The western boundary of this province is an artificial one; the eastern is very natural, both botanically and geographically; for Baffin's Bay and Davis' Straits (unlike Behring's Strait) have very deep water and different floras on their opposite shores." (p. 267). That portion of the province which is richest in plants lies between the Coppermine and Mackenzie rivers. From the whole area 370 species are recorded.

5) Arctic Greenland.— "In area Arctic Greenland exceeds any other arctic district except the Asiatic, but ranks lowest of all in number of contained species. In many respects it is the most remarkable of all the provinces, containing no peculiar species whatever, scarcely any peculiarly American ones, and but a scanty selection of European. A further peculiarity is that the flora of its temperate regions is extremely poor, and adds very few species to the whole flora, and, with few exceptions, only such as are arctic in Europe also. Being the only arctic land that contracts to the southward, forming a peninsula, which terminates in the

ocean in a high northern latitude, Greenland offers the key to the explanation of
most of the phenomena of arctic vegetation" (p. 270).

HOOKER accepts 207 species for the Phanerogamic flora of Greenland and of these he says only 11 are not European. He emphasizes as
"perhaps the most remarkable fact of all connected with the Greenland
flora" that the southern districts (*i.e.* those south of the Arctic Circle)
add no more than 74 species to its flora and these are almost all Arctic
European plants. He further states that "inasmuch as these additional
species increase the proportion of Monocotyledons to Dicotyledons of
the whole flora, Greenland as a whole is botanically more arctic in
vegetation than Arctic Greenland alone is!" The poverty of the
Greenland flora and its European affinities are restated in many paragraphs throughout the paper.

HOOKER concludes a section "On the Arctic Proportions of Species to Genera, Orders, and Classes" with the remarks "my object being
to show how little mutual dependence there is amongst the arctic
florulas. Each has profited but little through contiguity with its co-
terminous districts; though all bear the impress of being members of
one northern flora" (p. 276). His interpretation of some of his tabulated
figures in this section is not understood by the present writer.

In introducing his "Tabular view of the distribution of Arctic
plants," pp. 283–309, HOOKER explains at length his reasons for taking
a wide unit for comparative species. Thus he says:

"If the species thus treated conjointly really express affinities far closer
than those which exist between those treated separately, a certain amount of
definite information, useful for my purpose, is obtained; and it is a matter
of secondary importance to me whether the plants in question are to be
considered species or varieties" (p. 279). Again, "My thus grouping names
must not therefore be regarded as a committal of myself to the opinion that
the plants thus grouped are not to be held as distinct species; I simply treat
of them under one name, because for the purposes of this essay it appears to
me advisable to do so. Every reflecting botanist must acknowledge that
there is no more equivalence amongst species than there is amongst genera;
and I have elsewhere (Essay on the Australian Flora; introductory to the
Flora Tasmanica, p. v, &c.) endeavoured to show that, for all purposes of
classification, species must be treated as groups analogous to genera, differing in the number of distinguishable forms they include, and of individuals
to which these forms have given origin, and in the amount of affinity both
between forms and individuals. My main object is to show the affinities of
the polar plants, and I can best do this by keeping the specific idea comprehensive. It is always easier to indicate differences than to detect resemblances,
and if I were to adopt extreme views of specific difference, I should make
some of the polar areas appear to be botanically very dissimilar from others

with which they are really most intimately allied, and from which I believe them to have derived almost all their species" (p. 279).

Many details of his "lumping" and the reasons which led to his adopted nomenclature are given in the final part of the paper under the heading "Observations on the Species." The question of limitations of species is ever with us and HOOKER's views as expressed in the following sentences must still demand attention.

"I think I may safely affirm that the *specific term* has three different standard values, all current in descriptive botany, but each more or less confined to one class of observers, though more or less variable with all. With the general botanist it is a comprehensive term, and becomes more so with age and experience; with the monographer of large and widely diffused natural orders or genera its standard is contracted at first, but rapidly expands in successive revisions of his work; while the local botanist, or monographer of genera or orders with restricted ranges, begins with a rather broad standard, which rapidly contracts. This is no question of what is right or wrong as to the real value of the specific term. I believe each is right according to the standard he assumes as the specific; moreover, in the majority of cases all agree with regard to the absolute and undeniable distinctness of a moiety of the plants of every area (*see* Introd. Essay to Tasmanian Flora, p. v, for some ideas as to the objective and subjective values of the characters of species, and the division thereby of all species into groups); all agree with regard to the permanent distinctiveness of many of the subspecies, varieties, etc. of the other or variable moiety; and all agree with regard to the propriety and importance of tracing the characters and ranges of varieties as carefully as of species. Still the questions remain—Should the specific term ever be arbitrary? and if so, should it be broad or narrow? I believe it must often be arbitrarily defined, and that it should be broad, because the object of botanical nomenclature is defeated by an undue multiplication of names necessary to be borne in mind by the general botanist, whose convenience ought first to be considered, and also because the multiplication of specific names will demand a corresponding increase of generic ones; moreover the daily discovery of intermediate forms, or new or closely allied forms, is introducing an incessant change in the nomenclature of narrowly defined species" (p. 310).

HOOKER's paper was a bold first attempt to analyze the flora of the Arctic regions. Some of the conclusions he reached have stood the test of more recent work but others have been very much modified by subsequent investigators. This is not the place to attempt a summary of all the research carried out since the 21st of June 1860 (the date on which the paper was read) on the botany of the Arctic Region. It is, however, appropriate to refer briefly to some of the more modern views and the reasons for their divergencies from those of HOOKER.

It is essential in making comparisons to remember that HOOKER's essay is floristic and not ecological and he does not discuss in detail, as do many modern authors, the environmental factors limiting ranges and demarcating communities. As already noted, to HOOKER the Arctic Region was the area north of the Arctic Circle. Most modern writers, in theory, define the Arctic Region phytogeographically as having its southern boundary at the tree limit. In practice, however, as evidenced by maps, tabulated data, and textual discussions, there is by no means complete agreement in the practice of delimitation. BROCKMANN-JEROSCH (Baumgrenze und Klimacharakter, Zürich, 1919) gives numerous details and a coloured map which extends the "Kältewüsten (Frigorideserta)" rather generously to the south.

HOOKER's views were very much influenced by his bias in favour of "Scandinavian flora" as dominant in the flora of the Arctic. It seems probable that having given or accepted the term "Scandinavian" he more and more, perhaps unconsciously, argued that the flora originated in and spread from Scandinavia. In his Flora Tasmaniae 1, Introductory Essay, p. ciii. (1860), HOOKER refers to their being "a continuous current of vegetation (if I may so fancifully express myself) from Scandinavia to Tasmania ......the Scandinavian (assemblage) asserts his prerogative of ubiquity from Britain to beyond its antipodes." It has, however, to be acknowledged that though "Scandinavian" may be an unfortunate term, and even a misnomer, there still remain unsolved problems regarding widespread genera and species which are well represented in the North Temperate regions but extend beyond them.

Modern authors have the advantage over HOOKER in that during the last nine decades a very great deal has been learnt by new exploration and research regarding the flora and vegetation of the Arctic. HOOKER's relative lack of data is very evident for Greenland and Arctic North America. It is not that a great many more species have been recorded from these areas since he wrote but that so much more is known regarding the details of their distribution. The taxonomy is also much better known in many genera, thanks to monographic studies. In general, HOOKER used a species concept too wide to give the best results in phytogeographical studies, though extreme splitting may likewise sometimes prevent full and valid interpretations being obtained from distributional data. In phytogeographical research alternative classifications, even within the realm of taxonomy, are essential, so long as every classification is consistent and is logically constructed and used.

With regard to Greenland in particular, HOOKER was unaware of the theory of survival on nunataks and the importance of Norse introductions was unrealized.

We may consider some of the modern ideas regarding the subdivision of the Arctic as compared with the five major divisions of Hooker. All authors who have dealt with the Arctic flora as a whole agree that a very large proportion of the species occurring in the Arctic are more or less circumpolar. HULTÉN (1937), however, notes that there are very few *completely* circumpolar plants, that is species found without noticeable interruptions all round the North Pole. Nevertheless, the usual generic, and to a large degree the specific, similarity of the floras of any two large areas either north of the tree limit or north of the Arctic Circle makes natural subdivision a matter of difficulty. HOOKER's own map suggests clearly that land continuity is one explanation. The only significant gap is that between Scandinavia and Greenland. OSTENFELD has shown in several works (1925, 1926) that Baffin Bay and Davis Strait are not significant as barriers to the dispersal of Arctic plants whose disseminules can travel over the winter ice and snow. HOOKER considered that the flora of Greenland was essentially European and his previously mentioned conception of Scandinavia as the place of origin of all plants now occurring there and his deduction that the wide ranging species of the "Scandinavian flora" spread thence to many parts of the world, including Greenland, appears to have been partly responsible for this conclusion. WARMING (1888) discusses at some length the European and American relationships of the flora of Greenland and concludes that not the Davis Strait, as HOOKER thought, but the Denmark Strait, between Greenland and Iceland, forms the floristic boundary between Europe and America. NATHORST (1891) argued that the flora of the east coast came in the main from Europe and that of the west coast from America, the inland ice making the floristic boundary between Europe and America. OSTENFELD (1921, 1926) concludes, after a careful analysis, that the flora of Greenland is mainly composed of immigrants from America and that the whole island must be considered floristically as part of America. More recently some detailed studies on micro-species and paramorphs have suggested a partial return to HOOKER's views, NORDHAGEN (1935, quoted by NANNFELDT, 1940) believes that the West Arctic elements in the Scandinavian biota migrated from America to Europe previous to the Riss glaciation. NANNFELDT (1940) put forward the view that *Poa arctica caespitans* has spread "by a former land-bridge connecting Scandinavia, Novaya Zemlya, Spitsbergen, Greenland, and Ellesmereland." SAMUELSSON (1943) for microspecies of the *Alchemilla vulgaris* group concluded that they originated in European-Asiatic centres and spread over north and west Europe to Iceland, Greenland, and the extreme north of North America. RIKLI (1933) has also pointed out that for the Vascular Cryptogams, northern

Scandinavia is by far the chief mass centre, with 45 of the 54 species of "Polaris." These occur especially in northern Norway with its high air moisture and relatively mild winter.

KJELLMAN (1883) showed clearly that the Norwegian Polar Sea, with its temperatures higher than those of the rest of the Arctic Sea owing to its warm currents, was distinct in its marine algal flora. In Norwegian waters this, he believed, had been much enriched by immigration from the North Atlantic.

Arctic Asia, which, according to HOOKER, "contains by far the poorest flora of any on the globe," and which he extends from the Gulf of Obi eastwards to the Bering Strait, is still very incompletely known botanically and no complete critical survey of the flora and vegetation of this vast area appears to have been published since HOOKER wrote; except that the area is included in HULTÉN's and STEFFEN's studies referred to below. From numerous surveys of small (or relatively small) parts of the northern Eurasiatic coastlands it seems very probable that HOOKER was right in making a dividing line at the Gulf of Obi (or Ob). TOLMACHEV (1930), however, draws a line between Novaja Zemlja and Vaigatsch, the latter belonging to a West Siberian and the former being accepted as a small independent floristic region. HOOKER held that the Urals do not, in their northern part, make a floristic boundary. KJELLMAN (1882/83), however, separated, by a line drawn from the north point of the Urals to the Yugor (Jugor) Strait, a European district from a West Siberian one.

Arctic West America is, in HOOKER's sense, limited eastwards by the estuary of the Mackenzie river. Modern phytogeographers recognize that somewhat to the east of the Mackenzie Delta the southern limit of the Arctic (botanical) Region bends rapidly southwards so that while the Arctic belt is relatively narrow to the west it is very wide to the east (see map in Canada Year Book 1946, separate p. 13). Whether a phytogeographical boundary in the Arctic at the Mackenzie Delta is justifiable on HOOKER's evidence does not seem to have been discussed in recent years. The important works of HOLM and others in their Reports of the Canadian Arctic Expedition 1913–18 (1921–24) and of RAUP (1941), with their bibliographies should be consulted.

Arctic East America (exclusive of Greenland) has received much attention in the past two decades. It must suffice here to refer to the works already quoted, to SIMMONS's paper (1913), and to POLUNIN's taxonomic account (1940).

Modern views on the phytogeography of Greenland have been mentioned above in connection with the western boundary of Arctic Europe.

It remains to refer the reader to a few of the more general modern papers dealing with the Arctic flora as a whole or in large part. HULTÉN (1937) has considered together the Arctic and Boreal biota and classifies Arctic plants according to their total ranges on "the theory of equiformal progressive areas." RAUP (1941) has an adequate discussion of HULTÉN's views and here it is sufficient to remark that there are only about 8 groups of Arctic plants in his scheme. HULTÉN dealt especially with the plants of Eastern Asia and Western America and stated that neither wanderings of the Pole nor any displacements of the continents nor any land bridges (beyond one generally accepted across Bering Strait) were necessary to explain the distribution and history of the plants with which he was concerned. He particularly emphasized the importance of the Bering Strait area as a major centre of refuge of the Arctic and Boreal floras. STEFFEN (1924–1937, 1939) on the other hand accepts the Wegener-Köppen hypothesis of polar migrations and continental wanderings. HOOKER, of course, wrote long before the formulation of WEGENER's and KÖPPEN's views. It has been pointed out by various palaeobotanists that the distribution of the known records of fossil plants around the North Pole is difficult, if not impossible, to reconcile with the hypothesis of wanderings of the Pole during the Tertiary epoch.

Chapter III

SYRIA AND PALESTINE

HOOKER visited Syria and Palestine in the autumn of 1860 (*see* L. HUXLEY, Life and Letters of Sir J. D. HOOKER 1: 528–34, 1918). No full account of the expedition was prepared, so far as the present writer has been able to ascertain, but a number of articles of phytogeographical interest were published. In the Natural History Review 2: 11–8 (1862) there is one under the title "On the cedars of Lebanon, Taurus, Algeria, and India." In this, HOOKER describes the grove of *Cedrus libani* in the Kadisha valley at 6000 feet (1829 m). The cedars grew on a moraine immediately bordering a stream and nowhere else. They formed "one group, about 400 yards in diameter, with an outstanding tree, or two, not far from the rest, and appear as a black speck in the great area of the corry and its moraines, which contain no other arboreous vegetation, nor any shrubs, but a few small barberry and rose bushes, that form no feature in the landscape" (pp. 12–3). The trees numbered about 400 and were disposed in nine groups. Though of various sizes there was no tree of less than 18 inches girth and "no young trees, bushes, nor even seedlings of a second year's growth" were found. Calculating from annual rings in a branch HOOKER estimated "the youngest trees in Lebanon would average 100 years old, the oldest 2500, both estimates no doubt widely far from the mark." (p. 13). HOOKER doubts if cedar forests were ever extensive on the Lebanon though there is no doubt that the grove has, within the historic period, "increased and diminished in extent, owing to secular changes in the climate" (p. 15).

The characters and ranges of *Cedrus atlantica, C. deodara,* and *C. libani* are considered and HOOKER sums up as follows: "My own impression is that they should be regarded as three well-marked forms, which are usually very distinct, but which often graduate into one another, not as colours do by blending; but as members of a family do, by the presence in each of some characters common to most of the others, and which do not interfere with or obliterate all the individual features of their possessor." The discovery of moraines on the Lebanon led HOOKER to postulate that in the glacial period cedars occurred at lower levels on the Lebanon and were continuous with those on the Taurus; that these too were at lower altitudes and spread farther eastward and, via the Persian mountains, linked up with the range of the deodar in the Himalaya. Connection between the Algerian and Asiatic cedar forests may also have occurred in the glacial epoch. The present discontinuities in ranges may, therefore, be postulated as due to increasing warmth following the Ice Age, and "what may once have been three prevalent varieties in different parts of a continuous forest, became by isolation and extinction of intermediate forms in intermediate localities, three permanently distinct races or sub-species, which we now recognize as Lebanon, Algerian, and Deodar Cedars" (p. 18).

The notes and exact measurements made by HOOKER of the different cedar trees of the Kadisha grove were given in Woods and Forests for 7 May 1884, p. 328. A. HENRY in Country Life, 11 Dec. 1915, 810–3, may also be consulted. An excellent, and very readable account, of *Cedrus libani* is given in ELWES and HENRY: The trees of Great Britain and Ireland 3: 454–8 (1908). Other articles dealing with cedars in Lebanon are those of GADEAU in Bull. Soc. Dendrol. Fr. 1911: 125–33 and of YOUNÈS, *op. cit.* 1925: 38–43. In an article published in Gard. Chron. 3rd ser. 73: 325–6 (1923) HENRY quotes a statement (from Timber Trades Journal, 4 Sept. 1920, p. 615) that the cedar trees of the Kadisha grove "have been almost totally destroyed by the Turks during the war, the timber being used as fuel for the locomotives of the Palestine Railway." HENRY also describes the sacred nature of this grove. The latest published information regarding the grove that has been traced is a statement by DAVIS (*in* Journ. Roy. Hort. Soc. 72: 14, 1947) that the famous historical stand is now "reduced to a hundred trees which grow at 7000 feet on an old terminal moraine."

Reference should be made to a paper by EMBERGER (1938) in which he gives a general account of the distribution and ecology of the cedars—*Cedrus libani, C. deodara, C. atlantica,* and *C. brevifolia.* The last is endemic to Cyprus. EMBERGER puts forward the theory that the cedars are relicts owing their present limited ranges to contraction of these by age. The "geographical power" of a species does not depend solely on its own ecological characteristics but partly on those of species associated with it. To extend or even maintain its range a species must have a power of competition superior to that of other competitors. Plants, however, are evolving and an index of old age is the overtaking by other species in the power of competition. A species in course of extinction maintains its place only in such areas where not only is the climate suitable but where its competitive power dominates that of its competitors. This is the explanation, according to EMBERGER, of the present discontinuities of the genus *Cedrus.*

In the Trans. Linn. Soc. Lond. 23: 381–7 (1862) HOOKER published

2

a paper "On three oaks of Palestine". This is essentially taxonomic but contains references to his own field experiences. One quotation is given:

"*Q. pseudo-coccifera* is by far the most abundant tree throughout Syria, covering the rocky hills, of Palestine especially, with a dense brushwood of trees 8–12 feet high, branching from the base, thickly covered with small evergreen rigid leaves, and bearing acorns copiously. On Mount Carmel it forms nine-tenths of the shrubby vegetation, and it is almost equally abundant on the west flanks of the Antilebanon and many slopes and valleys of Lebanon. Even in localities where it is not now seen, its roots are found in the soil, and dug up for fuel, as in the valleys to the south of Bethlehem. Owing to the indiscriminate destruction of the forests in Syria, this oak rarely attains its full size. We saw but few very good trees, one of which is the famous oak of Mamre, called "Abraham's Oak", ... and I saw other good ones at Anturah on the Lebanon" (p. 382).

More recent publications on Abraham's oak are: W. J. BEAN, Abraham's Oak (Kew Bull. 1919: 233–6); M. PORTAL, Abraham's Oak (Kew Bull. 1920: 257–8); O. STAPF, The botanical history of 'Sindian' and the age of Abraham's oak (Kew Bull. 1920: 258–64); H. N. and A. L. MOLDENKE, Plants of the Bible (Waltham, Mass., 1952).

In SMITH's Dictionary of the Bible, 2 (1863), and in a series of later editions, the account of the botany of Syria and Palestine was by HOOKER. This is an excellent summary of what was then known of the flora and vegetation. The area is divided into three main phytogeographical areas: (*1*) Western Syria and Palestine, (*2*) Eastern Syria and Palestine, and (*3*) Middle and Upper Mountain Regions of Syria. The plant-life of these areas is described in clear language as is illustrated by the following extract.

"Nowhere can a better locality be found for showing the contrast between the vegetation of the eastern and western districts of Syria than in the neighbourhood of Jerusalem. To the west and south of that city the valleys are full of the dwarf oak, two kinds of *Pistacia*, besides *Smilax*, *Arbutus*, rose, Aleppo Pine, *Rhamnus*, *Phyllyraea*, bramble, and *Crataegus Aronia*. Of these the last alone is found on the Mount of Olives, beyond which, eastward to the Dead Sea, not one of these plants appear, nor are they replaced by any analogous ones. For the first few miles the olive groves continue, and here and there a carob and lentisk or sycamore recurs, but beyond Bethany these are scarcely seen. Naked rocks, or white chalky rounded hills, with bare open valleys, succeed, wholly destitute of copse, and sprinkled with sterile-looking shrubs of *Salsolas*, *Capparideae*, *Zygophyllum*, rues, *Fagonia*, *Polygonum*, *Zizyphus*, tamarisks, alhagi, and *Artemisia*. Herbaceous plants are still abundant, but do not form the continuous sward that they do in Judea. Amongst these, *Boragineae*, *Alsineae*, *Fagonia*, *Polygonum*, *Crozophora*, *Euphorbias*, and *Leguminosae* are the most frequent."

"On descending 1000 feet below the level of the sea to the valley of the Jordan, the subtropical desert vegetation of Arabia and West Asia is encountered in full force. Many plants wholly foreign to the western district suddenly appear, and the flora is that of the whole dry country as far east as the Panjab. The commonest plant is the *Zizyphus Spina-Christi*, or *nubk* of the Arabs, forming bushes or small trees. Scarcely less abundant, and as large, is the *Balanites Aegyptiaca*, whose fruit yields the oil called *zuk* by the Arabs, which is reputed to possess healing properties, and which may possibly be alluded to as Balm of Gilead. Tamarisks are most abundant, together with *Rhus* (Syriaca?), conspicuous for the bright green of its few small leaves, and its exact resemblance in foliage, bark, and habit to the true Balm of Gilead, the *Amyris Gileadensis* of Arabia. Other most abundant shrubs are *Ochradenus baccatus*, a tall, branching, almost leafless plant, with small white berries, and the twiggy, leafless broom called *Retama*. *Acacia Farnesiana* is very abundant, and celebrated for the delicious fragrance of its yellow flowers. It is chiefly upon it that the superb mistletoe, *Loranthus Acaciae*, grows, whose scarlet flowers·are brilliant ornaments of the desert during winter, giving the appearance of flame to the bushes. *Capparis spinosa*, the common caper-plant, flourishes everywhere in the Jordan valley, forming clumps in the very arid rocky bottoms, which are conspicuous for their pale-blue hue, when seen from a distance. *Alhagi maurorum* is extremely common; as is the prickly *Solanum Sodomaeum*, with purple flowers and globular yellow fruits, commonly known as the Dead Sea apple."

"On the banks of the Jordan itself the arboreous and shrubby vegetation chiefly consists of *Populus Euphratica* (a plant found all over Central Asia, but not known west of the Jordan), tamarisk, *Osyris alba*, *Periploca*, *Acacia vera*, *Prosopis Stephaniana*, *Arundo Donax*, *Lycium*, and *Capparis spinosa*. As the ground becomes saline, *Atriplex Halimus* and large *Statices* (sea-pinks) appear in vast abundance, with very many succulent shrubby *Salsolas*, *Salicornias*, *Suaedas*, and other allied plants to the number of at least a dozen, many of which are typical of the salt depressions of the Caspian and Central Asia."

"Other very tropical plants of this region are *Zygophyllum coccineum*, *Boerhavia*, *Indigofera*; several *Astragali*, *Cassias*, *Gymnocarpum*, and *Nitraria*. At the same time thoroughly European forms are common, especially in wet places; as dock, mint, *Veronica Anagallis*, and *Sium*. One remote and little-visited spot in this region is particularly celebrated for the tropical character of its vegetation. This is the small valley of Engedi (Ain-jidi), which is on the west shore of the Dead Sea, and where alone, it is said, the following tropical plants grow:— *Sida mutica* and *Asiatica*, *Calotropis procera* (whose bladdery fruits full of the silky coma of the seeds, have even been assumed to be the Apple of Sodom), *Amberboa*, *Batatas littoralis*, *Ærva Javanica*, *Pluchea Dioscorides*."

"It is here that the *Salvadora Persica*, supposed by some to be the mustard-tree of Scripture, grows: it is a small tree, found as far south as Abyssinia or Aden, and eastward to the peninsula of India, but is unknown west or north of the Dead Sea. The late Dr ROYLE—unaware, no doubt, how scarce and local it was, and arguing from the pungent,taste of its bark, which is used as horseradish in India—supposed that this tree was that alluded to in the parable of the mustard-tree; but not only is the pungent nature of the bark not generally known to the natives of Syria, but the plant itself is so scarce, local, and little known, that JESUS CHRIST could never have made it the subject of a parable that would reach the understanding of His hearers."

"The shores immediately around the Dead Sea present abundance of vegetation, though almost wholly of a saline character. *Juncus maritimus* is very common in large clumps, and a yellow-flowered groundsel-like plant, *Inula crithmoides* (also common on the rocky shores of Tyre, Sidon, etc.), *Spergularia maritima, Atriplex Halimus, Balanites Aegyptiaca,* several shrubby *Suaedas* and *Salicornias, Tamarix,* and a prickly-leaved grass (*Festuca*), all grow more or less close to the edge of the water; while of non-saline plants the *Solanum Sodomaeum, Tamarix, Centaurea,* and immense brakes of *Arundo Donax* may be seen all around."

"The most singular effect is however experienced in the re-ascent from the Dead Sea to the hills on its N.W. shore, which presents first a sudden steep rise, and then a series of vast water-worn terraces at the same level as the Mediterranean. During this ascent such familiar plants of the latter region are successively met with as *Poterium spinosum, Anchusa,* pink, *Hypericum, Inula viscosa,* etc.; but no trees are seen till the longitude of Jerusalem is approached."

* * *

A great deal has been added to our knowledge of the flora and vegetation of Syria and Palestine since HOOKER wrote. In particular the researches of EIG and his successors in Palestine itself have been most valuable. This work is still continuing and it would occupy undue space here to attempt even to outline at all completely the present position. The following references to papers, with the references they themselves contain, will enable students to evaluate the recent discoveries: – BOULOUMOY (1930), BOYKO (1945, 1947), EIG (1926, 1927 A and B, 1931–2, 1939), MONNET (1923), MONTERDE (1947), OPPENHEIMER (1938, 1940, 1949), PALACKY (1904), POST (1888, 1932–3), REIFENBERG (1938), ZOHARY (1937, 1940 A and B, 1942, 1944, 1945, 1947 A and B).

Comment is made only on EIG's 1931 paper. In this the flora of Palestine is analyzed into phytogeographical elements, of which the main groups are designated Saharo-Sindian, Irano-Turanian, Mediterranean, Sudano-Deccanian, and Eurosiberian-Boreo-American. Numerous "liaisons" between these groups are given. On p. 188 a text-figure map shows the phytogeographical territories of Palestine and southern Syria. The subdivision is very different from that proposed by HOOKER. The greater part of southern Syria and northern and Central Palestine is mapped as "Mediterranean." Southern Palestine with an extension northwards round the Dead Sea and up the Jordan

Valley is placed as Saharo-Sindian territory. Eastern Syria and much of Transjordania is said to be Irano-Turanian territory while at the southern end of the Dead Sea there is a small enclave of Sudano-Deccanian plants. Of Eig's phytogeographical groupings that needing most investigation and clarification from a general point of view is the Irano-Turanian. The boundaries of the ranges of elements so designated and of the territory as a whole, as well as the relationships of the group to the other accepted groups are by no means precise. For example, the present writer has previously suggested that a floristic boundary has probably to be drawn in eastern Persia between the southern and the northern portions, somewhere through the Grand Kavir. To the north of this the flora is more nearly related to that of the Aralo-Caspian basin and to the south to the widely ranging Saharo-Sindian flora. Another floristic boundary may run north and south about the longitude of Yezd.

Chapter IV

INDIA

Extracts from Diary:— In W. J. HOOKER's London Journal of Botany 6: 604–8 (1847) is an account of the plan for "Dr HOOKER's Botanical Mission to India," together with the information that he "embarked at Portsmouth on the 11th of November." It is worth noting that the original intention (of the First Lord of the Admiralty, Lord AUCKLAND) was that J. D. HOOKER should, after botanically exploring northern India, visit Borneo and ascend "if possible, the great mountain of Keeny Baloo, supposed to be 14,000 feet in height" (*see too* L. HUXLEY: Life and Letters of Sir J. D. HOOKER, 1: 216, 218, 329). This Bornean project fell through after the death of Lord AUCKLAND. Nevertheless, the journey to India not only led to the exploration of Sikkim and the enriching of Kew and other botanical institutions, but made the Indian flora the centre of HOOKER's studies for many years to come. HOOKER had returned from his Antarctic voyage in September 1843 and he sailed for India just over four years later. His experiences in India not only resulted in the publication of the "Himalayan Journals" but in a great amount of systematic research which culminated in the "Flora of British India", London, 1875–1897. In this great work of seven volumes the accounts of by far the greater number of families were prepared entirely by HOOKER.

Before dealing in some detail with the three major contributions of HOOKER to the plant geography of India, reference must be made to some other publications. In the Journ. Asiatic Soc. Bengal 17, 2: 355–411 (1848), there was published a paper under the title "Observations made when following the Grand Trunk Road across the hills of Upper Bengal, Parus Nath, in the Soane valley; and on the Kymaon branch of the Vindhya hills." This is very largely a printing of extracts from his diary. Thus, for leap-day 1848 he records:

"*February 29th.*— Being now nearly opposite the cliffs at Bidgegurh, where coal is reported to exist we again crossed the Soane, and for the last time. The ford is some three miles up the river, to which we marched through deep sand. On the banks saw a species of *Celtis* or *Sponia* covered with lac. This tree is said to produce it here in greatest abundance, as the *Butea* does at Burdwan and the Peepul in many parts of the country. I do not know which yields the best, nor whether the insects are different. The merchants do not distinguish the kinds. The bed of the river is about ¾ mile broad, and the rapid stream 50 or 60 yards, and breast-deep; the sand firm and silicious, with no mica; nodules of coal are said to be washed down here from the coal bed of Burdee, a good deal higher up, but we saw none."

"The cliffs come close to the river on the opposite side, their bases wooded and teeming with birds. The soil is richer and individual trees, especially of *Bombax*, *Pentapteris* and *Mahowa*, very fine; one tree of the *Hardwickia*, about 120 feet high, was as handsome a monarch of the forest as I ever saw, and it is not often that one sees trees in the tropics, which for a combination of beauty in outline, harmony of color, and arrangement of branches and foliage, would form so striking an addition to an English park."

"There is a large break in the Kymaon hills here, through which our route lay to Bidgegurh and the Ganges at Mirzapore, the cliffs leaving the river and trending to the N. in a continuous escarpment flanked with low ranges of rounded hills and terminating in an abrupt spur (Mungeza Peak) whose summit was covered with a ragged forest. Kunch, the village at which we halted is elevated 556 feet above the sea; four alligators basked in the river, like logs of wood at a distance, all of the short-nosed or Mager kind, dreaded by man and beast; I saw none of the sharp-snouted or Gharial, so common on the Ganges, where their long bills, with a garniture of teeth and prominent eyes peeping out the water, remind one of geological lectures and visions of *Ichthyosauri*."

"Botanized over the ridges near the river, but found little novelty. The *Mahowa*, *Ehretia*, *Hardwickia*, *Gmelina*, and especially *Diospyros* and *Terminalia* are the prevailing timber; the *Cochlospermum* on the very hottest and driest ridges, imitating the *Cistus* in habit; (and like the *C. Ladanum*), it is streaming with gum as was the *Mahowa* and *Olibanum*. *Catechu* and *Rhamneae* are ever present and ever troublesome to the pedestrian. *Phoenix acaulis* frequent, and in some places the woods appeared on fire from the bushes of *Butea frondosa* in full flower" (pp. 400–1).

Again, in Journ. Asiatic Soc. Bengal 18: 419–46 (1849) and, reprinted with corrections, in Journ. Hort. Soc. 7: 1–24 (1852), there was published a further extraction from HOOKER's diary for 1848 under the title "Notes, chiefly botanical, made during an excursion from Darjiling to Tongló, a lofty mountain on the confines of Sikkim and Nepal." The long entry for 21 May is worth quoting from the corrected reprint (1852): "*May 21st.*— Early this morning we proceeded upwards, our prospect more gloomy than ever. The road, still carried up steep ridges, was very slippery, owing to the rain upon the clayey soil, and was only passable from the hold afforded by interlacing roots of trees. At 8000 feet some enormous detached masses of micaceous gneiss rise abruptly from the ridge; these are covered with mosses, ferns, *Cyrtandreae* and *Begoniae*, and creeping *Urticeae*. Such masses occur on all the sharp ridges, and at all elevations; they project awkwardly through the soil, and are strangely confused and distorted in their stratification, down even to the ultimate lamination of the mica, felspar, and quartz. They are never *in situ*, and are generally strangely shattered, and evidently not the mere exposed top of any continuous rock forming the nucleus of the mountain. A uniformly dipping stratified rock of any extent would, if raised at the angle of the slopes of these hills, present a precipitous face somewhere; but the ranges of 4000 to 8000 feet ramify and inosculate in all imaginable directions, without presenting a bold face any where near Darjiling. The road cuttings from the plains to the Sanatarium, as well as the landslips, reveal highly inclined continuous strata, all variously distorted and much dislocated, but these are only at the foot of the hills. Above 4000 feet all appears a strangely piled mass of gneiss rocks, with no uniformity of dip. Amongst these the red clay lies deeper or shallower as the hollows retain it or otherwise."

"These rocks are scaled by means of the roots of trees, and from their summit (7000 feet) a good view of the surrounding vegetation is obtained. The mass of the forest is formed of (1) three species of oak, of which *Q. annulata?* with immense lamellated acorns, and leaves sometimes 16 inches long, is the tallest and the most abundant. (2) Chestnut. (3) *Laurineae*, of several species, beautiful forest trees, straight-boled and umbrageous above, chiefly *Tetranthera* and *Cinnamomum*. (4) *Magnoliaceae*; three species of *Michelia*. Other trees are *Pyrus*, *Saurauja* (both an erect and climbing species), *Olea*, cherry, birch, alder, maple (*Acer*), *Hydrangea*, and one species of fig; holly; several *Araliaceous* trees. Arborescent *Rhododendrons* commence here with the *R. arboreum*, which only occurs at one spot near Darjiling (Mr. Hodgson's grounds on Jillapahar, 7500 feet). *Helwingia* and brambles are the prevalent shrubs. Ferns were not yet fully expanded, and the upper limit of the tree ferns was passed. This is the region of pendulous mosses, lichens, and many herbaceous plants; of which latter, except *Arums*, few had yet appeared above ground. The pendulous mosses are chiefly species of *Hypnum*, *Neckera*, etc.; the lichens, *Borrera* and *Usnea*. Of *Arums*, *Arisaema speciosum* particularly affects this level, with some green spotted compound-leaved kinds, and the small *Remusatia* (*vivipara?*) on the rocks and trunks of trees. Neither *Pothos* (*Scindapsus*) *officinalis*, *decursiva*, nor *scandens* are found higher up the mountain; *Arum curvatum*, Roxb., and other species of *Arisaema*, are very frequent. *Calla*, *Colocasia*, and *Lasia* are confined to lower levels. Peppers reach this elevation, but no higher; whilst very prevalent shrubs are *Adamia cyanea*, *Pittosporum*; *Eurya* and *Camellia* in drier places; *Hypericum*; some species of *Vitis*; and several *Cucurbitaceae*, *Zanthoxylon* and *Sapindaceae*."

"Still ascending along very slippery paths, a considerable change is found in the vegetation of the following thousand feet, from 8000 to 9000. In the forest two gigantic species of *Magnolia* replace the *Michelias*, and were just past flowering. The *Quercus annulata* is less abundant. Chestnut disappears, with several *Lauri*; other kinds of maple are seen, and the *Rhododendron arboreum* is replaced by a much handsomer species, with capitula of very large white flowers and magnificent foliage, 16 inches long (*R. argenteum*). *Corneae*, *Viburnum*, *Lonicerae*, and *Aucuba* are frequent, with two or three *Hydrangeas*, many *Laurinae*, and some new oaks. *Helwingia* is still more abundant as a bush, with climbing and shrubby *Smilacineae*, epiphytical and other *Vaccinia*, and *Gaultheriae*. *Stauntonia* forms a handsome climber, with beautiful pendent clusters of lilac blossoms. The *Araliaceae* are chiefly scandent species and herbaceous, as *pseudo-ginseng*. *Symplocos*, *Limonia*, and *Celastrus* are common shrubs, and small trees. *Cissus capreolata* clothes the trees up to this height."

"At 9000 feet we arrived on a long flat spur or shelf of the mountain, covered with lofty trees, and a dense jungle of small bamboo. *Magnolias* here formed the majority of the trees, with a few oaks (*annulata* very rare). Great *Pyri* and two other species of *Rhododendron*, both attaining the height of 30 to 40 feet, *R. barbatum*, Wall., and *R. arboreum*, Wall., var. *roseum*, De C.; *Sphaerostemma*, a scandent *Araliacea*, and a *Saurauja* climb the loftiest trees: *Stauntonia* crawls round their base, or over lower bushes. *Limonia* and *Symplocos* are the common shrubs. A beautiful orchideous plant, with pale purple flowers (*Coelogyne Wallichii*), grows on the trunks of all the great trees, and attains a higher elevation than most other epiphytical species, for I have seen it at 10,000 feet. A very large, broadly cucullate spathed *Arisaema* first appears at 8000 feet, and is abundant thence to the top of the mountain, where smaller kinds are found at 10,000 feet."

"It is to be remarked that *Leguminosae* are all but unknown in this part of Sikkim above 6000 feet, except the *Parochetus communis*, which however I did not see on this ascent. This total absence of one of the largest and most ubiquitous natural orders through 4000 feet of elevation is most remarkable, and characterizes much of the Himalayan range of Sikkim. I know of no parallel case anywhere on the globe. In the equally humid forests of South Chili and Fuegia the order is extremely rare, but species do exist, and the whole flora of those countries is much poorer in numbers than this. Grasses are also extremely scarce above 4000 feet and below 10,000 feet, always excepting the bamboos, which by their giant dimensions may be fancifully supposed to compensate for the want of many herbaceous species: or it may perhaps be stated better thus:— where the proportion of trees is very great, both in number of species and of individuals, arboreous grasses replace the herbaceous species of less jungly regions."

"A loathsome tick infests the small bamboo, and a more hateful insect I never encountered. The traveller cannot avoid these coming on his person (sometimes in great numbers) as he brushes through the forest. They are often as large as the little finger nail, get inside one's dress, and insert the proboscis deeply without pain. Buried head and shoulders, and retained by a barbed lancet, it is only to be extracted by main force, which is very painful. I have devised many tortures, mechanical and chemical, to induce these disgusting intruders to withdraw the proboscis, but in vain."

"Leeches swarm at below 7000 feet; a small black species above 3000, a large yellow-brown solitary one below that. They are troublesome, but cause no irritation. In August and September these absolutely swarm, and are no less troublesome to man than to the feet of the ponies."

"The rain continuing heavily, we rested the men by some large pools on the flat. A small *Lobelia*, *Chrysosplenium*, *Procris*, and *Callitriche* formed a sward on the banks, amongst which some *Ranunculi* grew (*diffusus*, Wall., and a similar species). A large and handsome *Carex* flourished in the water."

"*Ranunculus*, though so common a genus literally almost everywhere else, is extremely scarce in the temperate and tropical zones of the Sikkim Himalaya; *R. sceleratus* abounds in the plains close to the foot of the hills, but between that elevation and 10,000 feet I have nowhere seen this or any other species. Here, and probably elsewhere in the Himalaya, the genus is very rare in these zones, though more abundant in the arctic zone above."

"*Cruciferae* is another natural order very frequent in the temperate and mountainous regions of all the world, except the Sikkim Himalaya. A variety of *Cardamine hirsuta*? is absolutely the only plant of this order occurring wild between the plains of India and the summit of Tongló."

"*Compositae* again are far from represented on the scale they are everywhere else. Though about Darjiling, where clearances have been effected, the amazing prevalence of *Gnaphalium* and *Anaphalis*, etc., gives an appearance of usual abundance of *Compositae*, these very species will be found elsewhere scarce in the temperate zone of southern Sikkim."

"*Labiatae* are also poorly represented, except in clearances."

"As far as I can guess, this paucity of representatives of orders for which the temperature of the Sikkim Himalaya is admirably adapted, can best be attributed—(1) to the uniform luxuriance of the arboreous vegetation, and the absence of either precipices or naked spots of any kind. (2) to the humid atmosphere; for some of these groups, as *Leguminosae*, are very rare in equally temperate climates which, in respect of humidity and equability of temperature, can be compared with Sikkim, namely, New Zealand and Fuegia. There, as here, *Cruciferae*, *Compositae*, *Ranunculi*, *Labiatae*, and above all, *Leguminosae* and grasses, are very rare in the forest region."

"Our ascent to the summit was by the bed of a watercourse, now a roaring torrent, from the heavy and incessant rain. A small *Anagallis* (like *tenella*) and a scapeless *Primula* grew by its bank also some small *Carices*, and *Hemiphragma*. The top of the mountain is another flat ridge, with depressions and broad pools or small lakes, in which grew an *Iris*. A square platform, raised by the Surveyor General (whose party were the only Europeans who had previously to ourselves visited this mountain), which had been cleared from jungle only eight months before, was already fast getting choked with bamboo and various trees."

"Upon the very top, though only about 500 feet above the flat, the number of additional species was great, and all betokened a rapid approach to the alpine region of the Himalaya, though large forest trees still abounded. In order of prevalence the trees are,—*Rhododendrons* of three species. (1) *R. arboreum*, var. *roseum*, in large bushy trees, 40 feet high. These ramify from the ground, the lower branches being long and spreading, and the apices of all loaded with a superb scarlet inflorescence. (2) *R. barbatum*, a tree of nearly

Plate 2. — SULTANGUNJE, APRIL 9, 1858. From HOOKER's Indian Sketches, at Kew.
Photo of original.

Plate 3. — Boodhist Monuments. *Pinus longifolia.* Great Rungeet.
Photo of Hooker's original field sketch.

the same height, but not so spreading; flowers as copious and beautiful, but foliage brighter, more luxuriant and handsomer. (3) *R. Falconeri*, in point of foliage the most superb of all the Himalayan species; trunks inclined, 30 feet high, branching but little, bark very smooth and papery. Branches naked, except at the apices, where clusters of white flowers are borne; the corollas are 10 cleft, and the stamens numerous. Leaves 18 inches long, very thick, above deep green, underneath wrinkled and covered with a rich, deep chesnut-brown tomentum. Next in abundance to *Rhododendrons* are shrubs of *Limonia*, *Symplocos*, and *Hydrangea*, forming small trees; there are still a few *Magnolias*, very large *Pyri*, of three species, and yew, the latter 18 feet in circumference; besides these, *Buddleia*, not in flower, *Pieris*, *Andromeda*, *Olea*, *Celastrus*, *Cerasus*, and *Daphne cannabina*. A white-flowered rose, *R. sericea*, was very abundant, growing erect, its numerous inodorous flowers pendent, apparently as a protection from the dashing rain. *Sphaerostemma*, *Sabia*, *Stauntonia*, and *Clematis montana?* were the prevailing climbers. I met with a cucurbitaceous plant at this great elevation, a *Smilax* and *Asclepiadeous* genus. A currant was common, always growing epiphytically on the trunks of large trees. Two or three species of *Berberis* and maple, I think, nearly complete the list of woody plants. Amongst the herbaceous and smaller shrubby plants were many of great interest, as a rhubarb, and *Aconitum palmatum*, a very pretty species, which, as well as various congeners, yields the "Bikh" poison of E. Nepal, Sikkim, and Bhotan. *Thalictrum*, one species. *Anemone vitifolia*, *Fumaria*, two *Violae*, *Stellaria*, *Hypericum*, *Geranium* two species, two Balsams, *Epilobium*, *Potentilla*, *Paris* (7000 to 10,000 feet), *Panax pseudo-ginseng*, and another species, *Meconopsis Nepalensis*, two species of *Gentiana*, *Ligularia*, and two *Crawfurdiae*, two species of *Arisaema*, *Anagallis*, *Hemiphragma*, and *Ajuga*, *Disporum*, and three *Convallariae*, one with verticillate leaves, whose root is called another "Bikh," and considered very virulent. *Gramineae* were very few in number, but a large *Carex* covered the ground, amongst the bamboo."

"Still the absence or rarity of several very large natural families at this elevation, which have numerous representatives at and much below the same level in the Western Himalaya, indicates a certain peculiarity in Sikkim. These are the following:— *Ranunculaceae*, *Fumariae*, *Cruciferae*, *Alsineae*, *Geraniae*, *Leguminosae*, *Potentilla*, *Epilobium*, *Crassulaceae*, *Saxifrageae*, *Umbelliferae*, *Lonicera*, *Valerianeae*, *Dipsaceae*, various genera of *Compositae*, *Campanulaceae*, *Lobeliaceae*, *Gentianae*, *Boragineae*, *Scrophularineae*, *Primulaceae*, *Gramineae*."

"All the above are genera of the north temperate and subarctic zones, which affect a much higher level in this part of Sikkim than in the Western Himálaya or Bhotan—the difference in this respect being very much greater than the small disparity of latitude will account for, or than the (if there be any) difference of mean temperature, for the snow-line is certainly very little different here from that of the N.W. Himalaya. On the other hand, certain tropical genera are more abundant in the temperate zone of the Sikkim mountains, and ascend much higher there than in the Western Himalaya. Of this fact I have cited conspicuous examples in the palms, plantains, and tree-fern ascending to nearly 7000 feet, and in the presence of many other orders at great elevations, as figs, peppers, *Lauri*, etc.; and to these could be added many others, none more remarkable than *Balanophora*, of which there are several species above 4000 and even 6000 feet, one ascending to 11,000 feet."

"This ascent and prevalence of tropical species is due to the humidity and the equability of the climate in this temperate zone, and is perhaps the direct consequence of these conditions. An application of the same laws accounts for the extension of similar features so far beyond the tropical limit in the Southern Ocean, where various natural orders which do not cross the 30th and 40th parallels of N. latitude, are extended to the 55th in Tasmania, New Zealand, the so-called Antarctic Islands south of that group, and to Cape Horn itself in Fuegia."

"The forest region, encroaching so far upon, and in fact covering the temperate zone of the Sikkim Himalaya, and the snow-level not being proportionally higher, it follows that, *caeteris paribus*, the belt occupied by upland alpine and Arctic species is more confined, and in all probability less prolific in species, than it is in the N.W. Himalaya. Of this the rarity of Pines (themselves indices of a severe drought in the air or soil) would appear to afford a proof; for between the level 2500, the upper limit of the *P. longifolia*, and the *Taxus*, 10,000, which also coincides with the lower limit of *Abies*, there is no coniferous tree whatever in Southern Sikkim, except on the mountain faces immediately subtending the perpetual snow; and there they descend 1000 feet lower. There are only six species of *Coniferae*, including *Taxus* and *Juniperus*, in this part of Sikkim, of which two are not common to the N.W. mountains, and none are by any means abundant."

"We encamped amongst the Rhododendron trees, on a spongy soil, of black vegetable matter, so oozy that it was difficult to keep the feet dry. The rain poured in torrents all the evening, and this, the calm, and wetness of the wood prevented our enjoying a fire. Except a transient view into Nepal, a few miles west of us, nothing was to be seen, the whole mountain being wrapped in dense masses of vapour. Gusts of wind, not felt in the forest, swept over the gnarled and naked tree tops; and though the temperature was 50°, this produced cold to the feelings on walking about, and being exposed to it."

"Our poor Lepchas were miserably off, but always happy under four posts and a bamboo-leaf thatch, and with no covering but a single thin cotton garment. They crouched on the sodden turf, joking with the Hindus of our party, who, though supplied with good clothing and shelter, were doleful companions."

"I made a shed for my instruments under a tree; BARNES, ever active and ready, floored the tent with logs of wood, and I laid a "corduroy road" of the same to my little observatory."

"During the night the rain did not abate; the tent-roof bagged and leaked in torrents, so that we had to throw pieces of waxcloth over our shoulders as we lay in bed" (pp. 12–9).

Further letters, probably these as printed are essentially extracts from HOOKER's diary, dealing with his Indian journey were published as follows: *in* W. J. HOOKER's London Journ. Bot. 7: 237–68, 297–321 (1848); HOOKER's Kew Journ. Bot. 1: 1–14; 41–56; 81–9; 113–20; 129–36; 161–75; 226–33; 274–82; 301–8; 331–36; 336–44; 361–70 (1849); *ibid.* 2: 11–23; 52–9; 88–91; 112–8; 145–51; 161–73; 213–8; 244–9 (1850); *ibid.* 3: 23–31 (1851). All the above "letters", or more probably the diary notes from which they were extracted, were used in the preparation of J. D. HOOKER's "Himalayan Journals". Since this work is dealt with very fully below there is no need to give further extracts from the "letters." Their value is that they can be referred to as a possible source for clearing up ambiguities in the earlier parts of the "Himalayan Journals" or for extending the information in this latter work. Unfortunately the main series of "letters" breaks off abruptly at p. 249 (1850). Some of the letters were published as a separate (London, 1848).

Himalayan Journals:— HOOKER's "Himalayan Journals", with the subtitle "Notes of a Naturalist in Bengal, the Sikkim and Nepal Himalayas, the Khasia Mountains, etc.", was first published in 1854 in two volumes. A second edition appeared in 1855 and another (Minerva Library, London) in one volume in 1891, with a re-issue in 1905. The quotations below are taken from the 1891 edition in which some revision of the text was made. The book is dedicated to CHARLES DARWIN.

"Himalayan Journals" holds a high place among travel books in the English language. It is written in a straight-forward descriptive style which is rather ponderous by modern standards. It contains very much botanical information, including descriptions of the flora and vegetation of northern India, of which much has not yet been superseded by more modern field studies.

The extracts given below have been selected mainly to illustrate HOOKER's abilities as a most observant field botanist but also to indicate his versatility. Besides the great additions to knowledge of the Indian flora that resulted from his travels there has to be added his exploration, surveying, and mapping of unknown territory and his description of the modes of life of little known peoples. Apart, then, from selections of particular phytogeographical value, the paragraphs below are somewhat diverse in subject matter.

Before his first Himalayan expedition proper HOOKER made a number of local excursions. The first was to Parasnath, the sacred mountain of the Jains, north-west of Calcutta. On the journey thither he made many records of the vegetation, and particularly of the cultivated plants.

"Though the botany of Paras-nath proved interesting, its elevation was not accompanied by such a change from the flora of its base as I had expected. This is no doubt due to its dry climate and sterile soil; characters which it shares with the extensive elevated area of which it forms a part, and upon which I could not detect above 300 species of plants during my journey. Yet, that the atmosphere at the summit is more damp as well as cooler than at the base, is proved as well by the observations as by the vegetation; and in some respects, as the increased proportion of ferns, additional epiphytal orchideous plants, *Begonias*, and other species showed, its top supported a more tropical flora than its base" (p. 17).

Excursions in the neighbourhood not only added to his botanical collections and data but resulted in many acute observations on very diverse subjects, as the following extracts show.

"At 10 a.m. the sun became uncomfortably hot, the thermometer being 77°, and the black-bulb thermometer 137°. I had lost my hat, and possessed no substitute but a silken nightcap; so I had to tie a handkerchief over my head, to the astonishment of the passers-by. Holding my head down, I had little source of amusement but reading the foot-marks on the road; and these were strangely diversified to an English eye. Those of the elephant, camel, buffalo and bullock, horse, ass, pony, dog, goat, sheep and kid, lizard, wild-cat and pigeon, with men, women, and children's feet, naked and shod, were all recognisable" (pp. 18–9).

"*Confervae* abound in the warm stream from the springs, and two species, one ochreous brown, and the other green, occur on the margins of the tanks themselves, and in the hottest water; the brown is the best Salamander, and forms a belt in deeper water than the green; both appear in broad luxuriant strata, wherever the temp. is cooled down to 168°, and as low as 90°. Of flowering plants, three showed in an eminent degree a constitution capable of resisting the heat, if not a predilection for it; these were all *Cyperaceae*, a *Cyperus* and an *Eleocharis*, having their roots in water of 100°, and where they are probably exposed to greater heat, and a *Fimbristylis* at 98°; all were very luxuriant. From the edges of the four hot springs I gathered sixteen species of flowering plants, and from the cold tank five, which did not grow in the hot. A water-beetle, *Colymbetes* (?) and *Notonecta*, abounded in water at 112°, with quantities of dead molluscs; frogs were very lively, with live molluscs, at 90°, and with various other water-beetles. Having no means of detecting the salts of this water, I bottled some for future analysis" (p. 20).

"In the woods I heard and saw the wild peacock for the first time. Its voice is not to be distinguished from that of the tame bird in England, a curious instance of the perpetuation of character under widely different circumstances, for the crow of the wild jungle-fowl does not rival that of the farmyard cock" (p. 22).

From the Parasnath district HOOKER travelled to the Soane (or Son) river. An interesting physiological observation is worth quoting:

Boodhist monuments.
Pinus longifolia

Plate 4. — Boodhist Monuments. *Pinus longifolia.* From Hooker's Indian
Sketches at Kew. Photo of plate as worked up for publication, probably by
Walter Fitch, from the field sketch of the previous photo.

Plate 5. — TAMBUR RIVER, E. Nepal, looking north, 23 Nov. 1848.
Photo of HOOKER's original field sketch at Kew.

"Finding the fresh milky juice of *Calotropis* to be only 72°, I was curious to ascertain at what depth this temperature was to be obtained in the sand of the river-bed, where the plant grew.

Surface.	104½°	3½ inches .	85°	Compact.
1 inch.	102	8 inches .	73	Wet.
2 inches.	94	15 inches .	72	Ditto."
2½ inches.	90			

"The power this plant exercises of maintaining a low temperature of 72°, though the main portion which is subterraneous is surrounded by a soiled heat to between 90° and 104°, is very remarkable, and no doubt proximately due to the rapidity of evaporation from the foliage, and consequent activity in the circulation. Its exposed leaves maintained a temperature of 80°, nearly 25° cooler than the similarly exposed sand and alluvium. On the same night the leaves were cooled down to 54°, when the sand had cooled to 51°. Before daylight the following morning the sand had cooled to 43°, and the leaves of the *Calotropis* to 45½°. I omitted to observe the temperature of the sap at the latter time; but the sand at the same depth (15 inches) as that at which its temperature and that of the plant agreed at mid-day was 68°. And assuming this to be the heat of the plant, we find that the leaves are heated by solar radiation during the day 8°, and cooled by nocturnal radiation 22½°" (p. 26).

After further exploration of the Soane valley he crossed to the Ganges and visited Benares and other towns till he reached Bhagulpore where preparations were made for his first Himalayan expedition. Sikkim, with extensions into western Nepal and just over the frontier into Tibet, was the scene of HOOKER's explorations in his Himalayan journeys. On the way to Darjeeling (Darjiling) he records the vegetation in clear descriptive language.

"After proceeding some six miles along the gradually ascending path, I came to a considerable stream, cutting its way through stratified gravel, with cliffs on each side fifteen to twenty feet high, here and there covered with ferns, the little *Oxalis sensitiva*, and other herbs. The road here suddenly ascends a steep gravelly hill, and opens out on a short flat, or spur, from which the Himalaya rise abruptly, clothed with forest from the base; the little bungalow of Punkabaree, my immediate destination, nestled in the woods, crowning a lateral knoll, above which, to east and west, as far as the eye could reach, were range after range of wooded mountains, 6,000 to 8,000 feet high. I here met with the India-rubber tree (*Ficus elastica*); it abounds in Assam, but this is its western limit."

"From this steppe, the ascent to Punkabaree is sudden and steep, and accompanied with a change in soil and vegetation. The mica slate and clay slate protrude everywhere, the former full of garnets. A giant forest replaces the stunted and bushy timber of the Terai Proper; of which the *Duabanga* and *Terminalias* form the prevailing trees, with *Cedrela* and *Gordonia Wallichii*.

Smaller timber and shrubs are innumerable; a succulent character pervades the bushes and herbs, occasioned by the prevalence of *Urticeae*. Large bamboos rather crest the hills than court the deeper shade, and of the latter there is abundance, for the torrents cut a straight, deep, and steep course down the hill flanks: the gulleys they traverse are choked with vegetation and bridged by fallen trees, whose trunks are richly clothed with *Dendrobium Pierardi* and other epiphytical Orchids, with pendulous *Lycopodia* and many ferns, *Hoya*, *Scitamineae*, and similar types of the hottest and dampest climates."

"The bungalow at Punkabaree was good—which was well, as my luggage-bearers had not come up, and there were no signs of them along the Terai road, which I saw winding below me. My scanty stock of paper being full of plants, I was reduced to the strait of botanising, and throwing away the specimens. The forest was truly magnificent along the steep mountain sides. The apparently large proportion of deciduous trees was far more considerable than I had expected; partly, probably, due to the abundance of the *Dillenia*, *Cassia*, and *Sterculia*, whose copious fruit was all the more conspicuous from the leafless condition of the plant. The white or lilac blossoms of the convolvulus-like *Thunbergia*, and other *Acanthaceae*, were the predominant features of the shrubby vegetation, and very handsome" (pp. 70–1).

"From Kursiong a very steep zigzag leads up the mountain, through a magnificent forest of chestnut, walnut, oaks, and laurels. It is difficult to conceive a grander mass of vegetation—the straight shafts of the timber-trees shooting aloft, some naked and clean, with grey, pale, or brown bark; others literally clothed for yards with a continuous garment of epiphytes, one mass of blossoms, especially the white Orchids *Coelogynes*, which bloom in a profuse manner, whitening their trunks like snow. More bulky trunks bore masses of interlacing climber, *Araliaceae*, *Leguminosae*, Vines, and *Menispermeae*, Hydrangea, and Peppers, enclosing a hollow, once filled by the now strangled supporting tree, which had long ago decayed away. From the sides and summit of these, supple branches hung forth, either leafy or naked; the latter resembling cables flung from one tree to another, swinging in the breeze, their rocking motion increased by the weight of great bunches of ferns or Orchids, which were perched aloft in the loops. Perpetual moisture nourishes this dripping forest, and pendulous mosses and lichens are met with in profusion" (p. 75).

Excursions were first made in the neighbourhood of Darjeeling and the ascent of Tongló proved especially interesting.

At 6000 feet, "we found great scandent trees twisting around the trunks of others, and strangling them: the latter gradually decay, leaving the sheath of climbers as one of the most remarkable vegetable phenomena of these mountains. These climbers belong to several orders, and may be roughly classified in two groups.— (1) Those whose stems merely twine, and by constricting certain parts of their support, induce death.— (2) Those which form a network round the trunk, by the coalescence of their lateral branches and aerial roots, etc.: these wholly envelop and often conceal the

Plate 6. — KANGLACHEM, 16,500 ft. Photo of plate as worked up, probably by WALTER FITCH, from HOOKER's field sketches.

Tibet Mastiff.

Plate 7. — TIBET MASTIFF. Photo of the original sketch at Kew, by C. JENYNS, Esq.

tree they enclose, whose branches appear rising far above those of its des-
troyer. To the first of these groups belong many natural orders, of which
the most prominent are—*Leguminosae*, ivies, hydrangea, vines, *Pothos*, etc.
The inosculating ones are almost all figs and *Wightia*: the latter is the most
remarkable, and I add a cut of its grasping roots, sketched at our encampment" (p. 114).

"At 8000 feet, some enormous detached masses of micaceous gneiss rose
abruptly from the ridge; they were covered with mosses and ferns, and from
their summit a good view of the surrounding vegetation is obtained. The
mass of the forest is formed of:— (*1*) Three species of oak, of which
Q. annulata? with immense lamellated acorns, and leaves sixteen inches
long, is the tallest and the most abundant.— (*2*) Chestnut.— (*3*) *Laurineae*
of several species, all beautiful forest-trees, straight-boled, and umbrageous
above.— (*4*) Magnolias.— (*5*) Arborescent rhododendrons, which com-
mense here with the *R. arboreum*. At 8,000 and 9,000 feet, a considerable
change is found in the vegetation; the gigantic purple *Magnolia Campbellii*
replacing the white; chesnut disappears, and several laurels: other kinds of
maple are seen, with *Rhododendron argenteum*, and *Stauntonia*, a handsome
climber, which has beautiful pendent clusters of lilac blossoms."

"At 9,000 feet we arrived on a long flat covered with lofty trees, chiefly
purple magnolias, with a few oaks, great *Pyri* and two rhododendrons,
thirty to forty feet high (*R. barbatum*, and *R. aboreum*, var. *roseum*): *Skimmia*
and *Symplocos* were the common shrubs. A beautiful orchid with purple
flowers (*Coelogyne Wallichii*) grew on the trunks of all the great trees, attaining a higher elevation than most other epiphytical species, for I have seen
it at 10,000 feet" (p. 115).

Relations with the Sikkim government authorities were, and
throughout HOOKER's stay in the country remained, very unsatisfactory. After much delay permission was obtained from the Rajah to
travel through Sikkim and the Rajah of Nepal granted the request to
visit the eastern districts of that country. Most of the difficulties
HOOKER encountered in both the first and second of his journeys in
Sikkim were due to the opposition and underhand behaviour of the
Dewan or minister of the Rajah of Sikkim.

The following extracts illustrate the numerous observations HOOKER
made in western Sikkim and eastern Nepal.

Towards the Tambur valley: "In some spots the vegetation was exceedingly fine, and several large trees occurred: I measured a Toon (*Cedrela*)
thirty feet in girth at five feet above the ground. The skirts of the forest were
adorned with numerous jungle flowers, rice crops, blue *Acanthaceae* and
Pavetta, wild cherry-trees covered with scarlet blossoms, and trees of the
purple and lilac *Bauhinia*; while *Thunbergia*, *Convolvulus*, and other climbers,
hung in graceful festoons from the boughs, and on the dry micaceous rocks
the *Luculia gratissima*, one of our green-house ornaments, grew in profusion,
its gorgeous heads of rose-coloured blossoms scenting the air" (pp. 134–5).

3

Numerous records were made of the peoples, their agriculture, and their domesticated animals.

"These motley groups of Tibetans are singularly picturesque, from the variety in their parti-coloured dresses, and their odd appearance. First comes a middle-aged man or woman, driving a little silky black yak, grunting under his load of 260 lb. of salt, besides pots, pans, and kettles, stools, churn, and bamboo vessels, keeping up a constant rattle, and perhaps, buried amongst all, a rosy-cheeked and lipped baby, sucking a lump of cheese-curd. The main body follow in due order, and you are soon entangled amidst sheep and goats, each with its two little bags of salt: beside these, stalks the huge, grave, bull-headed mastiff, loaded like the rest, his glorious bushy tail thrown over his back in a majestic sweep, and a thick collar of scarlet wool round his neck and shoulders, setting off his long silky coat to the best advantage; he is decidedly the noblest-looking of the party, especially if a fine and pure black one, for they are often very ragged, dun-coloured, sorry beasts. He seems rather out of place, neither guarding nor keeping the party together, but he knows that neither yaks, sheep, nor goats, require his attention; all are perfectly tame, so he takes his share of work as salt-carrier by day, and watches by night as well. The children bring up the rear, laughing and chatting together; they, too, have their loads, even to the youngest that can walk alone" (p. 142).

Much farther north, to Wallanchoon,

"The path lay north-west up the valley, which became thickly wooded with silver-fir and juniper; we gradually ascended, crossing many streams from lateral gulleys, and huge masses of boulders. Evergreen rhododendrons soon replaced the firs, growing in inconceivable profusion, especially on the slopes facing the south-east, and with no other shrubs of tree-vegetation, but scattered bushes of rose, *Spiraea*, dwarf juniper, stunted birch, willow, honey-suckle, berberry, and a mountain-ash (*Pyrus*). What surprised me more than the prevalence of rhododendron bushes, was the number of species of this genus, easily recognised by the shape of their capsules, the form and woolly covering of the leaves; none were in flower, but I reaped a rich harvest of seed. At 12,000 feet the valley was wild, open, and broad, with sloping mountains clothed for 1,000 feet with dark-green rhododendron bushes; the river ran rapidly, and was broken into falls here and there. Huge angular and detached masses of rock were scattered about, and to the right and left snowy peaks towered over the surrounding mountains, while amongst the latter narrow gulleys led up to blue patches of glacial ice, with trickling streams and shoots of stones. Dwarf rhododendrons with strongly-scented leaves (*R. anthopogon* and *setosum*), and abundance of a little *Andromeda*, exactly like ling, with woody stems and tufted branches, gave a heathery appearance to the hill-sides. The prevalence of lichens, common to this country and to Scotland (especially *L. geographicus*), which coloured the rocks, added an additional feature to the resemblance to Scotch Highland scenery. Along the narrow path I found the two commonest of all British

Plate 8. — YANGMA GUOLA, 9000 ft. 28 Nov. 1848. Photo of HOOKER'S original field sketch at Kew.

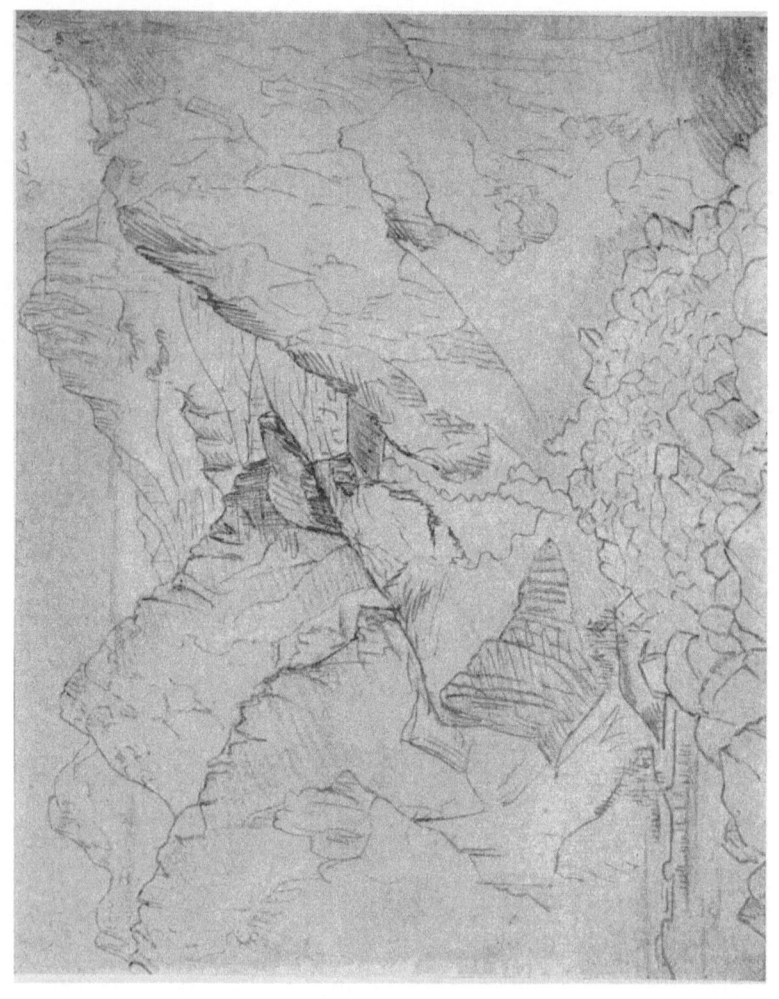

Plate 9. — ANCIENT MORAINE THROWN ACROSS THE YANGMA VALLEY, E. NEPAL, 11,000 ft. Photo of Hooker's original field sketch.

weeds, a grass (*Poa annua*), and the shepherd's purse! They had evidently
been imported by man and yaks, and as they do not occur in India, I could
not but regard these little wanderers from the north with the deepest interest" (pp. 152–3).

Discussions on more general problems of plant distribution are
sometimes included. Thus to his notes after ascending to Nango
mountain HOOKER adds comments as follows:

"We descended at first through rhododendron and juniper, then through
black silver-fir (*Abies Webbiana*), and below that, near the river, we came to
the Himalayan larch; a tree quite unknown, except from a notice in the
journals of Mr GRIFFITH, who found it in Bhotan. It is a small tree, twenty
to forty feet high, perfectly similar in general characters to a European
larch, but with larger cones, which are erect upon the very long, pensile,
whip-like branches; its leaves—now red—were falling, and covering the
rocky ground on which it grew, scattered amongst other trees. It is called
"Saar" by the Lepchas and Cis-himalayan Tibetans, and "Boarga-sella" by
the Nepalese, who say it is found as far west as the heads of the Cosi river:
it does not inhabit Central or West Nepal, nor the North-west Himalaya.
The distribution of the Himalayan pines is very remarkable. The Deodar
has not been seen east of Nepal, nor the *Pinus Gerardiana*, *Cupressus torulosa*,
or *Juniperus communis*. On the other hand, *Podocarpus* is confined to the east
of Katmandoo. *Abies Brunoniana* does not occur west of the Gogra, nor the
larch west of the Cosi, nor funereal cypress (an introduced plant, however)
west of the Teesta (in Sikkim). Of the twelve Sikkim and Bhotan *Coniferae*
(including yew, junipers, and *Podocarpus*) eight are common to the North-
west Himalaya (west of Nepal), and four are not: of the thirteen natives of
the north-west provinces, again, only five are not found in Sikkim, and
I have given their names below, because they show how European the absent ones are, either specifically or in affinity. I have stated that the Deodar
is possibly a variety of the Cedar of Lebanon. This now a prevalent opinion,
which is strengthened by the fact that so many more Himalayan plants are
now ascertained to be European than had been supposed before they were
compared with European specimens; such are the yew, *Juniperus communis*,
Berberis vulgaris, *Quercus Ballota*, *Populus alba* and *Euphratica*, etc. The cones
of the Deodar are identical with those of the Cedar of Lebanon: the Deodar
has generally longer and more pale bluish leaves and weeping branches, but
these characters seem to be unusually developed in our gardens; for several
gentlemen, well acquainted with the Deodar at Simla, when asked to point
it out in the Kew Gardens, have indicated the cedar of Lebanon, and when
shown the Deodar, declare that they never saw that plant in the Himalaya!"
(pp. 179–80).

Many passages recall HOOKER's interest in the genus *Rhododendron*—
an interest which led to his introduction of many species into cultivation at Kew and to the notes, descriptions, specimens, and drawings,
from which was prepared (under the editorship of W. J. HOOKER) the

fine folio volume "The Rhododendrons of Sikkim-Himalaya", London, 1849–51.

"At 9000 feet various shrubby rhododendrons prevailed, with mountain-ash, birch, and dwarf-bamboo; also *R. Falconeri*, which grew from forty to fifty feet high. The snow was deep and troublesome, so we encamped at 9,800 feet, or 800 feet below the top, in a wood of *Pyrus*, *Magnolia*, *Rhododendron*, and bamboo. As the ground was deeply covered with snow, we laid our beds on a thick layer of rhododendron twigs, bamboo, and masses of a pendent moss" (pp. 215–6).

HOOKER's descriptions of scenery and vegetation are often very vivid. This example relates to the ascent to Pemiongchi and the temple referred to is "the great Changachelling temple and monastery".

"The view of the snowy range from this temple is one of the finest in Sikkim; the eye surveying at one glance the vegetation of the Tropics and the Poles. Deep in the valleys the river-beds are but 3,000 feet above the sea, and are choked with fig trees, plantains, and palms; to these succeed laurels and magnolias, and higher up still, oaks, chestnuts, birches, etc.; there is, however, no marked line between the limits of these two last forests, which form the prevailing arboreous vegetation between 4,000 and 10,000 feet, and give a lurid hue to the mountains. Pine forests succeed for 2,000 feet higher, when they give place to a skirting of rhododendron and berberry. Among these appear black naked rocks, rising up in cliffs, between which are gulleys, down which the snow now (on the 1st January) descended to 12,000 feet. The mountain flanks are much more steep and rocky than those at similar heights on the outer ranges, and cataracts are very numerous, and of considerable height, though small in volume" (p. 231).

"On the following day we marched to Yoksun: the weather was fair, though it was evidently snowing on the mountains above. I halted at the Ratong river, at the foot of Mon Lepcha, where I found its elevation to be 7,150 feet; its edges were frozen, and the temperature of the water 36°; it is here a furious torrent flowing between gneiss rocks which dip south-south-east, and is flanked by flat-topped beds of boulders, gravel and sand, twelve to fourteen feet thick. Its vegetation resembles that of Darjeeling, but is more alpine, owing no doubt to the proximity of Kinchinjunga. The magnificent *Rhododendron argenteum* was growing on its banks. On the other hand, I was surprised to see a beautiful fern (a *Trichomanes*, very like the Irish one) which is not found at Darjeeling. The same day, at about the same elevation, I gathered sixty species of fern, many of very tropical forms. No doubt the range of such genera is extended in proportion to the extreme damp and equable climate, here, as about Darjeeling. Tree-ferns are, however, absent, and neither plantains, epiphytical *Orchideae*, nor palms, are so abundant, or ascend so high as on the outer ranges. About Yoksun itself, which occupies a very warm sheltered flat, many tropical genera occur, such as tall bamboos of two kinds, grasses allied to the sugar-cane, scarlet

Erythrina, and various *Araliaceae*, amongst which was one species whose pith was of so curious a structure, that I had no hesitation in considering the then unknown Chinese substance called rice paper to belong to a closely allied plant."

"The natives collect the leaves of many Aralias as fodder for cattle, for which purpose they are of the greatest service in a country where grass for pasture is so scarce; this is the more remarkable, since they belong to the natural family of ivy, which is usually poisonous; the use of this food, however, gives a peculiar taste to the butter. In other parts of Sikkim, fig-leaves are used for the same purpose, and branches of a bird-cherry (*Prunus*), a plant also of a very poisonous family, abounding in prussic acid"(pp. 251–2).

Kanchenjunga (Kinchinjunga), the great mountain to which HOOKER constantly refers in his accounts of Sikkim, and which he wrongly stated was the highest mountain in the world (it is now claimed to be the second highest), has not been climbed to its summit. It is situated on the Sikkim-Nepal border only a little south of Tibet. German-led expeditions in 1929, 1930, and 1931, were repulsed by the conditions on the mountain, in 1931 with loss of life. The mountain was photographed from the air in 1933.

The interval between the two main journeys was spent at Darjeeling, arranging and dispatching the collections already made (in 1848), "eighty loads" of specimens, and in an extended excursion to the lower land of the Terai to the south. The following extracts concern the more tropical vegetation he then examined.

"In the evening we walked to the skirts of the Sal forest. The great trunks of the trees were often scored by tigers' claws, this animal indulging in the catlike propensity of rising and stretching itself against such objects. Two species of *Dillenia* were common in the forest, with long grass, *Symplocos*, *Emblica*, and *Cassia Fistula*, now covered with long pods. Several parasitical air-plants grew on the dry trees, as *Oberonia, Vanda*, and *Ærides*" (p. 276).

"We walked to a stream, which flows at the base of the retiring sand-cliffs, and nourishes a dense and richly-varied jungle, producing many plants, as beautiful *Acanthaceae*, Indian horse-chestnut, loaded with white racemes of flowers, gay *Convolvuli*, laurels, terrestrial, and parasitic *Orchideae, Dillenia*, casting its enormous flowers as big as two fists, pepper, figs, and, in strange association with these, a hawthorn, and the yellow-flowered Indian strawberry, which ascends 7,500 feet on the mountains, and *Hodgsonia*, a new *Cucurbitaceous* genus, clinging in profusion to the trees, and also found 5,000 feet high on the mountains."

"In the evening we rode into the forest (which was dry and very unproductive), and thence along the river-banks, through *Acacia Catechu*, belted by *Sissoo*, which often fringes the stream, always occupying the lowest flats. The foliage at this season is brilliantly green; and as the evening advanced, a yellow convolvulus burst into flower like magic, adorning the bushes over which it climbed."

"It rained on the following morning; after which we left for the exit of the Teesta, proceeding northwards, sometimes through a dense forest of Sal timber, sometimes dipping into marshy depressions, or riding through grassy savannahs, breast-high. The coolness of the atmosphere was delicious and the beauty of the jungle seemed to increase the further we penetrated these primaeval forests."

"Eight miles from Rummai we came on a small river from the mountains, with a Cooch village close by, inhabited during the dry season by timber-cutters from Jeelpigoree. It is situated upon a very rich black soil, covered with *Saccharum* and various gigantic grasses, but no bamboo. These long grasses replace the Sal, of which we did not see one good tree" (p. 278).

"*Bombax*, *Erythrina*, and *Duabanga* (*Lagerstroemia grandiflora*) were in full flower, and with the profusion of *Bauhinia*, rendered the tree-jungle gay: the two former are leafless when flowering. The Duabanga is the pride of these forests. Its trunk, from eight to fifteen feet in girth, is generally forked from the base, and the long pendulous branches which clothe the trunk for 100 feet are thickly leafy, and terminated by racemes of immense white flowers, which, especially when in bud, smell most disagreeably of assafoetida. The magnificent Apocyneous climber, *Beaumontia*, was in full bloom, ascending the loftiest trees, and clothing their trunks with its splendid foliage and festoons of enormous funnelshaped white flowers" (pp. 282–3).

The second journey into Sikkim was made under greater difficulties than the first, because of the increasing opposition of the Dewan. This opposition was more to Dr. ARCHIBALD CAMPBELL, the Superintendent of Darjeeling, who was "likewise the Governor-General's agent or medium of communication between the British Government and the Sikkim Rajah", than to HOOKER personally. Not only did Dr. CAMPBELL make the necessary political arrangements for the journey and assist in many other ways, he also met HOOKER near Choongtam, about half-way between Darjeeling and the northernmost point reached on the second journey. The numerous inconveniences and insults the two travellers suffered culminated in the rough treatment of Dr. CAMPBELL and in their joint imprisonment (pp. 435 *seq.*) from which they were released only after prolonged negotiations with and threats from the British authorities.

The route this time was northwards from Darjeeling through eastern Sikkim up to and just over the ill-defined border with Tibet.

"From Temi the road descends to the Teesta, the course of which it afterwards follows. The valley was fearfully hot, and infested with mosquitos and peepsas. Many fine plants grew in it. I especially noticed *Aristolochia saccata*, which climbs the loftiest trees, bearing its curious pitcher-shaped flowers near the ground only; its leaves are said to be good food for cattle. *Houttuynia*, a curious herb allied to pepper grew on the banks, which, from the

Plate 10. — ANCIENT MORAINE THROWN ACROSS THE YANGMA VALLEY, E. NEPAL, 11,000 ft. Photo of the plate, as prepared for reproduction by WALTER FITCH, from HOOKER's field sketch.

Plate 11. — Ancient Moraine thrown across the Yangma Valley, E. Nepal, 11,000 ft. Photo of the plate reproduced. The series of three photos is particularly interesting as showing the degree of accuracy to be accepted in the final reproduction of field sketches.

profusion of its white flowers, resembled strawberry-beds; the leaves are eaten by the Lepchas. But the most magnificent plant of these jungles is *Hodgsonia*, (a genus I have dedicated to my friend, Mr. HODGSON), a gigantic climber allied to the gourd, bearing immense yellowish-white pendulous blossoms, whose petals have a fringe of buff-coloured curling threads, several inches long. The fruit is of a rich brown, like a small melon in form, and contains six large nuts, whose kernels (called "Katiorpot" by the Lepchas) are eaten. The stem, when cut, discharges water profusely from whichever end is held downwards. The "Took" (*Hydnocarpus*) is a beautiful evergreen tree, with tufts of yellow blossoms on the trunk; its fruit is as large as an orange, and is used to poison fish, while from the seeds an oil is expressed. Tropical oaks and Teaminalias are the giants of these low forests, the latter especially, having buttressed trunks, appear truly gigantic; one, of a kind called "Sung-lok" measured 47 feet in girth, at 5 feet, and 21 and 15 feet from the ground, and was fully 200 feet high. I could only procure the leaves by firing a ball into the crown. Some of their trunks lay smouldering on the ground, emitting a curious smell from the mineral matter in their ashes" (pp. 292–3).

Comparisons of a phytogeographical kind with other floras are frequently made:

"The vegetation in the neighbourhood of Lamteng is European and North American; that is to say, it unites the boreal and temperate floras of the east and west hemispheres; presenting also a few features peculiar to Asia. This is a subject of very great importance in physical geography; as a country combining the botanical characters of several others, affords materials for tracing the direction in which genera and species have migrated, the causes that favour their migrations, and the laws that determine the types or forms of one region, which represent those of another. A glance at the map will show that Sikkim is, geographically, peculiarly well situated for investigations of this kind, being centrally placed, whether as regards south-eastern Asia or the Himalayan chain. Again, the Lachen valley at this spot is nearly equi-distant from the tropical forests of the Terai and the sterile mountains of Tibet, for which reason representatives both of the dry central Asiatic and Siberian, and of the humid Malayan floras meet there.'

"The mean temperature of Lamteng (about 50°) is that of the isothermal which passes through Britain in lat. 52°, and east Europe in lat. 48°, cutting the parallel of 45° in Siberia (due north of Lamteng itself), descending to lat. 42° on the east coast of Asia, ascending to lat. 48° on the west of America, and descending to that of New York in the United States. This mean temperature is considerably increased by descending to the bed of the Lachen at 8,000 feet, and diminished by ascending Tukcham to 14,000 feet, which gives a range of 6,000 feet of elevation, and 20° of mean temperature. But as the climate and vegetation become arctic at 12,000 feet, it will be as well to confine my observations to the flora of 7,000 to 10,000 feet; of the mean temperature, namely, between 53° and 43° degrees, the isothermal lines corresponding to which embrace, on the surface of the globe, at the level of the sea, a space varying in different meridians from three to twelve degrees of latitude. At first sight it appears incredible that such a limited area, buried in the depths of the Himalaya, should present nearly all the types of the flora of the north temperate zone; not only, however, is this the case, but space is also found at Lamteng for the intercalation of types of a Malayan flora, otherwise wholly foreign to the north temperate region."

"A few examples will show this. Amongst trees the Conifers are conspicuous at Lamteng, and all are of genera typical both of Europe and North America: namely, silver fir, spruce, larch, and juniper, besides the yew: there are also species of birch, alder, ash, apple, oak, willow, cherry, bird-cherry, mountain-ash, thorn, walnut, hazel, maple, poplar, ivy, holly, Andromeda, *Rhamnus*. Of bushes: rose, berberry, bramble, rhododendron, elder, cornel, willow, honeysuckle, currant, *Spiraea, Viburnum, Cotoneaster, Hippophae*. Herbaceous plants are far too numerous to be enumerated, as a list would include most of the common genera of European and North American plants."

"Of North American genera, not found in Europe, were *Buddleia, Podophyllum, Magnolia, Sassafras? Tetranthera, Hydrangea, Diclytra, Aralia, Panax, Symplocos, Trillium*, and *Clintonia*. The absence of heaths is also equally a feature in the flora of North America. Of European genera, not found in North America, the Lachen valley has *Coriaria, Hypecoum*, and various *Cruciferae*. The Japanese and Chinese floras are represented in Sikkim by *Camellia, Deutzia, Stachyurus, Aucuba, Helwingia, Stauntonia, Hydrangea, Skimmia, Eurya, Anthogonium*, and *Enkianthus*. The Malayan by Magnolias, *Talauma*, many vacciniums and rhododendrons, *Kadsura, Goughia, Marlea*, both coriaceous and deciduous-leaved *Coelogyne, Oberonia, Cyrtosia, Calanthe*, and other orchids: *Ceropegia, Parochetus, Balanophora*, and many *Scitamineae*; and amongst trees, by *Engelhardtia, Goughia*, and various laurels" (pp. 313–5).

At Zemu Samdong Hooker records:

"On one occasion I ascended the steep hill at the fork; it was dry and rocky, and crowned with stunted pines. Stacks of different sorts of pine-wood were stored on the flat at its base, for export to Tibet, all thatched with the bark of *Abies Brunoniana*. Of these the larch (*Larix Griffithii*, "Sah"), splits well, and is the most durable of any; but the planks are small, soft, and white. The silver fir (*Abies Webbiana*, "Dunshing") also splits well; it is white, soft, and highly prized for durability. The wood of *Abies Brunoniana* ("Semadoong") is like the others in appearance, but is not durable; its bark is however very useful. The spruce (*Abies Smithiana*, "Seh") has also white wood, which is employed for posts and beams. These are the only pines whose woods are considered very useful; and it is a curious circumstance that none produced any quantity of resin, turpentine, or pitch; which may perhaps be accounted for by the humidity of the climate."

"*Pinus longifolia* (called by the Lepchas "Gniet-koong", and by the Boteeas "Teadong") only grows in low valleys, where better timber is abundant. The weeping blue juniper (*Juniperus recurva*, "Deschoo"), and the arboreous black one (called "Tchokpo") yield beautiful wood, like that of the

Plate 12. — SMALL TEMPLE, TASSIDING. Photo of HOOKER'S original sketch at Kew.

Plate 13. — 'THE BOTANIST IN SIKKIM.' From one of the copies made of WILLIAM TAYLER's sketch (1849) of J. D. HOOKER in the Sikkim Himalaya. This is apparently from FITCH's copy (with additions and modifications) of the original.

pencil cedar, but are comparatively scarce, as is the yew (*Taxus baccata*, "Tingschi"), whose timber is red. The "Tchenden," or funereal cypress, again, is valued only for the odour of its wood: *Pinus excelsa*, "Tongschi," though common in Bhotan, is, as I have elsewhere remarked, not found in east Nepal or Sikkim; the wood is admirable, being durable, close-grained, and so resinous as to be used for flambeaux and candles."

"On the flat were flowering a beautiful magnolia with globular sweet-scented flowers like snow-balls, several balsams, with species of *Convallaria*, *Cotoneaster*, *Gentian*, *Spiraea*, *Euphorbia*, *Pedicularis*, and honeysuckle. On the hill-side were creeping brambles, lovely yellow, purple, pink, and white primroses, white-flowered *Thalictrum* and *Anemone*, berberry, *Podophyllum*, white rose, fritillary, *Lloydia*, etc. On the flanks of Tukcham, in the bed of a torrent, I gathered many very alpine plants, at the comparatively low elevation of 10,000 feet, as dwarf willows, *Pinguicula*, (a genus not previously found in the Himalaya), *Oxyria*, *Androsace*, *Tofieldia*, *Arenaria*, saxifrages, and two dwarf heath-like *Andromedas*. The rocks were all of gneiss, with granite veins, tourmaline, and occasionally pieces of pure plumbago" (pp. 318–9).

He was now in the high Himalaya and his journals record the features of high mountain flora and vegetation.

"All my attempts to advance up the Zemu were fruitless, and a snow bridge by which I had hoped to cross to the opposite bank was carried away by the daily swelling river, while the continued bad weather prevented any excursions for days together. Botany was my only resource, and as vegetation was advancing rapidly under the influence of the southerly winds, I had a rich harvest: for though *Compositae*, *Pedicularis*, and a few more of the finer Himalayan plants flower later, June is still the most glorious month for show."

"Rhododendrons occupy the most prominent place, clothing the mountain slopes with a deep green mantle glowing with bells of brilliant colours; of the eight or ten species growing here, every bush was loaded with as great a profusion of blossoms as are their northern congeners in our English gardens. Primroses are next, both in beauty and abundance; and they are accompanied by yellow cowslips three feet high, purple polyanthus, and pink large-flowered dwarf kinds nestling in the rocks, and an exquisitely beautiful blue miniature species, whose blossoms sparkle like sapphires on the turf. Gentians begin to unfold their deep azure bells, aconites to rear their tall blue spikes, and fritillaries and *Meconopsis* burst into flower. On the black rocks the gigantic rhubarb (*Rheum nobile*) forms pale pyramidal towers a yard high, of inflated reflexed bracts, that conceal the flowers, and overlapping one another like tiles, protect them from the wind and rain: a whorl of broad green leaves edged with red spreads on the ground at the base of the plant, contrasting in colour with the transparent bracts, which are yellow, margined with pink. This is the handsomest herbaceous plant in Sikkim: it is called "Tchuka", and the acid stems are eaten both raw and boiled; they are hollow and full of pure water: the root resembles that of the medicinal rhubarb, but it is spongy and inert; it attains a length of four feet, and grows as thick as the arm. The dried leaves afford a substitute for to-bacco; a smaller kind of rhubarb is however more commonly used in Tibet for this purpose, it is called "Chula".".

"The elevation being 12,080 feet, I was above the limit of trees, and the ground was covered with many kinds of small-flowered honeysuckles, berberry, and white rose" (pp. 328–9).

"Above 11,000 feet the valley expands remarkably, the mountains recede, become less wooded, and more grassy, while the stream is suddenly less rapid, meandering in a broader bed, and bordered by marshes, covered with *Carex*, *Blysmus*, dwarf Tamarisk, and many kinds of yellow and red *Pedicularis*, both tall and beautiful. There are far fewer rhododendrons here than in the damper Zemu valley at equal elevations, and more Siberian, or dry country types of vegetation, as *Astragali* of several kinds, *Habenaria*, *Epipactis*, dandelion, and a caraway, whose stems (called in Tibet "Gzira") are much sought for as a condiment" (p. 334).

"Herbaceous plants are much more numerous here than in any other part of Sikkim; and sitting at my tent-door, I could, without rising from the ground, gather forty-three plants, of which all but two belonged to English genera. In the rich soil about the cottages were crops of dock, shepherd's-purse, *Thlaspi arvense*, *Cynoglossum* of two kinds (one used as a pot-herb), bal-sams, nettle, *Galeopsis*, mustard, radish, and turnip. On the neighbouring hills, which I explored up to 15,000 feet, I found many fine plants, partaking more or less of the Siberian type, of which *Corydalis*, *Leguminosae*, *Artemisia*, and *Pedicularis*, are familiar instances. I gathered upwards of 200 species, nearly all belonging to north European genera. Twenty-five were woody shrubs above three feet high, and six were ferns; sedges were in great profusion, amongst them three of British kinds: seven or eight were *Or-chideae*, including a fine *Cypripedium*" (p. 335).

In the Lachen valley towards Kongra Lama, the north slope of a contracted valley:

"Is covered with small trees and brushwood, rhododendron, birch, honey-suckle, and mountain-ash. These are the most northern shrubs in Sikkim, and I regarded them with deep interest, as being possibly the last of their kind to be met with in this meridian, for many degrees further north: perhaps even no similar shrubs occur between this and the Siberian Altai, a distance of 1,500 miles. The magnificent yellow cowslip (*Primula Sikki-mensis*) gilded the marshes, and *Caltha*, *Trollius*, *Anemone*, *Arenaria*, *Draba*, Saxifrages, Potentillas, Ranunculus, and other very alpine plants abounded" (p. 342).

The boundary between Sikkim and Tibet was reached at Kongra Lama.

"Isolated patches of vegetation appeared on the top of the pass, where I gathered forty kinds of plants, most of them being of a tufted habit charac-

Plate 14. — Black Juniper, 12,000 ft. From Hooker's Indian Sketches at Kew. This has probably been worked over by Walter Fitch.

Plate 15. — 'Living Bridge' formed of the aërial roots of the india-rubber and other kinds of figs. Photo of Hooker's original sketch at Kew.

teristic of an extreme climate; some (as species of *Caryophylleae*) forming hemispherical balls on the naked soil; others growing in matted tufts level with the ground. The greater portion had no woolly covering; nor did I find any of the cottony species of *Saussurea*, which are so common on the wetter mountains to the southward. Some most delicate-flowered plants even defy the biting winds of these exposed regions; such are a prickly *Meconopsis* with slender flower-stalks and four large blue poppylike petals, a *Cyananthus* with a membranous bell-shaped corolla, and a fritillary. Other curious plants were a little yellow saxifrage with long runners (very like the arctic *S. flagellaris*, of Spitzbergen and Melville Island), and the strong-scented spikenard (*Nardostachys*)" (p. 345).

Excursions and tours were made in various directions on both sides of the frontier.

"The Singtam Soubah being again laid up here from the consequences of leech-bites, I took the opportunity of visiting the Tunkra-lah pass, represented as the most snowy in Sikkim; which I found to be the case. The route lay over the moraines on the north flank of the Tunkrachoo, which are divided by narrow dry gullies, and composed of enormous blocks disintegrating into a deep layer of clay. All are clothed with luxuriant herbage and flowering shrubs, besides small larches and pines, rhododendrons and maples; with *Enkianthus*, *Pyrus*, cherry, *Pieris*, laurel, and *Goughia*. The musk-deer inhabits these woods, and at this season I have never seen it higher. Large monkeys are also found on the skirts of the pine-forests, and the *Ailurus ochraceus* (Hodgs.), a curious long-tailed animal peculiar to the Himalaya, something between a diminutive bear and a squirrel. In the dense and gigantic forest of *Abies Brunoniana* and silver fir, I measured one of the former trees, and found it twenty-eight feet in girth, and above 120 feet in height. The *Abies Webbiana* attains thirty-five feet in girth, with a trunk unbranched for forty feet."

"The path was narrow and difficult in the wood, and especially along the bed of the stream, where grew ugly trees of larch, eighty feet high and abundance of a new species of alpine strawberry with oblong fruit. At 11,560 feet elevation, I arrived at an immense rock of gneiss, buried in the forest. Here currant-bushes were plentiful, generally growing on the pine-trunks, in strange association with a small species of *Begonia*, a hothouse tribe of plants in England. Emerging from the forest, vast old moraines are crossed, in a shallow mountain valley, several miles long and broad, 12,000 feet above the sea, choked with rhododendron shrubs, and nearly encircled by snowy mountains. Magnificent gentians grew here, also *Senecio*, *Corydalis*, and the *Aconitum luridum* (n. sp.), whose root is said to be as virulent as *A. ferox* and *A. Napellus*. The plants were all fully a month behind those of the Lachen valley at the same elevation. Heavy rain fell in the afternoon, and we halted under some rocks: as I had brought no tent, my bed was placed beneath the shelter of one, near which the rest of the party burrowed. I supped off half a yak's kidney, an enormous organ in this animal."

"On the following morning we proceeded up the valley, towards a very steep rocky barrier, through which the river cut a narrow gorge, and beyond which rose lofty snowy mountains: the peak of Tunkra being to our left hand (north). Saxifrages grew here in profuse tufts of golden blossoms, and *Chrysosplenium*, rushes, mountain-sorrel (*Oxyria*), and the bladder-headed *Saussurea*, whose flowers are enclosed in inflated membranous bracts, and smell like putrid meat: there were also splendid primroses, the spikenard valerian, and golden Potentillas" (pp. 364–5).

"Very few plants grew amongst the stones at the top of the Tunkra pass, and those few were mostly quite different from those of Palung and Kongra Lama. A pink-flowered *Arenaria*, two kinds of *Corydalis*, the cottony *Saussurea*, and diminutive primroses, were the most conspicuous" (p. 367).

On the Donkia pass,

"One flowering plant ascends to the summit; the *Arenaria* one mentioned at p. 351 and 376. The Fescue grass, a little fern (*Woodsia*), and a *Saussurea* ascend very near the summit, and several lichens grow on the top, as *Cladonia vermicularis*, the yellow *Lecidea geographica*, and the orange *L. miniata*; also some barren mosses. At 18,300 feet, I found on one stone a fine Scotch lichen, a species of *Gyrophora*, the *"tripe de roche"* of Arctic voyagers, and the food of the Canadian hunters; it is also abundant on the Scotch alps" (p. 381).

An interesting interpolation on a subject that has much interested botanists, luminescence or phosphorescence in fungi (*see* the long list of literature quoted by WASSINK *in* Rec. trav. bot. néerland. 41: 150–212, 1948) may be quoted here:

"The phenomenon of phosphorescence is most conspicuous on stacks of fire-wood. At Darjeeling, during the damp, warm, summer months (May to October), at elevations of 5,000 to 8,000 feet, it may be witnessed every night by penetrating a few yards into the forest—at least it was so in 1848 and 1849; and during my stay there billets of decayed wood were repeatedly sent to me by residents, with inquiries as to the cause of their luminosity. It is no exaggeration to say that one does not need to move from the fireside to see this phenomenon, for if there is a partially decayed log amongst the fire-wood, it is almost sure to glow with a pale phosphoric light. A stack of fire-wood, collected near my host's (Mr. HODGSON) cottage, presented a beautiful spectacle for two months (in July and August), and on passing it at night, I had to quiet my pony, who was always alarmed by it. The phenomenon invariably accompanies decay, and is common on oak, laurel (*Tetranthera*), Birch, and probably other timbers; it equally appears on cut wood and on stumps, but is most frequent on branches lying close to the ground in the wet forests. I have reason to believe that it spreads with great rapidity from old surfaces to freshly cut ones. That it is a vital phenomenon, and due to the mycelium of a fungus, I do not in the least doubt, for I have observed it occasionally circumscribed by those black lines which are so often seen to bound mycelia on dead wood, and to precede a more rapid decay. I have often tried, but always in vain, to coax these mycelia into developing some fungus, by placing them in damp rooms, etc. When camping in the mountains, I frequently caused the natives to bring phosphorescent wood into my tent, for the pleasure of watching its soft undulating light, which appears to pale and glow with every motion of the atmosphere; but except in this difference of intensity, it presents no change of appearance night after night. Alcohol, heat and dryness soon dissipate it; electricity I never tried. It has no odour, and my dog, who had a fine sense of smell, paid no heed when it was laid under his nose" (pp. 396–7).

Mention has already been made of the imprisonment of CAMPBELL and HOOKER. After their release they returned to Darjeeling. Thence HOOKER made a trip to Calcutta and then an excursion to the Khasia mountains in Assam with Dr. THOMSON. He records that:

"The sub-tropical scenery of the lower and outer Sikkim Himalaya, though on a much more gigantic scale, is not comparable in beauty and luxuriance with the really tropical vegetation induced by the hot, damp, and insular climate of these perennially humid mountains. On the Himalaya forests of gigantic trees, many of them deciduous, appear from a distance as masses of dark gray foliage, clothing mountains 10,000 feet high: here the individual trees are smaller, more varied in kind, of a brilliant green, and contrast with gray limestone and red sandstone rocks and silvery cataracts. Palms are more numerous here; the cultivated *Areca* (betel-nut) especially, raising its graceful stem and feathery crown, "like an arrow shot down from heaven", in luxuriance and beauty above the verdant slopes. This difference is at once expressed to the Indian botanist by defining the Khasia flora as of Malayan character; by which is meant the prevalence of brilliant glossy-leaved evergreen tribes of trees (as *Euphorbiaceae* and *Urticeae*), especially figs, which abound in the hot gulleys, where the property of their roots, which inosculate and form natural grafts, is taken advantage of in bridging streams, and in constructing what are called living bridges, of the most picturesque forms. *Combretaceae*, oaks, oranges, *Garcinia* (gamboge), *Diospyros*, figs, Jacks, plantains, and *Pandanus*, are more frequent here, together with pinnated leaved *Leguminosae*, *Meliaceae*, vines and peppers, and above all palms, both climbing ones with pinnated shining leaves (as *Calamus* and *Plectocomia*), and erect ones with similar leaves (as cultivated cocoa-nut, *Areca* and *Arenga*), and the broader-leaved wild betel-nut, and beautiful *Caryota* or wine-palm, whose immense decompound leaves are twelve feet long. Laurels and wild nutmegs, with *Henslowia*, *Itea*, etc., were frequent in the forest, with the usual prevalence of parasites, mistleto, epiphytical *Orchideae*, *Æschynanthus*, ferns, mosses, and *Lycopodia*; and on the ground were *Rubiaceae*, *Scitamineae*, ferns, *Acanthaceae*, beautiful balsams, and herbaceous and shrubby nettles. Bamboos of many kinds are very abundant, and these hills further differ remarkably from those of Sikkim in the great number of species of grasses" (pp. 481–3).

"It is extremely difficult to give within the limits of this narrative any idea of the Khasia flora, which is, in extent and number of fine plants, the richest in India, and probably in all Asia. We collected upwards of 2,000 flowering plants within ten miles of the station of Churra, besides 150 ferns, and a profusion of mosses, lichens, and fungi. This extraordinary exuberance of species is not so much attributable to the elevation, for the whole Sikkim Himalaya (three times more elevated) does not contain 500 more flowering plants, and far fewer ferns, etc.; but to the variety of exposures; namely, (1) the Jheels, (2) the tropical jungles, both in deep, hot, and wet valleys, and on drier slopes; (3) the rocks; (4) the bleak table-lands and stony soils; (5) the moor-like uplands, naked and exposed, where many species of genera appear at 5,000 to 6,000 feet, which are not found on the outer ranges of Sikkim under 10,000. In fact, strange as it may appear, owing to this last cause, the temperate flora descends fully 4,000 feet lower in the latitude of Khasia (25° N.) than in that of Sikkim (27° N.), though the former is two degrees nearer the equator."

"The *Pandanus* alone forms a conspicuous feature in the immediate vicinity of Churra; while the small woods about Mamloo, Moosmai, and the Coal-pits, are composed of *Symplocos*, laurels, brambles, and jasmines, mixed with small oaks and *Photinia*, and many tropical genera of trees and shrubs."

"*Orchideae* are, perhaps, the largest natural order in the Khasia, where fully 250 kinds grow, chiefly on trees and rocks, but many are terrestrial, inhabiting damp woods and grassy slopes. I doubt whether in any other part of the globe the species of orchids outnumber those of any other natural order, or form so large a proportion of the flora. Balsams are next in relative abundance (about twenty-five), both tropical and temperate kinds, of great beauty and variety in colour, form, and size of blossom. Palms amount to fourteen, of which the *Chamaerops* and *Arenga* are the only genera not found in Sikkim. Of bamboos there are also fifteen, and of other grasses 150, which is an immense proportion, considering that the Indian flora (including those of Ceylon, Kashmir, and all the Himalaya) hardly contains 400. *Scitamineae* also are abundant, and extremely beautiful; we collected thirty-seven kinds."

"No rhododendron grows at Churra, but several species occur a little further north: there is but one pine (*P. Khasiana*) besides the yew, (and two *Podocarpi*), and that is only found in the drier interior regions. Singular to say, it is a species not seen in the Himalaya or elsewhere, but very nearly allied to *Pinus longifolia*, though more closely resembling the Scotch fir than that tree does."

"The natural orders whose rarity is most noticeable, are *Cruciferae*, represented by only three kinds, and *Caryophylleae*. Of *Ranunculaceae* there are six or seven species of *Clematis*, two of *Anemone*, one *Delphinium*, three of *Thalictrum*, and two *Ranunculi*. *Compositae* and *Leguminosae* are far more numerous than in Sikkim" (pp. 490–1).

"The vegetation is more alpine at Kala-panee (elevation, 5,000 feet); *Benthamia*, *Kadsura*, *Stauntonia*, *Illicium*, *Actinidia*, *Helwingia*, *Corylopsis*, and berberry—all Japan and Chinese, and most of them Darjeeling genera—appear here, with the English yew, two rhododendrons, and *Bucklandia*. There are no large trees, but a bright green jungle of small ones and bushes, many of which are very rare and curious. *Luculia Pinceana* makes a gorgeous show here in October" (p. 494).

The return was via Silhet to Chittagong. A deviation to the Burmese frontier took three days.

"The country continued a grassy level, with marshes and rice cultivation, to the first range of hills, beyond which the river is unnavigable; there also a forest commences, of oaks, figs, and the common trees of east Bengal. The road hence was a good one, cut by Sepoys across the dividing ranges, the first of which is not 500 feet high. On the ascent bamboos abound, of the kind called Tuldah or Dulloah, which has long very thin-walled joints; it attains no great size, but is remarkably gregarious. On the east side of the range, the road runs through soft shales and beds of clay, and conglomerates, descending to a broad valley covered with gigantic scattered timber-trees of jarool, acacia, *Diospyros*, *Urticeae*, and *Bauhiniae*, rearing their enormous trunks above the bamboo jungle: immense rattan-canes wound through the forest, and in the gullies were groves of two kinds of tree-fern, two of *Areca*, *Wallichia* palm, screw-pine, and *Dracaena*. Wild rice grew abundantly in the marshes, with tall grasses; and *Cardiopteris* covered the trees for upwards of sixty feet, like hops, with a mass of pale-green foliage, and dry white glistening seedvessels. This forest differed from those of the Silhet and Khasia mountains, especially in the abundance of bamboo jungle, which is, I believe, the prevalent feature of the low hills in Birmah, Ava, and Munnipore; also in the gigantic size of the rattans, larger palms, and different forest trees, and in the scanty undergrowth of herbs and bushes. I only saw, however, the skirts of the forest; the mountains further east, which I am told rise several thousand feet in limestone cliffs, are doubtless richer in herbaceous plants" (p. 528).

On the 28 January 1851 Calcutta was reached and ten days later HOOKER left India for England where he landed on 25 March.

Flora Indica:— Only volume 1 of the Flora Indica by J. D. HOOKER and T. THOMSON was published (London, 1855). In this volume there is an Introductory Essay of 280 pages. Throughout the Essay plural pronouns, "we", "our", etc., are used, but both the internal evidence of style and the theories enunciated as well as some external evidence gleaned from Hookerian correspondence, make it reasonably certain that HOOKER was largely responsible for its contents and for the conclusions reached. The Essay is of particular value for two reasons: it places on record what were, in 1855, "diplomatic" rather than advanced views on the nature of "species" and it includes a rather elaborate "sketch" of the "Physical Features and Vegetation of the Provinces of India".

Although outside the main thesis of this present work, it is not altogether irrelevant, and it is certainly of general historical interest, to

quote a few extracts which illustrate HOOKER's views on species at a
date shortly before the publication, in 1859, of DARWIN's "Origin of
Species".

"A most masterly view of the present state of the question will be
found in Sir C. LYELL's 'Principles of Geology', where the arguments
of LAMARCK and others are stated with great fairness, and answered by
the author, whose opinion is decided in favour of species being definite creations. In this we are disposed to agree, having seen no argument which is sufficient to alter the *a priori* conclusion to which facts
appear to point, that it is more probable that species should have been
created with a certain degree of variability, than that mutability should
be a part of the scheme of nature. This however is pre-eminently a
question for systematists. Long and patient observation in the field,
and much practice in sifting and examining the comparative value of
characters, can alone give the experience which will warrant the
expression of a decided opinion on a question of so much difficulty."

"It cannot be doubted that the general acceptance which the doctrine
of the mutability of species has met with amongst superficial naturalists, has originated in a reaction from early impressions of the absolute
fixity of characters. The student who is taught that species are definite
creations, constant and unchangeable, without being cautioned as to
their power of variation within certain limits, finds, when he begins to
observe for himself, that he has constant difficulty in determining their
limits, and that abler judges than himself are equally at fault. The more
books he consults, the greater are the discrepancies he meets with; if
he has recourse to gardens, he there finds species still more sportive;
and if he travels, he meets with a change of form under every climate;
till at last, perplexed and mortified, he gives up the study of specific
botany, and becomes a convert to the belief that species are the arbitrary creations of systematists. And such must be the result in the great
majority of instances, while each observer has to acquire for himself
that familiarity with the amount of variation to which organized beings
are subject, which alone will render him a sound systematist. For so
long as our early education does not teach us this important principle,
so long shall we find beginners refusing to accept the conclusions
arrived at by abler botanists."

"Even if we admit the hypothesis that the existence of species as definite
creations is inconsistent with facts, it does not necessarily follow that the
study of systematic botany is fruitless; for such a supposition involves the
operation of laws which govern the variations of plants, and in accordance
with which they remain fixed for a longer or shorter period; and such laws
it becomes the duty of the systematist to develop. The advocates for their
agency principally base their belief upon hybridity, and variability induced

by climatic influences; but we shall attempt to show, that all the legitimate conclusions which can be drawn from a study of these phenomena are opposed to the theory of universal mutability" (pp. 20–1).

The term "hybrid" was evidently limited to organisms which had originated from a cross between what were accepted as two distinct species—and HOOKER was, on modern standards, a decided "lumper".

"In the course of our extended wanderings, it has been our habit to acquaint ourselves with the plants as we gathered them, and so to observe their differential characters in the field, that we were never at a loss for the means of understanding one another when alluding to any particular species; yet we never met with a plant that suggested to us even a suspicion of hybridization" (p. 23).

"And now that we are on the subject of variation, it appears advisable to impress upon the Indian botanist the value of studying its phenomena in the field. We pledge our experience that he will find it the most profitable department of systematic botany he can pursue; and that the result of his investigations will be that he will take a wide and extended view of the variations of species, consistently with their still possessing certain definable limits" (p. 27).

"It is very much to be wished that the local botanist should commence his studies upon a diametrically opposite principle to that upon which he now proceeds, and that he should endeavour, by selecting good suites of specimens, produced under all variations of circumstances, to determine *how few*, not *how many* species are comprised in the flora of his district. The permanent differences will, he may depend upon it, soon force themselves upon his attention, whilst those which are non-essential will consecutively be eliminated. There is no better way of proving the validity of characters than by attempting to invalidate them. The unavoidable tendency of the human mind, when occupied with the pursuit of minute differences, is to seize on them with avidity, and to relinquish them with regret; hence the irresistible desire to rest contented with a character, however bad, so long as it is obtained with difficulty, and in the observer's opinion is tolerably constant. It is strange that local naturalists cannot see that the discovery of a form uniting two others they had previously thought distinct, is much more important than that of a totally new species, inasmuch as the correction of an error is a greater boon to science than is a step in advance" (pp. 35–6).

"The subject of geographical distribution leads to questions of practical importance, upon which we have a few remarks to offer, as eminently bearing upon all questions relating to the treatment of a systematic flora: these are,—(*1*) Its dependence on the doctrine of specific centres. (*2*) The power of migration as capable of effecting the present distribution. (*3*) The general effects of migration in producing a much wider dispersion and ubiquitous diffusion of species than is generally admitted by botanists who have not investigated tropical floras, and especially continental ones" (p. 39).

"There is a very curious theoretical point bearing upon the distribution of species, first enunciated, we believe, by a most accomplished observer, Dean

4

HERBERT, and which, we think, has never been sufficiently appreciated or followed out; it is, that species in general do not grow where they like best, but where they can best find room. Plants, in a state of nature, are always warring with one another, contending for the monopoly of the soil,—the stronger ejecting the weaker,—the more vigorous overgrowing and killing the more delicate. Every modification of climate, every disturbance of the soil, every interference with the existing vegetation of an area, favours some species at the expense of others. The life of a plant is as much one of strife as that of an animal, with this difference, that the contention is not intermittent, but continuous, though unheeded by the common observer. In the common course of events, therefore, the ground occupied by a widely-distributed plant is held on a very different tenure in different places; some individuals are obliged to grow in the shade, others in the sun; and they hence flower earlier in certain places: we say of such plants that they have a power of accommodating themselves to their altered conditions, or better, that they have the power of resisting the effects of the change. Now, this power we believe to be very much underrated, specific characters being too often founded on the differences in habit induced during a plant's migration over great areas, or brought about by the change of soil and climate and surrounding vegetation, to which individuals and their successors are subjected in different parts of one and the same area" (pp. 41-2).

In their general remarks on the vegetation of India the authors make numerous statements of fact and draw a variety of conclusions some of which may be questioned. The total number of Indian species is assumed to be 12-15000—"India" including the whole area inclusive from Afghanistan, Baluchistan, and Tibet to Ceylon and the whole of the Malay Peninsula (from 36° N. to the equator and from 62° to 105° E., "little less than two millions of square miles" of land surface). There are very few families (natural orders) peculiar to "India".

"Thus, *Aurantiaceae, Dipteraceae, Balsamineae, Ebenaceae, Jasmineae,* and *Cyrtandraceae* are the only orders which are largely developed in India, and sparingly elsewhere; and of these, few contain one hundred Indian species" (p. 91). "We believe that there is no part of the whole area included in our Flora where a radius of ten miles produces many more than 2000 species of flowering plants, and that this is very rare, confined to mountainous districts, and possibly to the Khasia" (p. 92). "There is almost a total absence of absolutely local plants in India, at least so far as our experience serves us" (p. 93).

The great influence of climate on the distribution of Indian plants is well summarized with reference to the tropical forests, hill and mountain vegetation, and the high mountain flora.

"The Tropical forests of India may be divided into those which inhabit perennially humid districts, and those which are confined to regions presenting contrasted seasons, of summer rain and winter drought."

"The perennially humid forests are uniformly characterized by the prevalence of Ferns, and, at elevations below 5000–7000 feet, by the immense number of epiphytal *Orchideae*, *Orontiaceae*, and *Scitamineae*: they contain a far greater amount of species than the drier forests, and are further characterized by *Zingiberaceae*, *Xyrideae*, Palms, *Pandaneae*, *Dracaena*, *Piper*, *Chloranthus*, *Urticaceae* (especially *Artocarpeae* and *Fici*), *Araliaceae*, *Apocyneae*, shrubby *Rubiaceae*, *Aurantiaceae*, *Garciniaceae*, *Anonaceae*, Nutmegs, and *Dipterocarpeae*."

"The drier tropical forests of the regions with contrasted seasons, are much modified in luxuriance and extension by the winter cold in those extratropical latitudes over which they spread. In the chapter upon the meteorology of India, it is shown that though the summer heat scarcely decreases with the increasing latitude till the 30th degree north, the cold of winter rapidly increases (see the map of Isothermals). Hence many tropical species, genera, and even families, which are sensitive to cold, are comparatively local when found beyond the tropic, as most Palms, *Cycas*, *Dipterocarpeae* (except *Vatica*), *Aurantiaceae*, *Connaraceae*, *Meliaceae*, *Myrtaceae*, *Rubiaceae*, *Ebenaceae*, and many more. Others are indifferent to the cold of winter, provided they experience a great summer heat; these advance far beyond the tropic, and lend a more or less tropical aspect to the Flora even of the base of the north-western Himalaya, in 33° north. Such are many *Leguminosae* (as *Bauhinia*, *Acacia*, *Erythrina*, *Butea*, *Dalbergia*, *Millettia*), *Bombax*, *Vatica*, *Nauclea*, *Combretaceae*, *Verbenaceae*, *Lagerstroemia*, *Grislea*, *Jasmineae*, and *Bignonia Indica*" (pp. 95–6).

"The transition from the tropical to the temperate Flora is more rapid in ascending above the level of the plains, than in advancing northward at the same level; the change of vegetation in a few thousand feet of ascent being much greater than in as many degrees of latitude as would compensate for the decrease of temperature experienced in that ascent" (p. 97).

"In the Himalaya the truly temperate vegetation supersedes the subtropical above 4000–6000 feet; and the elevation at which this change takes place corresponds roughly with that at which the winter is marked by an annual fall of snow" (p. 99).

"Lastly, the Alpine or Arctic Flora demands a few words here, though it forms comparatively so small a feature in the vegetation of all India, that its full discussion must be reserved to our remarks on the Alpine region of the Himalaya. This, which hardly reaches its extreme upper limit at 18,500 feet above the sea, commences (as we restrict it) above the limit of trees throughout a great part of the Himalaya; it partakes in its characteristic genera of the temperate Flora, and, though fully representing the Flora of the Polar regions, contains so many types that are foreign to them (as *Gentiana*, *Ephedra*, *Valerianeae*, *Corydalis*), and some which are even rare in Siberia, that it must rather be considered as a continuation of the Alpine Flora of Europe than a representation of that of the Arctic zone. It displays one remarkable feature throughout its whole extent, a comparative paucity of Cryptogamic plants; and it is especially poor in those luxuriant mosses of tall growth and succulent habit, which form vivid and broad green tufts, loaded with rich brown capsules, and which abound both in the Alps and Polar regions. This is no doubt indirectly due to the elevation of the region, and directly to the sudden accessions of great heat and drought, which are the effects of a highly rarefied atmosphere, and which, though strongly enough marked to check the development of Mosses and Hepaticae, are not of sufficient duration to affect phaenogamic vegetation in the same degree" (pp. 100–101).

The following "geographical alliances or affinities" of the Indian flora are considered to be well established: the Australian type, the Malay Archipelago type, the China and Japan type, the Siberian type, the European type (including 222 British plants which extend into India), the Egyptian type, and the Tropical African type.

"India" is divided by the authors into four primary divisions: "(1) *Hindostan*, in the widest sense of that term, including the whole Western (Madras) Peninsula, and the Gangetic plain to the base of the Himalaya. (2) The *Himalaya*, a mountain chain which rises abruptly from the Gangetic plain, and is connected with a still loftier mountain mass (of Tibet) to the north, and beyond India: (3) *Eastern India* (India ultra Gangem), including the kingdom of Ava and the Eastern or Malayan Peninsula. (4) *Afghanistan*" (pp. 115–6).

Hindostan is further divided into 18 provinces, the Himalaya into 16 (and Tibet into 9), and Eastern India into 9. It is not possible here adequately to illustrate the wealth of geographical and botanical information condensed in 140 pages. Two extracts are given below, one dealing with Sind (or Sindh), as an example of the treatment of an area not personally known to HOOKER, and the other with Sikkim, which province he so thoroughly explored.

"14. *Sindh*.— The province of Sindh extends from the sea on the south to the borders of the Panjab on the north. Westward it is bounded by the mountains of Beluchistan, and on the east it is continuous with the desert of Marwar. Sindh is an alluvial plain watered by the various branches of the Indus. For the most part it is perfectly level, but a few low hills (spurs from the Beluch mountains) here and there, as at Rori, Hyderabad, and Karachi, advance close to the Indus."

"The climate of Sindh is perfectly arid, little or no rain falling at any period of the year. Now and then, however, exceptional seasons occur, when heavy showers fall at intervals, especially at the commencement of the south-west monsoon, at which time there is a considerable rain-fall in the mountains of Beluchistan and Afghanistan. The average rain-fall of Sindh is not more than four or five inches, but occasionally upwards of twenty inches of rain have been registered. Even with this amount of rain, however, the climate is so dry that the air does not remain humid for any length of time, the storms being transitory in duration. The heat is therefore very great, and the mean temperature probably as high as anywhere in India."

"Though extremely fertile where irrigation is practicable, Sindh is, in consequence of the great dryness of the air, naturally sterile. There is no forest of large trees; and though extensive tracts near the river are covered with dense jungle, chiefly of *Acacia Arabica* and *Prosopis spicigera*, the greater part of the surface is barren of vegetation, and the driest parts are an absolute desert. In the lower part of the delta, within reach of the tides, a low jungle of mangroves occupies the swampy islets."

"The vegetation of Sindh was first made known to science by GRIFFITH, who traversed the upper part of the province on his way to Afghanistan, and has recorded in his private journals and literary notes the most characteristic plants which he observed. It has also been explored by Major VICARY, who has published in the Asiatic Society's Journal a list of its plants. For our very complete knowledge of its flora we are, however, mainly indebted to the late Dr. STOCKS, whose labours in this interesting province throw much light on Indian botany. Dr. STOCKS' collections amount to little more than four hundred species, so that the flora is a very poor one. No doubt, as he has himself stated, a careful exploration of the hilly districts would considerably increase this number; but we feel confident that the novelties would be almost if not entirely western forms, and, would therefore increase the proportion, already great, which these bear to forms characteristic of Eastern India vegetation."

"More than nine-tenths of the Sindh vegetation, on a rough estimate, consist of plants which are indigenous in Africa. At least one-half of these are common Nubian or Egyptian plants, but which, from being indifferent to moisture, are diffused over all parts of India. As examples we may mention *Gynandropsis pentaphylla, Abutilon Indicum, Tribulus terrestris, Tephrosia purpurea, Glinus lotoides, Grangea Maderaspatana, Trichodesma Indicum, Lippia nodiflora, Solanum Jacquini, Ærua lanata, Achyranthes aspera*. A smaller number but still considerable, are tropical African, which are also widely diffused over India. Among these are many *Convolvulaceae*, as *Batatas pentaphylla, Pharbitis Nil, Ipomoea muricata* and *reptans* and many of the commonest Indian weeds, such as *Peristrophe bicaliculata* and several species of *Corchorus* and *Triumfetta*. A considerable proportion (perhaps one-sixth of the whole) consists of common Egyptian plants, which are too intolerant of moisture to withstand the climate of the more humid parts of India, but which extend along the Arabian and Persian coasts to Sindh, and thence to the Panjab and the drier parts of the Gangetic plain, and some even to the Dekhan and Mysore. Such are *Peganum Harmala, Cocculus Leaeba, Capparis aphylla, Fagonia Arabica, Alhagi Maurorum, Acacia Arabica, Prosopis spicigera, Zizyphus Lotus*, and *Calotropis procera*, all of which extend to the drier parts of the peninsula; and *Malcolmia Africana, Corchorus depressus, Cucumis Colocynthis, Berthelotia lanceolata, Heliotropium undulatum, Salvia Ægyptiaca, Lycium Europaeum, Cometes Surattensis*, several *Chenopodiaceae*, and *Crypsis schoenoides*, which are confined to northern India. With these there occur also a few central European plants, though far fewer than in the northern Panjab, as for example *Ranunculus sceleratus, Convolvulus arvensis, Heliotropium Europaeum, Rumex obtusifolius, Asphodelus fistulosus*, and *Potamogeton pectinatus* and *natans*."

"Sindh also contains a considerable number of species which have not been met with elsewhere in India, but which are Arabian or Nubian plants. Such are, *Zygophyllum album* and *simplex*, *Balsamodendron*, *Neurada procumbens*, *Aizoon Canariense*, *Seddera latifolia*, *Trichodesma Africanum*, *Acanthodium hirtum*, and several *Barleriae*. A few Persian and Mesopotamian plants not yet known further west, such as *Populus Euphratica* and *Gaillonia*, occur also in the list. *Puneeria coagulans*, Stocks, is confined to Sindh, and the neighbouring province of Beluchistan. Eastern species which find their western limit in Sindh are almost entirely wanting. The following are all that are contained in Dr. Stocks' catalogue, excluding plants manifestly cultivated (such as *Tamarindus*), *Rhus Mysorensis*, *Zizyphus Jujuba*, *Hedyotis aspera*, *Coldenia procumbens*, *Salvia plebeia* (a New Holland plant), *Clerodendron phlomoides*, *Aristolochia bracteata*, and *Zeuxine sulcata*. There are, however, a considerable number of species which have not been met with in Egypt or Arabia, but which belong to genera characteristic of those countries, and are very closely related to Egyptian species. Instances of this kind are *Crotalaria Burhia*, *Dicoma lanuginosa*, *Leptadenia Jacquemontiana*, *Oxystelma esculentum*, *Linaria ramosissima*, *Streptium asperum*, *Solanum gracilipes*, *Chamaerops Ritchiana*. If we add to this enumeration the coast flora of *Sonneratia*, *Rhizophora*, *Ceriops*, *Scaevola*, *Ægiceras*, *Ipomaea Pes-caprae*, and *Avicennia*, a good general idea is given to the nature of the flora of Sindh". (pp. 151–4).

"3. *Sikkim*:— The province of Sikkim, though of very limited extent, is now the best known part of the central or eastern Himalaya, and presents many features of much interest. It consists entirely of the basin of the river Tista, which, with its tributaries, drain the whole country. The course of this river is for the most part meridional, that is, perpendicular to the plains; and the same may be said of its great tributary the Rangit river, which joins it from the west, flowing for a short distance parallel to the plains, through a deep ravine not 1000 feet above the sea, to the north of a transverse range elevated 7–8000 feet."

"The position of Sikkim, opposite to the opening of the Gangetic valley between the mountains of Bahar on the one hand, and those of Khasia on the other, exposes it to the full force of the monsoon; its rains are therefore heavy and almost uninterrupted, and are accompanied by dense fogs and a saturated atmosphere. This weather indeed prevails throughout the year, as there are frequent winter rains, which are generally accompanied by cold fogs, and alternate with frost and snow. March and April are the driest months, and in fine seasons are often bright and clear, but the rains commence in May, to continue with little intermission till October. The bounding mountain-chains are very lofty, and snow-clad throughout a great part of their extent, but the central range which separates the Rangit from the Tista is depressed till very far in the interior. The river-valleys are also considerably depressed, but less markedly so than those of western Bhotan. The rainy winds have thus free access to the heart of the province, and sweep almost without interruption up to the base of Kanchinjanga (28,178 feet), the loftiest mountain and most enormous mass of snow in the world. The snow-level is here about 16,000 feet. Between the two principal sources of the

Tista, however, the Lachen and the Lachung, a lofty snowy range is projec-
ted; and as this chain has a southwest direction, and is moreover sheltered
to a considerable extent by the boundary chain between Sikkim and the
Tibetan valley of Chumbi, we have in these valleys a rapid diminution of
the rain-fall and an equally rapid transition to the Tibetan climate, while
the level of perpetual snow rises to above 18,000 feet."

"From the level of the sea to an elevation of 12,000 feet Sikkim is covered
with a dense forest, only interrupted where village clearances have bared
the slopes for the purposes of cultivation; and there the encroachment of the
forest is with difficulty prevented by frequent fires and the incessant labour
of the villagers. The forest consists everywhere of tall umbrageous trees;
with little underwood on the drier slopes, but often dense grass jungle;
more commonly however it is accompanied by a luxuriant undergrowth of
shrubs, which renders it almost impenetrable. In the tropical zone large Figs
abound, with *Terminalia*, *Vatica*, *Myrtaceae*, Laurels, *Euphorbiaceae*, *Meliaceae*,
Bauhinia, *Bombax*, *Morus*, *Artocarpus*, and other *Urticaceae*, and many
Leguminosae; and the undergrowth consists of *Acanthaceae*, Bamboos, several *Calami*, two dwarf *Arecae*, *Wallichia*, and *Caryota urens*. Plantains and tree-
ferns, as well as *Pandanus*, are common; and, as in all moist tropical countries, ferns, orchids, *Scitamineae*, and *Pothos* are extremely abundant. Few
oaks are found at the base of the mountains, and the only conifers are a
species of *Podocarpus* and *Pinus longifolia*, which frequents the drier slopes of
hot valleys as low as 1000 feet above the level of the sea, and entirely avoids
the temperate zone. The other tropical Gymnosperms are *Cycas pectinata* and
Gnetum scandens, genera which find their north-western limits in Sikkim."

"The rarity of oaks at the base of the mountains must be ascribed to the
great dryness and winter's cold of that part of the chain, for we miss also
other eastern types which abound in the equable and moist climate of the
Malayan archipelago and peninsula, such as *Liquidambar* and nutmegs;
whilst *Dipterocarpeae*, and especially *Anonaceae*, are exceedingly few in
number. *Liquidambar* is common in the Assam jungles, and indicates their
greater humidity. The same inference may be drawn with regard to the tropical
belt of the Khasia, from the occurrence there of two nutmegs and numerous
Anonaceae."

"Oaks, of which (including chesnuts) there are upwards of eleven species
in Sikkim, become abundant at about 4000 feet, and at 5000 feet the temperate zone commences, the vegetation varying with the degree of humidity.
On the outermost ranges, and on northern exposures, there is a dripping
forest of cherry, laurels, oaks and chesnuts, *Magnolia*, *Andromeda*, *Styrax*,
Pyrus, maple and birch, with an undergrowth of *Araliaceae*, *Hollböllia*,
Limonia, *Daphne*, *Ardisia*, *Myrsine*, *Symplocos*, *Rubi*, and a prodigious variety
of ferns."

"*Plectocomia* and *Musa* ascend to 7000 feet. On drier exposures bamboo
and tall grasses form the underwood. Rhododendrons appear below 6000
feet, at which elevation snow falls occasionally. From 6–12,000 feet there
is no apparent diminution of the humidity, the air being near saturation
during a great part of the year; but the decrease of temperature effects a

marked change in the vegetation. Between 6000 and 8000 feet epiphytical orchids are extremely abundant, and they do not entirely disappear till a height of 10,000 feet has been attained. Rhododendrons become abundant at 8000 feet, and from 10,000 to 14,000 feet they form in many places the mass of the shrubby vegetation. *Vaccinia*, of which there are ten species, almost all epiphytical, do not ascend so high, and are most abundant at elevations of from 5000 to 8000 feet."

"The flora of the temperate zone presents a remarkable resemblance to that of Japan, in the mountains of which island we have a very similar climate, both being cold and damp. *Helwingia, Aucuba, Stachyurus*, and *Enkianthus* may be cited as conspicuous instances of this similarity, which is the more interesting because Japan is the nearest cold damp climate to Sikkim with whose vegetation we are acquainted. At 10,000 feet (on the summit of Tonglo) yew makes its appearance, but no other conifer except those of the tropical belt is found nearer the plains than the mountain Phalút, north of Tongló, on which *Picea Webbiana* is found, at levels above 10,000 feet. *Abies Brunoniana* is first met with at 9000 feet in the Rangit valley, at Mon Lepcha, and *A. Smithiana* and *Brunoniana*, and the larch, are found everywhere in the valleys of the Lachen and Lachung rivers, above 8000 feet. The Pines are thus specifically the same as those of Bhotan, except *Pinus excelsa*, which occurs nowhere in Sikkim."

"A subtropical vegetation penetrates far into the interior of the country along the banks of the great rivers; rattans, tree-ferns, plantains, screwpines, and other tropical plants occurring in the Ratong valley, almost at the foot of Kanchinjanga, and 5000 feet above the level of the sea. With the pines, however, in the temperate zone, a very different kind of vegetation presents itself. Here those great European families which are almost entirely wanting in the outer temperate zone become common, and the flora approximates in character to that of Europe, though not to the same extent as that of the western Himalaya does. Shrubby *Leguminosae*, such as *Indigofera* and *Desmodium, Ranunculaceae (Thalictrum, Anemone, Delphinium, Aconitum*, etc.), *Umbelliferae, Caryophylleae, Labiatae*, and *Gramineae*, increase in numbers as we advance into the interior. The air becomes drier, and from the increased action of the sun the temperature does not diminish in proportion to the elevation, the summers being warmer, though the winters are colder. The forests at the same time become more open, and are spread less uniformly over the surface, the drier slopes being bare of trees, and covered with a luxuriant herbaceous vegetation. It is only in the upper part of the valley of the Tista, however, above the junction of the Lachen with the Lachung, that this change becomes marked; and from the rapidly increasing elevation not only of the surrounding mountains, but of the floors of the valleys, it proceeds with great rapidity, and the temperate soon gives place to an alpine flora."

"The subalpine zone in Sikkim scarcely begins below 13,000 feet, at which elevation a dense rhododendron scrub occupies the slopes of the mountains, filling up the valleys so as to render them impenetrable. Here the summer is short, the ground not being free of snow till the middle of

June. It is, however, comparatively dry, and the alpine flora very much resembles that of the western Himalaya and (in generic types at least) the alps of Europe and western Asia; while as we advance towards the Tibetan region we have a great increase of dryness, so that a Siberian flora is rapidly developed, which at last entirely supersedes that of the subalpine zone, and ascends above 18,000 feet."

"A small herbarium of Darjiling plants was, we believe, formed by collectors sent by GRIFFITH while in charge of the Calcutta Botanic Garden, but our knowledge of the vegetation of Sikkim is entirely derived from our own collections, which we believe to be very complete. These consist of about 2770 species of flowering plants and 150 ferns, of which the majority inhabit the temperate zone; fewer are tropical, and still fewer alpine. The prevailing natural orders are:—

Ranunculaceae	55	Gentianeae	38
Papaveraceae	25	Asclepiadeae } Apocyneae	45
Fumariaceae	16		
Magnoliaceae	7	Scrophularineae	70
Malvaceae } Bombaceae } Tiliaceae } Byttneriaceae	30	Labiatae	90
		Cyrtandreae	27
		Myrsineae	12
		Primulaceae	36
Ternstroemiaceae	11	Boragineae	18
Aurantiaceae	12	Acanthaceae	35
Caryophylleae	30	Polygoneae	45
Cruciferae	30	Euphorbiaceae	35
Vitaceae	20	Urticeae	110
Balsamineae	18	Amentaceae	15
Acerineae	6	Coniferae	10
Leguminosae	100	Laurineae	30
Rosaceae	80	Aroideae	16
Umbelliferae	50	Orchideae	150
Araliaceae	26	Scitamineae	24
Melastomaceae	10	Palmeae	10
Cucurbitaceae	20	Smilaceae } Liliaceae	40
Rubiaceae	80		
Crassulaceae	16	Junceae	25
Compositae	170	Gramineae	180
Ericeae } Vaccinieae	60	Cyperaceae	106"

(pp. 178–83)

Sketch of the Flora of British India:— HOOKER's masterly essay "A sketch of the flora of British India", written as a chapter in the descriptive volume of "The Indian Empire" in the Imperial Gazetteer of India (new ed. Oxford, 1907) and published separately in 1904 (London: Eyre and Spottiswoode) has not yet been replaced by a more up-to-date summary of the flora of a vast area. The title needs some explanation. Not only are the modern Dominions of India and Pakistan included but also Nepal, Bhutan, Ceylon, Burma, and the Malay Peninsula. The Kuram Valley and British Baluchistan are treated in appendices. On the other hand, only the vascular plants are considered. HOOKER states that the flora of this large slice of Asia "is more varied than that of any other country of equal area in the Eastern hemisphere, if not in the globe". It is, however, doubtful if "India" be richer in a number of genera and species than any other area of equal dimensions and "it is certainly far poorer in endemic genera and species than many others" (p. 2).

The principal elements are Malayan (dominant), European-Oriental, African, Tibetan-Siberian (all but confined to the Himalaya), and Sino-Japanese. No natural order (family) of vascular plants is peculiar to "India". The endemic genera are few and mostly local and restricted to few species. "It may hence be affirmed that in a large sense there is no Indian Flora proper" (p. 3).

"The British Indian Flora, though so various as to its elements, presents few anomalies in a phytogeographic point of view. The most remarkable instances of such anomalies are the presence in it of one or a few species of what are very large and all but endemic genera in Australia, namely *Baeckia*, *Leptospermum*, *Melaleuca*, *Leucopogon*, *Stylidium*, *Helicia* and *Casuarina*. Others are *Oxybaphus himalaicus*, the solitary extra-American species of the genus; *Pyrularia edulis*, the only congeners of which are a Javan and a North American; and *Vogelia*, which is limited to three species, an Indian, South African and Socotran. Of absentee Natural Orders of the Old World, the most notable are *Myoporineae*, which though mainly Australian, has Chinese, Japanese, and Mascarene species; *Empetraceae*, one species of which girdles the globe in the north temperate hemisphere and re-appears in Chili (the rarity of bog land in the Himalaya must be the cause of its absence); and *Cistineae*, an Order containing upwards of 100 European and Oriental species, of which one only (a Persian) reaches independent Baluchistan. The absence of any indigenous Lime (*Tilia*) or Beech (*Fagus*) or Chestnut (*Castanea*) in the temperate Himalaya is remarkable, all three being European, Oriental, and Japanese genera" (p. 3).

There are in "India" few assemblages of species of peculiar or conspicuous plants giving a character to the landscape, except the Rhododendron belt in the high Eastern Himalaya. Indigenous palms are comparatively few. Bamboos are important, whether as clumps growing in the open, or forming in association all but impenetrable jungles; the taller kinds monopolise large areas in the hot lower regions, and the smaller clothe mountain slopes up to 10,000 feet in the Himalaya. Tree-ferns frequent the deepest forests of the Eastern Himalaya, Burma, Malabar, the Malay Peninsula, and Ceylon.

"Of shrubs that form a feature in the landscape from their gregarious habits, the most conspicuous examples are the Rhododendrons of the temperate regions of the Himalaya, and the genus *Strobilanthes* in the Western hills of the Peninsula; many species of the latter genus do not flower till they have arrived at a certain period of growth, and then, after simultaneously flowering, seed profusely and die. Some Bamboos, also gregarious, display the same habit, which they retain under cultivation in Europe. Three local, all but stemless Palms are eminently gregarious; *Phoenix farinifera* of the Coromandel coast, *Nannorhops Ritchieana* of extreme North West India, and *Nipa fructicans* of the Sundarbans. Amongst the herbaceous plants the beautiful genus *Impatiens* takes the first place, from abounding in all humid districts except the Malay Pensinula, and from its numerous species being (with hardly an exception) endemic; added to which is the fact that, though profuse in individuals, the species are remarkably local, those of the Eastern Himalaya differing from those of the Western, these again from the Burmese, and all from those of the Western Peninsula and Ceylon; and most of these last from one another" (pp. 4–5).

HOOKER estimates the number of species of flowering plants in "India" at approaching 17,000, in 176 families (natural orders), and the vascular cryptogams at 600 species.

"The ten dominant Orders of Flowering Plants in all British India are in numerical sequence:—

1. Orchideae.	6. Acanthaceae.
2. Leguminosae.	7. Compositae.
3. Gramineae.	8. Cyperaceae.
4. Rubiaceae.	9. Labiatae.
5. Euphorbiaceae.	10. Urticaceae."

"The proportion of Monocotyledons to Dictotyledons is approximately as 1 to 2.3; of genera to species as about 1 to 7. Of Palms there are more than 220 recorded species; of Bamboos, 120; of Conifers, only 22; of Cycadeae, 5. Of genera with 100 or more species there are 10, of which 4 are Orchids, headed by 200 of *Dendrobium*; the others are *Impatiens, Eugenia, Pedicularis, Strobilanthes, Ficus, Bulbophyllum, Eria, Habenaria*, and *Carex*."

"British India is primarily divisible into three Botanical areas or regions, a Himalayan, an Eastern and a Western. The two latter are roughly limited by a line drawn meridionally from the Himalaya to the Bay of Bengal. The prominent characters of the three are, that the Himalayan presents a rich tropical, temperate and alpine Flora, with forests of Conifers, many Oaks, and a profusion of Orchids; the Eastern has no alpine Flora, a very restricted temperate one, few Conifers, many Oaks and Palms, and a great preponderance of Orchids; the Western has only one (very local) Conifer, no Oaks, few Palms, and comparatively few Orchids. Further, the Himalayan Flora abounds in European genera; the Eastern in Chinese and Malayan; the

Western in European, Oriental and African. These three Botanical regions or areas are divisible into nine Botanical Provinces, for the determination of which I have, after long deliberation, resorted to the number of species of the ten largest Natural Orders in each Province as the leading exponent of their botanical differences" (p. 7).

A brief summary of the more important features recorded for his nine provinces may be given as follows.

I. THE EASTERN HIMALAYAN PROVINCE

The only portion of this province botanically at all well known in 1904 was Sikkim, explored by HOOKER himself. Sikkim is the most humid district in the whole range of the Himalaya. *Orchidaceae* is the largest family. Three zones are distinguished. The tropical zone extends upwards to 6,500 feet and has (or had) at its base the unhealthy Terai covered with a loose forest especially of leguminous trees and Sal (*Shorea robusta*); a rich undergrowth of shrubs, coarse grasses, and the herbs of the Gangetic Plain. The foot-hills and spurs of the tropical zone are (or were) clothed with a dense forest, Malayan in general character.

"There are in this zone about 850 trees and shrubs, many of them timber-trees, amongst the most conspicuous of which are species of *Magnoliaceae, Anonaceae, Guttiferae, Sterculiaceae, Tiliaceae, Meliaceae, Sapindaceae, Anacardiaceae, Leguminosae, Combretaceae, Myrtaceae, Lythraceae, Rubiaceae, Bignoniaceae, Laurineae, Euphorbiaceae, Urticaceae,* and *Myristicaceae*."

"There are four species of Oak, as many of *Castanopsis* (a genus allied to *Castanea*), a Poplar, a Willow (*Salix tetrasperma*), one Pine (*P. longifolia*), a *Cycas* and two species of *Musa*, 18 indigenous Palms, a dwarf and a tall *Pandanus*, and 12 arborescent or frutescent Bamboos. Ferns abound. Of shrubs *Acanthaceae* and *Melastomaceae*, together with others of the above arboreous Orders, are amongst the most frequent. Of climbers there are many species of *Ampelideae, Cucurbitaceae, Convolvulaceae, Apocyneae, Asclepiadeae, Smilax, Dioscorea,* and *Aroideae*. Herbaceous plants are well represented by *Malvaceae*, Balsams, Orchids and *Scitamineae*, together with many species of other ubiquitous tropical Indian Orders" (pp. 12–3).

Secondly, there is the temperate zone of Sikkim, from 6,500 to 11,500 feet. It is roughly divisible into a lower non-coniferous and an upper coniferous and Rhododendron belt. The orchids in this zone are strongly Malayan in character, but the other important families are mostly European, Central Asian, Japanese, or Chinese.

"The most conspicuous trees are *Magnoliaceae* (five species), of which one, *Magnolia Campbellii*, before the destruction of the forests, clothed the slopes around Darjeeling, starring them in spring, when still leafless, with its magnificent flowers. Other conspicuous trees of this region are Oaks, Laurels, Maples, Birches, Alder, *Bucklandia*, *Pyrus* and Conifers. Of these the Conifers are chiefly confined to a belt from 9,000 to 12,000 feet in elevation. The monarch and most common of them is a Silver fir (*Abies Webbiana*), which is also the most gregarious; others are the English Yew, a Spruce (*Picea Morinda*), a Larch (*Larix Griffithii*, the only deciduous Conifer in the Himalaya), the weeping *Tsuga Brunoniana*, and two species of Juniper, both of which, in dwarf forms, ascend high in the Alpine zone. The absence of any true Pine or Cypress in the forests of this region of the Himalaya is notable, in contrast with similar elevations in the Western Himalaya. Of shrubs the most conspicuous are the Rhododendrons (25 species), which abound between 9,000 and 12,000 feet elevation, some of them forming impenetrable thickets; a few of these are arboreous, though never attaining any considerable height. Other shrubs are species of *Clematis, Ternstroemiaceae, Berberideae, Ilex, Rosa, Rubus, Cotoneaster, Spiraea, Hydrangea, Aucuba, Lonicera, Leycesteria, Osmanthus, Osbeckia, Luculia, Buddleia, Vacciniaceae* (some epiphytic), *Ericaceae*, Elder, *Viburnum, Polygonum*, Ivy, &c. Beautiful herbaceous plants abound — Anemones, Aconites, Violets, many species of Balsam, *Potentilla, Fragaria, Gentianeae, Campanulaceae, Gesneraceae, Scrophularineae, Orchideae* (*Coelogyne*, 8 species), *Cypripedium, Polygonatum, Smilacina, Lilium, Fritillaria, Arisaema*. Only two Palms inhabit this zone, a scandent rattan (*Plectocomia himalaica*), and very rarely a Fan-palm (*Trachycarpus Martiana*). Dwarf bamboos, of which there are six species, abound, some of them forming impervious thickets infested with leeches and large ticks. Ferns are characteristic of this zone" (pp. 13–4).

Thirdly, the alpine zone of Sikkim has a lower or outer humid and an upper or inner dry Tibetan climate and conforming vegetation. *Compositae* make the largest family.

"The largest genera are *Pedicularis, Primula, Corydalis* and *Saxifraga*. The low position of *Cyperaceae* and *Gramineae* in the decad is notable, remarkably so in contrast to the Western Himalayan decad; but future herborizations may bring them up higher. The few trees to be found on the lower skirts only of this zone are scattered Birches and Pyri. The principal bushes are Rhododendrons (of which several species attain 14,000 feet elevation, and three dwarf ones 16,000 feet), two junipers, species of *Ephedra, Berberis, Lonicera, Caragana, Rosa, Cotoneaster, Spiraea* and dwarf Willows. Of ferns there are very few. About 30 species reach 18,000 feet elevation, some of them a little higher. The highest recorded plant is a *Festuca* (not found in flower) at about 18,300 feet. In drier valleys above 15,000 feet elevation several species of *Arenaria* occur, which form hard, hemispheric or globose white balls and are a characteristic feature in the desolate landscape. By far the most striking plants of this zone are the species of *Meconopsis, Rheum nobile*, the Edelweiss, many Primulas, *Tanacetum gossypinum, Saussurea obvallata* and *gossypifera*, and the odorous *Rhododendron Anthopogon*" (pp. 14–5).

II. THE WESTERN HIMALAYAN PROVINCE

The largest family is *Gramineae* and *Orchidaceae* takes seventh place. Three altitudinal zones are again recognized. The tropical zone is perhaps 1000 feet lower at its upper limit than in the eastern Himalaya. Epiphytic orchids are few.

"Among the most interesting tropical and sub-tropical trees and shrubs of the Western Himalaya that are absent in the Eastern are *Cocculus laurifolius*, *Boswellia thurifera*, *Pistacia integerrima*, the Pomegranate and Oleander, *Roylea elegans*, *Engelhardtia Colebrookeana*, and *Holoptelea integrifolia*, most of which are also Oriental" (p. 17).

The temperate zone of the western Himalaya, in Kumaun and Gharwal is estimated as 1000 feet lower than in Sikkim. The vegetation of the outer ranges is in character on the whole like that of Sikkim but there is a greatly increased number of European genera.

"All the Conifers of the Eastern Himalya are present, except the Larch, and to these are added forests of Deodar, *Pinus longifolia*, (a comparatively rare plant in Sikkim and there only tropical) *Abies Pindrow*, *Cupressus torulosa*, *Juniperus macropoda* and (in dry regions) *Pinus Gerardiana*; the first and three last giving, where found, a character to the landscape. Of Oaks there are six, four of them Sikkim species, two only being western, of which one is the European Holm Oak (*Quercus Ilex*), which extends eastward to Kumaun at 3,000 to 8,500 feet elevation, and westward to Spain. The Indian Horse-Chestnut, *Æsculus indica*, represents the eastern *Æ. punduana*. Two Birches and two Hornbeams are common to both regions, but the eastern nut, *Corylus ferox*, is replaced by the Oriental *C. Colurna*."

"Amongst other shrubs and small trees peculiar to the Western Himalaya are the Indian Bladder-nut and Lilac (*Staphylea Emodi* and *Syringa Emodi*), *Rosa Webbiana*, *moschata* and *Eglanteria*, *Parrottia Jacquemontiana*, the Mountain Ash (*Pyrus Aucuparia*), the Bullace (*Prunus insititia*), and the common Hawthorn. On the other hand, the most striking difference between the temperate Floras of Eastern and Western Himalaya is the paucity of species of *Rhododendron* in the latter, where only four are found, all common in Sikkim *R. Anthopogon*, *barbatum*, *campanulatum*, and *arboreum*; the latter also inhabits the mountains of southern Malabar, Burma and Ceylon. Of European herbaceous plants there occur several hundreds unknown in Sikkim, as *Nymphaea alba*, *Lythrum Salicaria*, *Caltha palustris*, *Ranunculus aquatilis*, and *R. Lingua*, which all occur in the lake of Kashmir; also species of *Aquilegia* and *Paeonia*, *Parnassia palustris*, *Adoxa Moschatellina*, *Polemonium caeruleum*, *Eriophorum vaginatum*, and many Grasses, Rushes and *Carices*. The genus *Impatiens* abounds in the temperate zone, at all elevations, except the highest, the species being with few exceptions endemic. The Orchids of this zone are almost uniformly terrestrial; they include several European species unknown in the Eastern Himalaya, as *Corallorhiza innata*, *Epipogum aphyllum*,

Listera ovata and *cordata*, *Orchis latifolia*, and *Habenaria viridis*. The only Palm is *Trachycarpus Martiana*, confined to and rare in Kumaun and Garhwal, but also a Sikkim plant. Of Bamboos there are four, all dwarf and gregarious" (pp. 18–9).

The alpine zone of the western Himalaya is assigned the lower limit of 11–12,000 feet in the outer ranges and 18,000 feet is accepted as the normal upper limit of flowering plants. The western alpine flora is much richer than the eastern, in correlation with the greater area it covers. The increment consists mainly of small European genera, and not in additions to the large genera common to both floras.

"Tibetan valleys of the Western Himalaya.— Ascending the Indus river a few tropical plants extend up to Gilgit (alt. 4–5,000 feet). At greater elevations the full effects are experienced of a dry climate, great cold alternating with fierce sun-heat, and consequent aridity. Between 12,000 and 14,000 feet the principal indigenous trees are *Populus euphratica* and *P. balsamifera*, and of shrubs or small trees *Ulmus parvifolia* and species of *Tamarix*, *Caragana*, *Rosa*, *Lonicera*, *Hippophae*, *Myricaria*, *Elaeagnus*, and *Salix*. The cultivated trees are the fruit-bearing European ones, with *Populus alba* and *P. nigra*. Above 14,000 feet and up to 18,000, is a region of alpine perennials of European, Oriental, and Central Asian Orders and Genera, as *Fumariaceae*, *Leguminosae* (*Astragalus* especially), *Compositae*, *Labiatae* and *Stipaceae*. The only Orchids are a few species of *Orchis* and *Herminium*. Above 17,000 feet, 25 genera are recorded, all (except *Biebersteinia*) European, and many of them British, as *Potentilla Sibbaldi* and *anserina*, *Saxifraga cernua*, *Lloydia serotina*. The most typical plant of this region is *Arenaria rupicola*, which forms hard white cushions or balls a foot in diameter, apparently the growth of centuries. The genera *Astragalus*, *Saussurea*, *Artemisia*, *Tanacetum*, and *Allardia* have many endemic species at these elevations. Where saline soils occur *Chenopodiaceae* abound, with two endemic Crucifers (*Dilophia* and *Christolea*), also *Sonchus maritimus*, *Glaux maritima* and *Triglochin maritimum*. The Fresh-water plants of this region include *Ranunculus aquatilis*, *Hippuris vulgaris*, *Limosella aquatica*, and species of *Utricularia*, *Potamogeton* and *Zannichellia*. Ferns are all but absent" (pp. 20–1).

III. THE INDUS-PLAIN PROVINCE

"Whether proceeding across this Province in a S.W. direction from the Himalaya to Sind, or in a S.E. from the Afghan border to Western Rajputana, vegetation rapidly diminishes, approaching extinction in the Indian desert. Over the whole province a low, chiefly herbaceous, vegetation of plants common to most parts of India, mixed with Oriental, African and European types, is spread, with thickets of shrubs and a few trees, the latter most luxuriant along the banks of the rivers. With few exceptions all are leafless and the herbaceous species dry up in the hot season" (p. 21).

The chief arboreous vegetation consists of isolated groups of trees on the outskirts of the Western Himalayan Province, on the banks of rivers, and on the western flank of the Aravalli hills. Palms and ferns are rare.

IV. THE GANGETIC PLAIN PROVINCE

This is divided by HOOKER into three sub-provinces: the upper Gangetic plain, Bengal proper, and the Sundarbans.

The indigenous vegetation of the upper part of the first sub-province is that of a dry country, the trees in the dry season being leafless (for the most part) and the grasses and other herbs burnt up, but to the eastward by far the greater part of the land is under cultivation. In the extreme west, the flora is continuous with that of the Indus Plain. The principal forest is that of Ajmir. "Considerable areas of this Sub-province are occupied by the Usar, or Reh-lands, which, being impregnated with alkalis, converted into swamps in the rainy season and into deserts in the dry, are as unfavourable to a native as they are to an introduced vegetation. *Salvadora persica* is said to be the only tree that will succeed on the most saline of them, and of herbaceous plants a few perennial-rooted grasses are the only ones which thrive" (p. 25).

Bengal proper is humid and has a luxuriant evergreen vegetation. There are many kinds of trees, both indigenous and introduced. In the east, the overflow of the great rivers during the rains provides conditions for many kinds of marsh plants.

"*The Sundarbans*:— The estuarial Floras of India are notable, inasmuch as that, considering the limited areas they occupy, they contain more local species than do any other botanical regions in India. This is due to the saline properties of their waters, and to tidal action on the land. The islets of the Sundarbans are in great part clothed with a dense evergreen forest of trees and shrubs, amongst which the various Mangroves hold the first place, with an undergrowth of climbers and herbaceous plants, together with *Typhaceae*, *Gramineae*, and *Cyperaceae*. Two gregarious Palms form conspicuous features, the stemless *Nipa fruticans* in the swamps and river banks with leaves 30 feet long, and the elegant *Phoenix paludosa* in drier localities; as do the cultivated Coco-nut and Betel-nut palms. The principal exceptions to these forest-clad tracts are the sand hills occurring at intervals along the coasts facing the sea, the vegetation of which differs from that of the inland muddy islets and grassy savannahs which become more frequent in advancing eastward towards the mouth of the Megna" (p. 26).

"A most remarkable character of the estuarian vegetation is the habit of several of the endemic species to send up from their subterranean roots a multitude of aerial root-suckers, in some cases several feet long, which act as respiratory organs. These suckers occur in species of *Avicennia*, *Carapa*, *Heritiera*, *Amoora*, *Sonneratia*, and in *Phoenix paludosa*" (p. 28).

V. MALABAR PROVINCE

This is almost throughout a hilly or mountainous country and is (except in the north) of excessive humidity.

"Its abrupt western face is clothed with a luxuriant forest vegetation of Malayan type, except towards the north, where, with the drier climate, the elements of the Deccan and Indus-Plain Floras compete with that of Malabar. The eastern face slopes gradually into the elevated plateau of the Deccan, but it is varied by many spurs being thrown off which extend far to the eastward, often as above stated enclosing valleys with a Malabar flora" (pp. 30–1).

"The Nilgiri Hills form a noeud of the Western Ghats, where they attain their greatest elevation, viz. 8,760 feet. They rise precipitously from the west to extensive grassy downs and table lands seamed with densely-wooded gorges (Sholas). These grassy downs possess in parts a rich shrubby and herbaceous Flora. Amongst the shrubs some of the most characteristic are *Strobilanthes Kunthianus, Berberis aristata, Hypericum mysorense*, many *Leguminosae*, as the common Gorse (introduced), *Sophora glauca* and *Crotalaria formosa, Rhododendron arboreum*, species of *Rubus, Osbeckia, Myrtaceae, Hedyotis, Helichrysum, Gaultheria*. Amongst the herbaceous plants, are species of *Senecio, Anaphalis, Ceropegia, Pedicularis* and *Cyanotis*. Most conspicuous of all is the tall *Lobelia excelsa*."

"But the richest assemblage is found in the Sholas which, commencing at about 5,000 feet, ascend to the summits of the range. They are filled with an evergeen forest of tall, usually roundheaded trees with a rich undergrowth. Of the trees, some of the most conspicuous are *Michelia nilagirica, Ternstroemia japonica, Gordonia obtusa*, species of *Ilex, Meliosma, Microtropis, Euonymus, Photinia, Viburnum bebanthum, Eugenia* (three species), and several of *Symplocos, Glochidion, Araliaceae*, and *Laurineae*" (p. 32).

"Peat bogs, which are of the rarest occurrence in India, are found in depressions of the Nilgiri Hills at about 7,000 feet elevation. Their chief constituents are the debris of grasses, sedges, mosses and rushes. The curious *Hedyotis verticillata*, found elsewhere only in Ceylon, is characteristic of these bogs, whose surface is covered with a herbaceous vegetation of species of *Utricularia, Scrophularineae, Eriocaulon, Xyris, Exacum, Commelynaceae, Lysimachia*, etc." (p. 33).

The Laccadive Archipelago, of coral islets fringed with coconut palms has a Malayan vegetation with no endemic species. The trees have mostly been introduced by man as have also many of the herbs.

VI. THE DECCAN PROVINCE

Sub-provinces can only be rather vaguely indicated though the plateau is distinguished from the low coast land of Coromandel. Deciduous forests are the most conspicuous features on the plateau, and comparatively evergreen ones on the coasts and slopes with an eastern aspect.

"Much of the open country presents a jungle of small trees and shrubs, together with a herbaceous vegetation which is leafless or burnt up in the dry season. In the large river valleys and those of the higher hills, types of the Malabar Flora penetrate far to the east. Of forest trees there are several hundred species, amongst which *Sterculiaceae, Meliaceae, Leguminosae, Combretaceae, Bignoniaceae,* and *Urticaceae*" (p. 35).

Acanthaceae are notable amongst the herbs. The rich black cotton soil that prevails over large areas in the Deccan has a peculiar assemblage of the indigenous plants of the Province.

The Coromandel sub-province has a vegetation mainly of the Deccan type with a reduced number of species. Thickets of thorny evergreens and deciduous trees and shrubs abound, belonging especially to the genera *Flacourtia, Randia, Scutia, Diospyros, Mimusops, Garcinia, Sapindus, Pterospermum,* etc.

VII.   THE CEYLON PROVINCE

The botanical features of Ceylon coincide with its physical, the moist mountainous southern and south-western districts having a flora of the Malabar type, and the hot dry northern districts one of the Coromandel type. It differs from the Malabar flora in having many more Malayan types.

"Ceylon possesses no fewer than 23 endemic genera, of which 10, comprising 46 species (all but two endemic), belong to the typical Malayan Order of *Dipterocarpeae,* which is represented by only 12 species in the Peninsula. The principal Orders containing very many endemic species are *Orchideae* 74, *Rubiaceae* 72, *Euphorbiaceae* 53, *Melastomaceae* 38, and *Myrtaceae* 26; and of genera *Strobilanthes, Eugenia, Memecylon, Phyllanthus,* and *Hedyotis.* The genus *Impatiens* abounds; upwards of 21 species are recorded, nearly all of them endemic. Of other conspicuous Orders, Orchideae contains 160 species, more than half of them endemic. Of Palms eight genera are endemic, and there are 18 endemic species, exclusive of the introduced Betel-nut, *Borassus* and Coco-nut. The Talipot (*Corypha umbraculifera*) is one of the most imposing of the Order. The *Nipa* occurs rarely, Ceylon being its western limit, Australia its eastern. *Cycas circinalis* is common in the forests. Of *Bambuseae* there are five genera and 10 species (of which four are endemic). At elevations above 6,000 feet a few temperate northern genera appear, fewer than might have been expected in mountains that attain heights of upwards of 7,000 and 8,000 feet. Of these genera, *Agrimonia, Crawfurdia,* and *Poterium* are not Pensinsular. The following are also Peninsular: *Anemone, Thalictrum, Berberis, Cardamine, Viola, Cerastium, Geranium, Rubus, Potentilla, Alchemilla, Sanicula, Pimpinella, Peucedanum, Galium, Valeriana, Dipsacus, Artemisia, Vaccinium, Gaultheria, Rhododendron, Gentiana, Swertia, Calamintha, Teucrium, Allium.* Of Peninsular temperate genera that are absent in Ceylon, *Fragaria* and *Rosa* are two the occurrence of which might have been expected, both being Nilgiri genera."

"Remarkable features in the vegetation of Ceylon are the *Patanas*, grass- or shrub-covered stretches of country, most prevalent in the south-east of the island, from the sea to 5,000 feet altitude. They are partly natural and partly due to the destruction of the forests. A peculiar, endemic, pale green Bamboo covers hundreds of square miles of these Patanas, *Ochlandra stridula*, so called from the crackling noise caused by treading on the broken stems. In grassy places *Imperata arundinacea* prevails; and in scrub-forests such tropical trees occur as *Pterocarpus Marsupium*, *Careya arborea*, *Phyllanthus Emblica*, *Terminalia Belerica* and *T. Chebula*. At higher levels *Rhododendron arboreum* appears. In moist districts a fern (*Gleichenia linearis*) occupies the ground" (pp. 39–40).

The Maldive Archipelago of coral islets has few flowering plants other than coconut palms, littoral shrubs, and weeds of cultivation.

VIII. THE BURMESE PROVINCE

Burma is botanically by far the richest of the Provinces distinguished here by HOOKER but, to him, was the least known. KURZ's classification of the forests is outlined (KURZ 1877: Forest Flora of British Burma, Calcutta). Four divisions of the Province are made: Northern, Western, Eastern, and Central.

"Northern Burma" is a mountainous country, belonging politically to Assam. The climate is of maximum humidity, and the vegetation approximates to that of the Eastern Himalaya, differing conspicuously in the absence of an alpine zone and of any species of *Picea*, *Abies*, *Tsuga*, *Larix*, or *Juniperus*, and in the presence of *Pinus khasya* and *Nepenthes*. From that of Central and Southern Burma it differs in the absence of teak, in the paucity of Dipterocarps, in the presence of sal which, in the Garo Hills, finds its eastern limit, and of *Pinus khasya* which is replaced farther south by *P. merkusii*. In the valleys and lower elevations the vegetation of the tropical zone of the Himalaya prevails, but at elevations above 4000 feet temperate genera and species mainly replace them, many of them identical with the Himalayan, though maintaining a lower level by 3000 feet or more. In the western districts, a conspicuous feature is the open unforested character of elevations above 4000 feet. *Nepenthes khasiana* in the Jaintia Hills alone, growing prostrate amongst wet grass and stones at 4000 feet, marks the northern limit of the genus.

Western and Southern Burma is humid and has a vegetation of dense evergreen forest where Dipterocarps, oaks, and bamboos are conspicuous features. Ferns, scandent palms, and orchids abound.

Eastern Burma had been relatively little explored when HOOKER

wrote and his account consists of summaries of collections made by
POTTINGER (Kachin Hills), COLLETT (Shan States), PARISH, and LOBB
(Tenasserim Mountains).

Central Burma is dry in its northern and humid in its southern part.
The character of the upper Central vegetation is largely, if not typically, that of the Deccan Province. Two species of teak (*Tectona grandis*
and *T. hamiltoniana*) occur, with many leguminous trees, *Acacia catechu*
often forming forests, and more rarely *Dipterocarpus tuberculatus*.

The Andaman Islands are mostly clad with mixed evergreen and
deciduous forests which are typically Burmese. Their most remarkable
feature is the apparently total absence of *Quercus* and *Castanopsis*, of the
first of which there are 40 species on the neighbouring continent, and
of the second 11.

The Nicobar Islands are even less known botanically than the Anda-
mans, and it is questionable whether they belong to the Burmese or to
the Malay Peninsular flora.

IX. THE MALAYAN PENINSULA PROVINCE

Except when cultivation interferes, the whole Peninsula is clothed
with an evergreen vegetation, that of the shore being esturial. HOOKER's
account is meagre.

Appendices to the main essay outline the flora of the Kurum Valley
by means of a short summary of AITCHISON's investigations (Journ.
Linn. Soc. 18: 1–113, 1880 and 19: 139–200, 1881) and of British
Baluchistan, based mainly on LACE's researches (Journ. Linn. Soc.
28: 288–327, 1891).

India: more recent investigations:— HOOKER's phytogeographical studies on India can be most conveniently considered under
two main headings: (1) those concerned with his own fieldwork and (2)
those of wider scope and more general import. We will deal first with
the area covered by HOOKER in his travels and described particularly in
the Himalayan Journals.

Before setting out on his main Sikkim expeditions HOOKER visited
Bihar and crossed the Ganges Valley. An important excursion was to
Parasnath. HAINES (1921–5, p. 45) notes that now "The flora of Parasnath has been more carefully investigated than that of any other
portion of our flora." He refers to it under the heading "The Mixed
Forests of The Central Tract". ANDERSON (1863) and BISWAS (1934)
have described aspects of the flora and vegetation. The Son (Soane)
area is also included in HAINES's flora.

HOOKER's explorations in Sikkim and eastern Nepal resulted in one
of the botanically least known areas of India becoming one of the best

known. Since the publication of the "Himalayan Journals" HOOKER's results for Sikkim and eastern Nepal have been extended but not yet superseded. The following works may be consulted in proof of this statement: BANERJI (1948), BURKILL (1906, 1907, 1908, 1916), CLARKE (1876, 1885), COWAN (1929), GAMBLE (1875), GAMMIE (1894a and b), HAY (1934), LACAITA (1916), PARKER (1931), SCOTT (1874), SMITH (1913), and SMITH and CAVE (1911). For Assam, the following are references: BOR (1938b, c, 1942a and b), CLARKE (1886), DAS (1942), GAMMIE (1895), and KANJILAL, KANJILAL, DAS, DE, and BOR (1934). Two general accounts are worth noting: L. A. WADDELL's book "Among the Himalayas" (Westminster 1899), is an exceedingly read-able book of travels covering much of the area traversed by HOOKER, and a paper by D. W. FRESHFIELD, "The Sikkim Himalaya" in Scot. Geogr. Magn. 21: 173–82 (1905). Most of the above mentioned authors refer, often again and again, to HOOKER's "Himalayan Journals", most frequently to substantiate his observations. There are, however, extensions to HOOKER's pioneer work as well as some valuable generalizations. These may be illustrated by some extracts from and comments upon several of the papers.

C. B. CLARKE (1876) notes that the "dripping forest of Sikkim extends from about 5500 ft. to 9000 ft. (1676 m. to 2743 m.); above 9000 ft. the rhododendrons come in; below 5500 ft. is cultivation. This forest is specially characterized by the prevalence of oaks, magnolias, laurels, and araliads, and beneath these trees by members of the Urticaceae, and Cyrtandraceae, species of Impatiens and Rubi, and small bamboos. Noteworthy absentees are species of Leguminosae and Malvaceae". CLARKE stated that for the top of Tongló, which he visited in September, some families and genera, as Umbelliferae, Geraniaceae, Fumariaceae, Epilobium, and Potentilla were "very fairly represented", though HOOKER, who visited Tongló in May, said they were conspicuously absent. This illustrates well the need for visits at all seasons to, or better still a residential study of, an area before its flora and vegetation can be said to be properly known. The tremendous development of rhododendrons in Sikkim is commented upon by CLARKE (1886), by WADDELL, and other authors, as by HOOKER previously. Attention may be called to the remarkable picture of rhododendron trees (presumably Rhododendron arboreum) on p. 319 of WADDELL's book and one of the giant wild rhubarb (Rheum nobile), so often mentioned by HOOKER, on p. 185.

GAMMIE's papers (1894) partly summarized and partly extended HOOKER's work. It is noted that HOOKER's collection of Sikkim plants amounted to 2920 species and that the total flora of Sikkim probably numbers 4000 species (presumably of vascular plants). Owing to the humid climate of Sikkim, and the absence of excessive cold at any season of the year over the greater part of it, the prevailing vegetation is of an evergreen character, though some species are exceptional in being deciduous.

SMITH and CAVE (1911) dealt with the vegetation of the Zemu and Llonakh Valleys in the north-western corner of the Sikkim Himalaya. These valleys are difficult of access on account of rhododendron forest or high passes. HOOKER had failed to reach the upper part of these valleys. The Zemu area may be divided botanically into three regions: a temperate forest from 8–11,000 ft. (2438–3353 m.), a "subalpine" shrub region from 11–14,000 ft. (3353–4267 m.), and an "alpine" region from 14–17,000 ft. (4267–5182 m.). "Taken as a whole the valley is undoubtedly a transition from the moist prolific area such as prevails to the south and is typified by Jongri, to the dry area of Tibetan Sikkim lying beyond the Thé La. Though much poorer in species than the Jongri area it has on the whole more affinities with it than the Llonakh. The lack of epiphytic ferns, the comparative scarcity of ferns, mosses, lichens, show no approximation, however, to what we found in the dry area of Llonakh."

In another paper SMITH (1913) dealt with the high mountain vegetation of south-east Sikkim. The limit of vegetation is lower than in other parts of Sikkim. Thus at 15,000 ft. (4572 m.) in the Changu area there is a crest of bare rocks while at this altitude in west and north Sikkim there is a conspicuous and varied flora. In the south-east there is exposure to rains and the snow lies long giving a very short season for the high mountain flora. The most tenacious phanerogams on the upper rocks and sometimes the only ones were *Primula muscoides* and *Chrysosplenium carnosum*. The flora of the Chola Range between 10,000 and 15,000 ft. (3048 and 4572 m.) is very homogeneous and only in the northern area is there a gradual transition to the flora of a drier region. The short vegetative season is one of mist and rain. This in conjunction with the low temperature is conducive to the formation of an acid soil. The result is a vegetation which in many respects is xeromorphic, with rhododendrons as a striking example. In the "subalpine" tract they form a great part of a forest with a general vegetation level of 20–30 ft., becoming lower at greater altitudes and finally (at 4572 m.) making a heath of prostrate forms which do not usually arise above 2 ft. The preponderance of rhododendrons induced by the climatic factors is no doubt a reason for the comparative absence of variety in herbaceous plants. Shrubs of such genera as *Berberis*, *Pyrus*, and *Salix* are much less prominent and rarely succeed in monopolising even a small area. Yet rhododendrons are slow in recolonizing an area in which they have been destroyed. The flora of Sikkim is remarkable for the isolation of many of the species. "The deep valleys and the sterile mountain ridges are the chief cause of this. HOOKER has pointed out in his Himalayan Journals the isolation of even distinct floras such as the temperate flora of the Lachen-Lachung area. The broad belt of rhododendrons no doubt plays its part in keeping the areas distinct." The transition to the Tibetan flora as one proceeds northward is a gradual one in the Chola-Range. In October, the chief impression was the rapidity with which the alpine vegetation prepared for winter.

COWAN (1929) has dealt with the ecology of the forests of Kalimpong, east of Darjeeling. Altitude, configuration of the ground, geological formation, soil, and rainfall are the principal factors influencing the vegetation. Three vegetation zones are recognized: the tropical or lower hill zone from the plains up to 914 m. Forests which are mainly deciduous and come within SCHIMPER's monsoon forest type form the climax vegetation. Four forest associations occur in the zone. The subtropical or middle hill zone occupies the area from 914 to 1828 m. Much of the land is now cleared for cultivation. Most of the trees of the four associations recognized are evergreen. The temperate zone occurs from 1828 to 3170 m. There is here great humidity and the trees are covered with mosses and lichens. Again, four associations are recognized. COWAN (1942) deals with the rhododendrons of India, a group to which HOOKER paid so much attention in Sikkim.

BURKILL (1906) describes a journey into Nepal. He deals with the bhavàr or great forest, the pinewoods of the south face of the hills, of the top of the foothills, plants of sal forests, downs in the cultivated belt, and the temperate forest belt. Very few plants of the temperate forest belt reach the plains but a few reappear on Parasnath and in the Chittagong hills. The details given concern the vegetation between Raksál and the Himalaya of Central Nepal as far back as 35 miles from the plains and not higher than 2134 m. The outstanding result is an expression of the easternness of the vegetation. It is much more like that of the Darjeeling district than that of the N.W. Himalaya. This emphasizes HOOKER's classification of Central Nepal with Sikkim.

Turning now to recent work on the flora and vegetation of Assam, with special reference to areas visited by HOOKER, or more or less contiguous to such, we will first consider two papers by Bor (1938b, 1942). In the former there is recorded a synecological study of the area "which is loosely known as the Aka Hills in Assam" and lies against the northern frontier of the province. This is north of the area in which HOOKER collected but contains so much of interest as extending HOOKER's researches that a reference must be made to it. Corresponding to areas of different climates four formations are recognized. (1) The tropical, *Laurus-Melia, Mesua* formation has the following characters: high evergreen forest with an admixture of deciduous species where the rainfall is 200–300 cm.; the underwood and shrubby undergrowth are all evergreen; a tall impenetrable forest with cane brakes, patches of bamboo with a profuse growth of climbers; many of the dominants have plank-buttresses. (2) Montane subtropical, *Phoebe-Beilschmiedia-Engelhardtia* formation is characterized as follows: high evergreen forest in which lauraceous species predominate; the formation occupies a belt lying between 1000 and 2000 m.; the underwood and undergrowth are evergreen; very few dominants are buttressed. (3) Montane temperate, *Quercus-Rhododendron-Tsuga-Abies* formation, which is found between 2000 and 3750 m., has the characters: a high forest of immense trees with umbrageous crowns at the lower altitudes dwindling to forest scarcely 10 m. high where *Rhododendron* forms the tree growth, with patches of tall *Tsuga* and *Abies* at the higher altitudes; undergrowth is often absent or is represented by large stretches of *Arundinaria* and *Lomaria*; mosses are very conspicuous. (4) Transmontane, *Quercus-Pinus-Cupressus* formation composed of rather open high forest of which pine and oak are the most conspicuous dominants; a pure proclimax of *Cupressus* is found in a limestone habitat in the area. Various seral units are also described. Grassland covers enormous areas in the plains. It is concluded that "In the grassland of the plains we have a case of succession-retardation due to fire and the browsing of wild animals. In the plains of Assam, grassland, although covering hundreds of square miles, is not a climax community and could not persist were it not for the aid it receives from fire."

In another paper BOR (1942) dealt with the relict vegetation of the Shillong Plateau, Assam. This is within the area of the Khasi (or Khasia) Hills so fully explored botanically by HOOKER who recorded the floristic richness of the area. BOR notes that HOOKER and GRIFFITH have left behind their impressions of the scenery and that they were struck by the bareness of the country. He comments that this is quite understandable in the neighbourhood of Charrapunji where the rainfall is so heavy as to have removed completely the soil covering of the horizontally stratified sandstone. Farther to the north, however, where the rainfall is very much less, there are grassy rolling downs with patches of pinewood and, still rarer, clumps of evergreen forest. The grassland and pine forest are continually being interfered with by human activities.

The largest portion of the Shillong plateau is occupied by fields and grasslands, the latter almost, if not entirely, devoid of trees. No doubt the community is very stable indeed and GRIFFITH (in 1837) and HOOKER (in 1849) found the hills very much as they are to-day. The climate is a forest one and the annual burning, in March, April, or May, or even earlier, is the most important controlling factor in the maintenance of the grasslands. The grasses, belonging to the *Andropogoneae*, are more or less tall, coarse, and able to resist fire. They are perennials with rhizomes or thick rootstocks. The grassland reaches its full luxuriance in the autumn and the majority of the grasses flower at the end of the rains or during the cold weather. By spring the grass has completely dried up and fire raging everywhere, rapidly reduces the hills to smoking black wastes. There are numerous associates of the grasses — especially herbaceous and shrubby species of *Leguminosae* and *Melastomataceae*.

Pinus insularis (*P. khasya* Royle is now considered a synonym) covers large areas in Assam (Khasia Hills, Naga Hills, Manipur, Lushai Hills) and Burma. In the Khasia Hills it ranges from 870 m. to 1950 m. in scattered clumps of trees. Its seeds germinate well often where there is a minimum of soil. Where protected from fire the undergrowth is impenetrable owing to the multitude of evergreen species similar to those in the evergreen forest.

Small woods of evergreen species are to be seen here and there on the plateau especially in folds in the downs or on slopes with a northern exposure, though HOOKER's statement that all woods are found on northern exposures is not true. The groves are surrounded by grassland or pinewood and the line of demarcation between them is precise. The tallest trees in the dark and gloomy wood are about 20 m. high. A small tree layer can sometimes be distinguished but is not at all distinct. The shrubby and herbaceous layers are in places well marked but in others are almost non-existent. The soil is covered with mosses and *Selaginella* and is deeply stained with humus. The evergreen forest is the climatic climax since with protection from human interference, particularly fire, grassland will become colonized by *Pinus insularis* and, in due time, the pine needle covering on the floor of a pine forest will prevent development of pine seedlings. Should the forest then become more open, so that the layer of pine needles can decay more rapidly, pine seedlings do not colonize the ground but an undergrowth of evergreen species appears developing into the evergreen climatic climax.

U. N. KANJILAL (*in* Flora of Assam by U. N. KANJILAL and others, 1: vi–xi, 1934) gives "A brief ecological sketch of the botany of Assam". Four main ecological groups are recognized: evergreen forests (including pine forests), deciduous forests, swamp forests, and grasslands. It is noted that besides the pine areas there is in the Khasia Hills a very interesting type of forest known as sacred forests. These generally occupy hill-tops and cool aspects as a rule above the pine zone. They are primeval forests and till recently there was no felling in them for fear of annoying the sylvan gods. It is, indeed, to these fascinating groves that the Khasia Hills owe the reputation they enjoy of being the richest botanical area, not only in India but perhaps in the world. Predominating arboreous species belong to the *Magnoliaceae, Fagaceae,* and *Lauraceae* and the trees are often of enormous size. Other characteristic trees are: *Dendropanax japonicum, Randia wallichii, Croton laevigatus, Myrsine capitellata, Taxus baccata, Podocarpus neriifolia, Daphniphyllum himalayense, Eriobotrya bengalensis,* and many others. There are also numerous interesting shrubs, undershrubs, and herbs.

HOOKER's general review of the flora of India has not yet been superseded so far as India proper (modern India and Pakistan) is concerned but is unsatisfactory for other areas, particularly Burma and the Malay Peninsula. An exhaustive bibliography of Indian botany from 1902 to the present is not attempted here but the following references and those given above would form a basis for this with the references included in them: AGHARKAR (1920), AIYAR (1932), BAMBER (1916), BISWAS (1933), BLATTER and McCANN (1929), BOR (1938*a*), BOURDILLON (1908), BRANDIS (1906), BURKILL (1924), CHAMPION (1936), CHATTERJEE (1940), CHENGAPA (1944), COLLETT (1902), COOKE (1903–8), COVENTRY (1929), COWAN (1928, 1929*a* and *b*), DUDGEON (1920), DUDGEON and KENOYER (1925), DUTHIE (1903–29), FISCHER (1918, 1921), FYSON (1915–20, 1932), GAMBLE and FISCHER (1918–35), GORRI (1933), GRIFFITH (1946), GUPTA (1939), HAINES (1910), HOWARD (1928), KASHYAP (1925), KASHYAP and JOSHI (1936), KENOYER (1923), LESTER-GARLAND (1927), MAHABALE and KHARADI (1946), MEEBOLD (1909), MISRA (1944), MOONEY (1938, 1941, 1942*a* and *b*), OSMASTON (1922, 1927), PARKER (1918, 1942), PARKINSON (1923), PIROTTA E CORTESI (1912), PRAIN (1903*a* and *b*, 1905), PRING (1947), PURI (1947), RAMASWAMI (1914), RUTTLEDGE (1933), SAXTON (1922, 1924), SAXTON and SEDGWICK (1918), SMYTHE (1938), SMYTHIES (1921), SRIVASTAVA (1944), STEBBING (1922), TROLL (1939), TROUP (1921, 1926), WARD (1927, 1930*a* and *b*, 1931), and WRIGHT (1931).

The subdivisions of "India" accepted by HOOKER (A sketch of the Flora of British India) are obviously (as is acknowledged, *l.c.* footnote p. 3 and p. 9) based upon the paper by CLARKE *in* Journ. Linn. Soc. Bot. 34: 1–146 (1898), with the title "On the subareas of British India, illustrated by the detailed distribution of the *Cyperaceae* in that Empire" HOOKER records (Sketch, p. 9):

"The principal differences between his Sub-areas and my Provinces lies in his inclusion of Central Nepal in the Eastern Himalayan Province, and of the Afghan E. boundary mountains, all Baluchistan, S. E. Rajputana and Central India in the Indus-plain Province; in his treatment of N. and N.E.

Burma with the Assam Valley as a separate sub-area (Assam); of eastern and southern Burma as another (Pegu); and of his inclusion of all Ceylon in the Deccan Province".

The only modern attempts, known to the present writer, which in a broad general sense suggest modifications (and some of them very minor ones) of HOOKER's scheme of botanical provinces (whose boundaries, of course, do not necessarily coincide with the boundaries of political divisions) are those of CALDER (1937), CHATTERJEE (1940), and BISWAS (1943). CALDER has six main divisions: the North Western Himalaya, the Eastern Himalaya, the Indus Plain, the Gangetic Plain, the Deccan (with án eastern subprovince), and Malabar. The flora and vegetation of these are described in some detail. The provincial boundaries are very similar to those of HOOKER.

BISWAS (1943) has a map (p. 110) "showing the Phytogeographical Divisions of India with the explored, unexplored, and insufficiently explored regions." Unfortunately the printing of this map leaves much to be desired and there is no detailed explanation of it in the text. Apparently "India", including Burma but excluding Ceylon and Malaya, is divided into fifteen "phytogeographical divisions". These show considerable divergence from those proposed by earlier botanists and by CHATTERJEE.

CHATTERJEE (1940) divides India, with Burma but excluding Ceylon and Malaya, into ten main divisions: (1) Deccan; (2) Malabar; (3) Indus Plain; (4) Gangetic Plain; (5) Assam; (6) Eastern Himalaya; (7) Central Himalaya; (8) Western Himalaya; (9) Upper Burma; (10) Lower Burma. CHATTERJEE was particularly interested in problems of endemism. He deals with the dicotyledons whose genera he gives as 1831 and species as 11,124. Of these species, 4274 (38.5%) are endemic. The ranges of the endemics are: common generally in India 533 (4.9 %); Continental India 2045 (18.2 %); Himalaya with Assam 3169 (28.8 %); and Burma 1071 (9.6 %). The majority of the endemics appear to be new (young) species produced from stocks capable of change and, not having had opportunity or time for migration, are restricted to more or less narrow ranges. The non-endemics belong to three categories: (1) species chiefly tropical and subtropical of fairly wide distribution in Asia, and sometimes beyond; (2) a considerable number of species of limited distribution occurring also just beyond the boundaries of our area, e.g. in south-west China, Siam, Tibet, and Afghanistan; (3) species associated with cultivation. Details are given of endemism in dicotyledonous families.

Attention must be called to the important work of CHAMPION (1936). This deals with the classification and distribution of forest types in

India and Burma. Since forests form the natural climaxes over the greater part of India and Burma their taxonomy and ecology must be important criteria for any more modern phytogeographical subdivision of central southern Asia. CHAMPION's main concern is the forest types and their classification and not the determination of phytogeographical boundaries. His coloured map of the distribution of climatic types (diagrammatic) is, however, most instructive. About half the total area is covered by only two of the 14 types mapped: tropical dry deciduous forest and tropical thorn forest. There are patchwork concentrations of types in western peninsular India and in Burma and Assam, and long approximately parallel ranges of types in the Himalaya. The major divisions accepted are: moist tropical forests (including tropical wet evergreen forests, tropical semi-evergreen forests, tropical moist deciduous forests, subsidiary edaphic types of moist tropical forests, moist tropical seral types); dry tropical forests (including tropical dry deciduous forests, tropical thorn forests, tropical dry evergreen forests, dry tropical seral types); montane subtropical forests; (montane) temperate forests; and alpine forests (including alpine seral types, alpine scrub). A list of types occurring in every (political) province is provided. TROUP (1926) and HOWARD (1928) give useful concise accounts of forest ecology in India. RIDLEY (1942) also discusses distribution areas of the Indian flora.

The flora of Ceylon, an island dominion included as a distinct province by HOOKER (Sketch, 1904, pp. 37–40), has been better investigated than any other of the "Indian" provinces. The great flora by TRIMEN (1893–1900), with ALSTON's supplement (1931) has been the taxonomic basis for much detailed ecological research. Readers are referred to papers by ABEYESUNDERE and DE ROSAYRO (1939), ALSTON (1938), BAKER (1938), CHAPMAN (1947), HOLMES (1945), LEWIS (1920, 1926), PARKIN and PEARSON (1903), PEARSON (1899), ROSAYRO (1942, 1945–6), TANSLEY and FRITSCH (1905), WILLIS (1915, 1916), and WRIGHT (1905).

The forests and waste lands of Ceylon are described by A. F. BROUN in TRIMEN, Part. 5: 355–63 (1900). Papers by BAKER and DE ROSAYRO (1942) are modern accounts of forest ecology. Several of the papers, notably those of PEARSON, PARKIN and PEARSON, and DE ROSAYRO (1945–6) deal with the Ceylon patanas or hill grasslands. TANSLEY and FRITSCH consider in detail the flora and vegetation of the littoral.

HOOKER (Sketch, 1904, p. 40) notes that "Burma is botanically by far the richest Province of British India, and at the same time, as such, the least known". More modern researches confirm the floristic richness of the now independent state of Burma and, in addition, emphasize the great phytogeographical interest particularly of the northern parts. Reference should be made to the following works: ATKINSON (1948), BARRINGTON (1931), BISWAS (1932), FISCHER (1938), GAGE (1904), HANDEL-MAZZETTI (1927), MERRILL (1941), MOREHEAD (1944), STAMP (1923, 1924, 1925a and b), TROUP (1911), WARD (1921, 1923, 1924a, b and c, 1930, 1932a and b, 1933, 1941, 1942, 1944-5, 1946a, b, c, and d). The best general account of the vegetation of Burma is probably that by STAMP (1925). The following types of vegetation are recognized and described: I. Mountain vegetation (above 3000 ft. = 914 m., or the frost line), with oak forest, pine forest, rhododendron forest, bracken brake, mountain bamboo brake, and mountain grassland; II. Lowland vegetation (below 3000 ft.), with evergreen dipterocarp forest, wet evergreen forest (northern type), mixed cane brake, bamboo brake, pyinkado or semi-evergreen forest, moist teak forest, dry teak forest, pyinma or plains forest, semi-indaing, indaing (dry dipterocarp forest), *Diospyros* forest, dry deciduous forest without teak, than-dahat forest (*Tectona-Terminalia* forest), sha-dahat thorn forest, sha thorn scrub, *Zizyphus* thorn scrub, *Euphorbia* semi-desert, valley swamp forest, kanazo forest (*Heritiera* forest), mixed delta scrub, true mangrove forest, beach or dune forests, salt-marsh vegetation, freshwater swamp vegetation, and lake vegetation; III. seral communities, with kaing grassland (sandbanks), *Combretum* hedgerow community, riverside and village parkland, and ponzos (fallow returning to jungle). It would appear that some of the communities classified under II. above are also seral communities. In MOREHEAD's (1944) useful summary of the forests of Burma the classification adopted is into: tidal forests, beach and dune forests, swamp forests, tropical evergreen forests, mixed deciduous forests, moist upper mixed deciduous forest, dry upper mixed deciduous forests, lower mixed deciduous forests, dry forest (with subtypes thandahat and thorn forests), deciduous dipterocarp or indaing forests, and subtropical and temperate evergreen forests.

KINGDON WARD's numerous books and papers indicate the extent of his exploration of northern Burma and contiguous areas. He has shown that three distinct floral regions have contributed to make up the bulk of the flora of northern Burma. These are: Indo-Malayan from the south, Eastern Asiatic from the east, and Sino-Himalayan from the north and northwest. There are also a certain number of species which might be regarded as north temperate, or cosmopolitan, others pantropical, and some endemic (so far as is known). This great mixture of phytogeographical types is connected with the great range in altitude, the orientation of the mountain chains, and the geological history. Especially important has been the effects of the oncoming of the Ice Age and the glacial and interglacial periods. A very full account of the botany and geography of North Burma is given by WARD (1944). The probable history of the flora is well sketched by WARD in an earlier paper (1932*a*). Any extensive eastward migration could only have occurred after the Oligocene-Miocene Himalayan uplift. A much later event, the Ice Age, has not received the attention it deserves. During the last maximum advance of the ice, a vast area in the Himalayan and Sino-Malayan mountains, as well as in western China and Tibet, was covered by ice, and the effects were tremendous. The ice-sheet of the last great fluctuation had a two-fold effect: (*1*) it isolated the entire Indo-Malayan region from the Central Assam, and likewise cut off the East Asiatic region from the Central Asian, though it did not prevent a mingling of East Asiatic and Indo-Malayan forms in the southeast; (*2*) it extinguished or drove south the alpine flora which had already established itself. The southern drive of the flora must have had a telescoping effect, crowding the alpine on to the subalpine, the subalpine on to the warm temperate, and so forth. On the return of the flora there was plenty of room; plants did not compete with one another so much as with external conditions.

The last of HOOKER's provinces is the Malayan Peninsula Province. He deals with this rather briefly. The following references will serve as a guide to recent researches; BURKILL and HOLTTUM (1923), CORNER (1940), CUBITT (1920, 1924), FOXWORTHY (1910), HENDERSON (1930, 1939), HOLTTUM (1924), MEAD (1912), RIDLEY (1900, 1901, 1906, 1909, 1910, 1922–25), WATSON (1928). CUBITT divides the forests into the following types: A. Littoral forests, including the mangrove swamps and dry forests (*Casuarina* belts and other coast forests). B. Inland forests, including freshwater swamp forests and dry forests (lowland forests up to about 2000 ft. = 610 m. and high hill forests over 2000 ft.). No recent complete and detailed phytogeographical account of the Malay Peninsula is known to the present writer. RIDLEY, in the Introduction to his Flora of the Malay Peninsula, p. xi, says that the area of the Peninsula falls botanically into two divisions lying north and south respectively of the mouth of the Kedah River. The northern division has a flora with close relations with that of Tenasserim and Mergui, upwards of forty genera being represented in this division but not in the southern one. The greatest part of the southern division is or was at one time dense forest of large trees of tropical rain forest type up to about 3000 ft. = 914 m.

Chapter V

AFRICA

Morocco:— In 1871, HOOKER visited Morocco in the company of JOHN BALL and GEORGE MAW. The tour in Morocco lasted from early April to early June. A full account of their journey was published under the title "Journal of a tour in Marocco and the Great Atlas", London, 1878, under the names of HOOKER and BALL. The general use of the plural pronoun "we" (and also "our") throughout the main part of the book indicates clearly that (apart from the Appendices) joint authorship is acknowledged. It is, however, somewhat difficult to determine exactly what this implies. In the Preface to the work it is stated: "Sir JOSEPH D. HOOKER, who made careful notes throughout the journey, hoped to complete the work without delay, and actually wrote the greater part of the first two chapters; but the constant demands upon his time ... ... so far interfered with the completion of the original design as to compel him to request his fellow-traveller, Mr. BALL, to undertake the completion of the work." In L. HUXLEY: "The Life and letters of Sir JOSEPH DALTON HOOKER" 2: 95 (1918) we read: "the book was ultimately brought out by BALL in 1878, using HOOKER's journal and fragment of narrative as well as his own and MAW's journal." Three of the appendices (two of them phytogeographical) are by HOOKER alone. BALL was the sole author of the "Spicilegium Florae Maroccanae", published in the Journ. Linn. Soc. 16: 281–742 (1877–8), which is based mainly on the collections made during this expedition, and of the new species described in Journ. Bot. 11: 267–73, 296–307, 332–5, 364–74(1873). On the other hand HOOKER published accounts of "The ascent of the Great Atlas" in Proc. Roy. Geogr. Soc. 15: 212–21 (1871) and Rep. Brit. Assoc. 1871, 179–80.

With these facts and uncertainties in mind it has been decided that quotations should here be limited to three from the "Journal" proper and others from contributions that were certainly HOOKER's alone.

"After standing the fire of some harmless 'chaff' from the Jew and Moorish boys that loitered about the city gate, we soon got clear of the enclosures near the town, and descended through cultivated land into a little grassy valley that lies below the hilly range of the Djebel Kebir. Bright spring annuals—blue and yellow lupen, crimson Adonis, a deep orange marigold (*Calendula suffruticosa*), blue pimpernel, and other less conspicuous flowers—enlivened the tillage ground; but the northern botanist is more struck by the perennial species that hold their ground on the large portion of the soil which the plough has not touched. Predominant among these, as elsewhere throughout a large part of the Mediterranean region, is the palmetto, or dwarf palm (*Chamaerops humilis*). Where unmolested by animals, and protected from the periodic fires that the native herdsmen renew for the sake of getting herbage for their cattle, it forms a thick trunk, ten or twelve feet in height, which probably takes a long time to attain its full size; but in the open places it is commonly stemless, and covers the ground with its radiating tufts of stiff fan-shaped leaves. Many plants of the lily tribe abound; but in this mild climate most of them had flowered in winter, and few now showed more than their tufts of large root-leaves. Most conspicuous is the large maritime squill (*Scilla maritima* of LINNAEUS). The flowers are not large or showy, and do not correspond with the size of the bulb which often equals that of a man's head. Another species of the same genus (*Scilla hemisphaerica*) is more ornamental, as are the two common asphodels. The slender iris (*I. Sisyrhynchium* of LINNAEUS), whose delicate flower lasts only a few hours—opening one at a time on successive days, appearing about mid-day and withering in the afternoon—is very abundant."

"On reaching the hollow ground, where a slender stream runs through damp meadows, we were charmed by the delicate tint of a pale blue daisy that enamels the green turf. It is merely a slight variety of the little annual daisy (*Bellis annua*), so common in many parts of Southern Europe; but the blue tint does not seem to have been noticed elsewhere. The larger blue daisy, afterwards seen as one of the ornaments of the mountain region of the Great Atlas, was at first supposed to belong to the same species; but, besides that this is perennial, it shows other less obvious differences."

"It was on the slopes of the Djebel Kebir, where the stony ground is almost exclusively occupied by a dense mass of small shrubs, few of them rising more than three or four feet from the ground, but nearly all covered with brilliant flowers, that we first began to seize the really characteristic features of the North Marocco flora. A great variety and abundance of flowering perennials of shrubby habit is, indeed, a distinguishing feature of the whole Mediterranean region; but very little observation was needed to show that we were in that well marked division that includes Southern Portugal, South-western Spain, and the opposite corner of Africa. This may be called for distinction the Cistus and Heath region; for though most of the same kinds of Cistus and Helianthemum extend as far as the south of France, and many species of heath inhabit the Atlantic coasts of Europe as far north as Connemara, it is only here that both these tribes flourish together, and give a prevailing character to the vegetation. Most conspicuous of all is the gum-cistus (*C. ladaniferus*), which in the Sierra Morena and the adjoining parts of Spain and Portugal obtains such predominance that for twenty miles together one may ride through a continuous thicket where the peculiar scent of the gum that covers the leaves and young branches is never absent."

"About Tangier the rich purple spot that usually adorns the base of the

large petals is wanting, and the flowers show unmixed snowy white. Of the
tribe, besides several true *Cisti*, there are many species of *Helianthemum*. Of
heaths, along with the commoner kind (*Erica arborea* and *E. scoparia*), we
saw in abundance the rarer and more characteristic forms, *E. australis* and
E. umbellata. *E. ciliata*, one of our English rarities, is here very scarce,
though it grows on the opposite side of the Strait. Our common heather
(*Calluna vulgaris*) still holds its ground, but in a poor and stunted condition.
The rhododendron of the East (*Rh. ponticum*), that is at home in the mountain region of Asia Minor and Syria, and which strangely reappears here
and there among the low hills between Tarifa and Algeciras, on the north
side of the Straits, has not been found on the African shore; but until the
coast between Tangier and Ceuta has become more accessible, it will not
be safe to assume that it is wanting. Among the many shrubby leguminous
plants whose flowers give the prevailing golden tint to the hill sides, two
of the Broom tribe (*Genista triacanthos* and *Cytisus tridentatus*), plants of very
peculiar aspect and characteristic of this region, attracted our attention. It
is impossible to omit another ornament of the hills—a plant rather widely
diffused but nowhere common (*Lithospermum fruticosum*), whose azure blue
flowers formed a charming contrast with the surrounding masses of golden
colour."

"The botanical district to which the northern corner of Marocco belongs
has been already called that of the Cistus and Heath, but no single species
of those tribes exactly conforms to the limits above pointed out. There are,
however, several less conspicuous plants whose distribution more closely
agrees with those limits. The most singular of these is the *Drosophyllum lusi-
tanicum*, a plant of the sun-dew tribe, whose branched stem bears several
large yellow flowers. The numerous slender strap-shaped root-leaves, nearly
a foot in length, that are gradually contracted to the thickness of whip-
cord, are beset with pellucid ruby-tipped glands, and present a peculiarity
that appears to be unique in the vegetable kingdom. Any one who has
remarked the growth of ferns must have seen that in the young state the
leaves are rolled or curled inwards, so that in the process of unfolding the
face or upper side of the leaf, which was at first concealed, is gradually opened
and turned to the light. A similar process occurs in many other plants; but
in *Drosophyllum* alone, so far as we know, the young leaf is rolled or curled
the reverse way, so that the upper side of the leaf is that turned outwards.
It appears to grow in many parts of Southern Portugal; reappears on the
north side of the Straits of Gibraltar near Tarifa and Algeciras, and on the
southern side about Cape Spartel and on the hills above Tetuan, where it
commands a view of the opening of the Mediterranean, but extends no
farther eastwards. Very similar is the distribution in Europe of two ferns
whose natural home seems to be in the Canary Islands—the graceful *Davallia
canariensis*, and the *Asplenium Hemionitis* of LINNAEUS. Both occur here and
there in shady spots, from the rock of Lisbon to Algeciras and Tangier, but
are unable to travel eastward beyond the Pillars of Hercules."

"The scarcity of trees in this country is mainly due to the mischievous
interference of man. The same ignorant greed of the herdsman, who to

6

procure a little meagre herbage for goats sets fire to wide tracts of brushwood, that has reduced whole provinces of Spain to a nearly desert condition, has been equally busy and equally effectual in Marocco. The evergreen
oak, which might produce much valuable timber, is the chief indigenous tree
of this country; but, except on the rocky western declivity of the hill above
Cape Spartel, few here arrive at a moderate growth, and the same is true of
the Portuguese oak (*Quercus lusitanica*). The latter, indeed, never attains a
considerable stature; but, where preserved from damage, it forms thickets
some twenty or thirty feet in height, and, if duly protected, would help
to preserve the hilly districts of this region from being annually parched by
the summer sun. One of the shrubby evergreen oaks of this country (*Quercus coccifera*, L.), whose dark green spiny leaves are more like those of a
holly than of an ordinary oak, might perhaps be successfully introduced in
the south-western parts of the British islands. Its very dense foliage would
make it valuable as a screen, and it produces a good effect when mixed
with other shrubs." (Journal, pp. 16–20).

"Our increased acquaintance with the flora of the Great Atlas did not
much modify our first impressions. Making due allowance for the earliness
of the season, and for the adverse conditions that may have concealed from
us some species inhabiting the higher zone, it was clear that the vegetation
here differs very much from that of all the lofty mountain masses of Southern Europe and Western Asia, and especially in the absence of those
families that elsewhere form the chief ornaments of the higher mountain
zone, and which we are accustomed to associate with the glories of the
Alpine flora. There was here to be seen no gentian, no primrose or *Androsace*,
no rhododendron, no anemone, no potentilla, and none but lowland forms
of saxifrage and ranunculus."

"Our first impression had been that the flora is absolutely very poor;
but this was due mainly to the fact that so large a proportion of the plants
have inconspicuous flowers. Comparing the produce of our day's work
with that of high mountain excursions made elsewhere, the species are not
deficient in variety, but show a singularly small proportion of showy
flowers. As regards novelty, we had nothing to complain of; for, in the
upper part of this valley, out of 151 species collected, 31 are described as
new; and, so far as we know, are peculiar to the Great Atlas chain. This
gives about the same proportion of endemic species as the Sierra Nevada of
Granada, always regarded as a singularly rich botanical district."

"The most remarkable feature of the flora of this region is, undoubtedly,
the very large proportion of common plants of the colder temperate region
(Central and North-Western Europe), here found associated with species of
very different type. Nearly one-half (70 out of 151) of the species found in
the upper zone, belong to this category, and the proportion is here actually
larger than it is in the higher mountains of Southern Spain. It was further
remarkable that several of these northern species, such as the wild goose-
berry, are plants that do not extend to the South of Spain, although climatal
conditions must be at least equally favourable, and whose nearest known
habitat is six or seven hundred miles distant. Especially to be noted was

the fact that, with the doubtful exception of *Sagina Linnaei* (the *Spergula saginoides* of the older botanists), not one of the plants in question is characteristically an Alpine species, or typical of the Arctic or glacial flora. Combining this with the almost complete absence of rushes and sedges, we are forced to conclude that, whatever agencies may have contributed to make up the existing flora of the Great Atlas, transport by floating ice during the last glacial period cannot have been amongst them. If such ice-rafts were ever borne to what was then probably a long western peninsula of Northern Africa, they must either have foundered at sea with all their vegetable crew, or, if cast ashore, must have found an inhospitable region where the voyagers were starved, and left no descendants."

"As was to be expected, from the habitual dryness of the climate, ferns were here deficient in number and variety. In the uppper region we found very sparingly six of the common species of Northern Europe; and lower down, in the middle part of the valley, we were able to add to our lists but two southern forms."

"About one-third only of the species found in the upper region could be described as properly belonging to the Mediterranean flora; most of these being widely-spread plants, while a few are exclusively confined to the nearest neighbouring mountain regions—the Lesser Atlas of Algeria or the mountains of Southern Spain. But there was little in the general aspect of the vegetation to suggest any special connection with either; and several of the conspicuous plants have been hitherto known only in very distant regions. A bright-flowered *Veronica* appeared to be no more than a large variety of a species peculiar to Asia Minor; *Medicago suffruticosa* had hitherto been seen only in the Pyrenees; and *Evax Heldreichii* had been detected nowhere nearer than the mountains of Sicily and Eastern Algeria. Our original expectation of finding some connecting links between the special flora of the Canary Islands and that of North Africa was so far completely negatived, and we saw nothing to suggest their existence."

"The most prominent characteristics of the mountain flora of the Great Atlas were found to be of a negative character. If asked to point out the positive features that most struck us, we should in the first place note the prevalence of *Cruciferae* and *Caryophylleae*, the former reckoning one-ninth and the latter one-tenth of the whole number of flowering plants. Of conspicuous genera we had especially remarked *Chrysanthemum*, *Galium*, and *Linaria*. Of the first of these we found two new species, one of which, from its remarkable buff-coloured rays and large scarious involucres, has been named *Chrysanthemum Catananche*" (Journal, pp. 230–2).

"But in the pursuits of a naturalist there are abundant sources of satisfaction not suspected by the uninitiated. These are not merely derived from the objects themselves, suggesting as they often do interesting trains of thought and speculation; there are further springs of keen enjoyment in the countless impressions with which they are linked by the subtle influence of association. Much of the pleasure that an artist, however unskilled, derives from travel, arises from the power of each sketch to bring back again to the mind the original scene of which it is but the imperfect transcript. If he be active and industrious, he may preserve a dozen such keys to the impressions of each day's journey. But to the botanist almost every specimen is indissolubly linked in the memory with the spot where it was collected; and as he goes through the produce of his day's work, every minute detail is vividly presented to the mind, along with the wider background that lay behind the original picture. The wonder and awe that dwell around the mountain fastnesses, the consolation of the forest glade, the indefinable grandeur of the desert plain, nay, even the bleak solitariness of northern moorland and morass—these dominant impressions suggested by the aspects of nature are varied and enriched for the naturalist by the myriad phases of beauty that are disclosed to the eye of the observer. The glory of colour in the gentian and saxifrage and golden *Alyssum*, and the other bright creatures that haunt the mountain tops; the tender grace of the delicate ferns that dwell in the rocky clefts; the teeming life of the warm woodland; the strange beauty of the unaccustomed forms that spring up in the desert solitudes; the purple glow of the heath relieving the sombreness of the leaden sky, and the delicate structures of the *Drosera* and *Menyanthes*, and bog-asphodel, and many another inhabitant of our northern bogs—these and countless other images are instantaneously revived by contact with the specimen that grew beside them. Strangest of all is, perhaps, the enduring nature of this connection. Often does it happen, as many a botanist can testify, that after a lapse of a quarter, nay, even half, a century, the sight of a specimen will bring back the picture seemingly effaced long ago, of its original home" (Journal, pp. 273–4).

In Appendix E to the Journal (pp. 404–21) HOOKER has an essay, "On the Canarian Flora as compared with the Maroccan", which, from the vantage point of seventy years later, can only be described as bold. He had to rely almost entirely on two taxonomic accounts for his comparisons: BALL's "Spicilegium" (*see above*) for the Moroccan flora and WEBB and BERTHELOT's "Histoire naturelle des îles Canaries" (Paris, 1835–60) for the flora of the Canary Islands. There has been much botanical exploration in both areas since 1878. In particular, HOOKER did not realize the richness of the flora of Morocco. Yet there is much in this essay that repays study.

He notes that the flora of the Canary Islands, "though consisting, like the Maroccan, for the most part of Mediterranean species, yet differs from that of Marocco, in containing many plants that may be classed under" five categories: (*1*) non-Moroccan plants obviously introduced by man; (*2*) apparently indigenous plants which are not Moroccan but are widely distributed elsewhere; (*3*) a few peculiar plants more closely allied to endemic species of Morocco than to those of any other country and that may have been derived from species that originally were transported from that country; (*4*) a large class of plants hitherto not found in Morocco and more closely allied to Mediterranean species than to any others; (*5*) many species that are representatives of floras more distant than those of Morocco or Western Europe and are not found in those countries—the Orient, America, and Tropical and South Africa. HOOKER gave a list of 34 genera and 1 section of a genus (the genera are broad ones) with a tabulation of the numbers of their species in the Canary Islands and Morocco and of the endemics in each area. He comments:

"It is most remarkable; the number of endemic species being in the Canaries three-fourths of the whole and in Marocco only one-sixth; and were the peculiar genera of the Canaries added, the disproportion would of course be increased." 54 Macronesian genera are named which were not known from Morocco. In Morocco, out of a total of 517 genera, 202, in 67 families, and with more than 300 Moroccan species, had no known indigenous species in the Canaries or Madeira. Only 16 species are named as exclusively common to Macronesia and Morocco.

"The wonderful development in the Canaries of endemic species belonging for the most part to Mediterranean types, points to the very early introduction of the parent forms of these, and the long isolation both of the Archipelago and its separate islets. It is in accordance with generally accepted views, to assume that the endemic species of each genus have been derived from parent forms originally introduced into one or more of the islets; and that as the descendants of these species spread over the Archipelago they were exposed to different conditions in each islet, resulting in their varying, and in the segregation and conservation of different local varieties each in its own insular birth-place; a supposition which is in accordance with the fact that those endemic species are really very local, many being confined to a single islet. In Marocco the parent forms of its Flora would be exposed to no such diverse conditions, and the areas in which varieties occurred, not being isolated, would be exposed both to invasion on all sides by other plants, and to destruction by agencies that affected the whole surrounding country, as drought, floods, insects, and birds."

"The tropical types in the Canaries, with the exception of the Egypto-Arabian and the trees mentioned under V.c., are chiefly weeds of wide distribution, which have not reached Marocco, because of its want of ports and its limited commerce."

"Finally the *Dracaena*, together with the tropical trees of *Myrsineae*, *Sapotaceae* (in Madeira), and *Laurineae*, and the Egypto-Arabian types, suggest the hypothesis that at a very remote period these and many other plants of warmer and damper regions flourished in the area included in North-West Africa and its adjacent islands, and that they have been expelled from the continent by altered conditions of climate, but have been preserved in the more equable climate and more protected area of the Atlantic-Islands."

"BALL, who has given me valuable aid on many points discussed in this article, directs my attention to the important differences that exist between the vegetation of the eastern group of the Canary Islands—Fuertaventura, Lanzarote, and the adjacent islets—the 'Purpurariae' of authors, and the western group, including Teneriffe, Grand Canary, etc."

"In the first place, nearly all the characteristic Canarian types are absent in the eastern group. Out of fifty-four genera above enumerated as present in the Canaries but wanting in Marocco, two are in the Canaries confined to the eastern islands: one of these, *Traganum*, is an African desert type, probably to be found in South Marocco; the other, *Melianthus*, a Southern African plant, and scarcely indigenous. Of the remainder *Plocama* alone is certainly present, and three other generic types probably exist in that group; while forty-eight genera, including eight out of nine peculiar to the Canaries, are apparently absent. In the next place several characteristic desert plants, such as *Oligomeris subulata, Ononis vaginalis, Convolvulus Hystrix*, and *Traganum nudatum*, are present in the 'Purpurariae', but absent from the western islands."

"Although the Flora of the Purpurariae is incompletely known, and our acquaintance with that of the neighbouring African coast between the rivers Sous and Draha is extremely imperfect, these facts tend to prove that there is a closer botanical relationship between the eastern islands and the adjoining continent than there is between them and the western portion of the Canarian Archipelago. Such relationship might be brought about in three different ways.

1) The greater dryness and heat of the eastern islands may have favoured the immigration of African forms, and at the same time led to the destruction, or weeding out, of the characteristic Canarian types. In this case the cause would be of a purely local and climatic character.

2) We may believe in the trans-oceanic migration of some African species to the nearer islands, along with the transport of some Canarian species (those enumerated in p. 416, and others which may be hereafter found) to the neighbouring continent.

3) An ancient extension of the continent to the Purpurariae, leaving the other islands separated by deep sea."

"It is an objection to the latter hypothesis that a profoundly deep ocean bed lies between the lines of 100 fathom soundings that girdle the islands and the African coast respectively; and that while the 100 fathom line extends about thirty miles from the coast of the continent, it is never more than five miles, rarely more than one or two, from those of the islands."

"In favour of the hypothesis of trans-oceanic transport it may be remarked that the distance between the African coast and Fuertaventura is not more than seventy miles, and that a moderate change of level of about 600 feet would reduce that distance by one-half, while it would but slightly affect the interval that separates the Purpurariae from the other islands" (Journal, pp. 417–9).

LYELL is quoted as against the view of a general previous continuity with the African continent and of the islands one with another.

"It remains a point of some nicety to decide whether the Macronesian islands should be regarded as a Botanical province apart from the Mediterranean, or a sub-division of the latter. The assemblage of American and Oriental genera which their Flora contains, together with the arboreous representatives of tropical *Laurineae*, all so entirely foreign to the European Flora, would give it a title to be called a Botanical province; and to this as a further title is the prevalence of a considerable proportion of North European plants, in the Northern Archipelago especially. On the other hand, fully two-thirds of the species are typical of the Mediterranean Flora, and by far the majority of the remainder are derivative species of the same origin; so that, on the whole, I am disposed to regard it as a very distinct sub-division of the Mediterranean province, which owes its peculiarities partly to the conservation of types once common to West Europe and North Africa, but which have been eliminated in those regions, and partly to the effect of isolation and climate on the progeny of species still existing in those regions" (Journal, pp. 420–1).

The title of the short Appendix F is "Comparison of the Maroccan flora with that of the mountains of Tropical Africa" (Journal, pp. 421–3). The African mountains referred to are the peak of Fernando Po and Cameroons Mountain. The temperate vegetation of these West African tropical mountains is stated to be notable for its poverty, the preponderance of Abyssinian genera and species, the considerable proportion of European plants, the paucity of South African genera and species, the great rarity of new genera, and the absence of St. Helena and Canarian types.

"I find no other evidence of relationship between the Biafran mountain Flora and that of Marocco than what is afforded by the European species common to both. In most other respects the Floras differ totally, the other mountain plants of Biafra being Abyssinian or Cape types, or more nearly related to tropical African ones."

From the article in the Proc. Roy. Geogr. Soc. 15 (1871) two extracts may be given:

"Our camp is in an open valley, about midway, roughly speaking, between the Plain of Marocco and the axis of the chain. We are surrounded with olive and walnut groves, and there are, besides, abundance of fig-trees, prickly pear (*Opuntia*), vines, mulberries, and almonds. The native trees are poplar, ash, two junipers, willow, and *Callitris* (the famous Thuja of the Romans), of which such beautiful articles are made in Algeria. The bushes are lentisks, honeysuckle, cistus, elder, rose, alaternus, phillyraea, ivy, bramble, balloot oak, colutea, and shrubs allied to the broom. The climate is temperate, and scenery rather pretty than grand or mountainous, except up the valley, which is backed by the rugged black but snow-streaked crest of the range" (pp. 214–5).

"Botanically, the upper (alpine) region is as bare as the middle region is rich; we found no trace of alpine plants, and remarkably few of any kind; few grassy slopes or mossy rocks, not a gentian, primrose, anemone, or any representative of those lovely alpine genera; not even the mountain sorrel

(*Oxyria reniformis*), which I found on the Lebanon, and I regard as the typical representative of an Arctic and N. Alpine flora. Considering the nature
of the rocks – granite and very hard porphyries – and the climate, I was not
much surprised at the absence of alpine plants, and indeed expected a very
poor flora. It is true that the higher points may present exceptions, but it
is not likely, as we examined carefully the banks of the streams, which in
all alpine regions carry down specimens of the plants that grow at higher
levels. The rocks also were singularly bare of tufts of vegetation, such as
alpine rocks should present, and this even on their shaded sides. Another
proof of the dryness of the climate is the great rarity of ferns at all elevations,
even in the narrow, fertile, well watered valleys. We have found some half-
dozen in all—most of them very sparsely indeed; and all but two, I think,
are British."

"The flora of the middle region again is Spanish in type, and is rich,
varied, and beautiful; and we shall carry home a splendid herbarium, in
point of interest and, I hope, novelty. As presenting the southern limit of
the Mediterranean and indeed North Temperate Flora, the Atlas has a
special interest, for here many prevalent European types die out, as it were—
the ivy, oak, ash, bramble, and some hundred of other British genera, probably exist nowhere to the southward of the Atlas in this longitude; a few
may possibly lurk in the valleys of the mountain-range that runs parallel to
and south of this, on the opposite side of the Sous Valley, but it is unlikely
that many, if any, do so."

"In the poverty of this alpine region the Atlas appears to me to be quite
exceptional; the Abyssinian mountain-tops, though equally barren, being
lower and less extended. Another curious fact in this range, is the exceptionally upward extent of lowland plants; at between 10,000 and 11,000
feet we found many plants of the foot of the range, and even some inhabitants of the hot plains of Marocco. In short, in every point of view the Flora
of the great Atlas is excessively interesting; its general nature has long been
a botanical desideratum, and I hope we shall clear it up" (pp. 216–7).

* * *

Since the visit of Hooker, very great changes have taken place in
the political status of Morocco and consequently in its accessibility and
in our botanical knowledge of the country. Thanks mainly to the
energy of French botanists there is now available a mass of new facts
regarding both the flora and the vegetation. The flora is known to be
richer than it was judged to be by Hooker as is shown by the following figures compiled from Jahandiez and Maire (1931–4): total
species (of vascular plants) 3480; endemics 496 = 14.2%. In addition
there are some hundreds of endemic subspecies and varieties.

In Jahandiez and Maire (1931–4) there is a useful bibliography
which, with the supplement, gives many of the important papers and
books published on Moroccan botany up to 1933. A selected list is
also given *in* Rikli, Pflanzenkleid der Mittelmeerländer 3: 1231–1242,

Bern, 1948. The following, including more recent publications are important: ANDREÁNSZKY, B. G. (1932), ANDREÁNSZKY, J. (1939) BRAUN-BLANQUET et MAIRE (1921), CERECEDA (1916), EMBERGER (1927, 1932, 1934, 1936, 1938), EMBERGER, FONT-QUER, et MAIRE (1928), EMBERGER et MAIRE (1927), FONT-QUER (1929), HESS (1925), HURÉ (1945–6), JAHANDIEZ (1923), JAHANDIEZ et MAIRE (1931–4), JACCARD (1926), JOUBERT (1932, 1933), MAIRE (1921,1939), PITTARD (1913), RIKLI (1933–5), and RÜBEL and LÜDI (1939).

HOOKER's own published accounts of the botany of Morocco being phytogeographical comparisons, it is not within the general terms of reference to the present work to discuss the flora and vegetation, as now known, in any detail. It is agreed that the flora is basically of Mediterranean type, thus confirming HOOKER's conclusion. MAIRE recognized that the major part of the species comprising it has been derived from species of the plains, but in this basis there are types of four other floral regions: holarctic, Indo-African, Neo-intertropical, and southern (S. African and Australian), witnessing to some geographical connections now broken. It can be assumed that the plants of tropical and austral affinities are descendants of the Tertiary flora of Morocco. The glacial period affected Morocco but the Mediterranean considerably impeded the invasion of Morocco and the remainder of North Africa by boreal species. The Appennine-Sicilian route was the most favourable, for the east-west orientated mountains of Spain made a barrier rather than a bridge. The glacial period in Morocco was represented by abundant rains, glaciers only forming in high mountains. Tropical elements migrated to the south. The mountains—the Great and Middle Atlas—played a part similar to the Mediterranean: they barred the route to the return of tropical species and were an obstacle to the reconquest of Morocco during the xerothermic periods. This enforced isolation of Morocco was favourable to the preservation of old types, now endemics, and to the development of neo-endemism.

HOOKER's essay comparing the Moroccan and Canarian floras is in many ways still surprisingly sound. The following are important publications on or referring to the flora and vegetation of the Canary Islands: ENGLER, in ENGLER und DRUDE, Veg. der Erde 9, I, 2 (1910), pp. 822–866, BANNERMAN (1922), BOLLE (1893), BØRGESEN (1924), BURCHARD (1929), CHRIST (1885, 1886), GUPPY (1921), HIRMER (1924–6), KNOCHE (1923), LINDINGER (1926), MATTICK (1933–5), PITARD and PROUST (1908?), RIKLI (2, 1946, pp. 930–5), SCHRÖTER (1909), and SPRAGUE and HUTCHINSON (1913).

RIKLI (1946) accepts the calculation that the endemic species compose 50% of the flora from the sea-coast to the summit of the Pico de

Teyde (3716 m.) in Tenerife. The very high degree of endemism is generally accepted as the most characteristic feature of the Canarian flora and this endemism includes that of varieties, species, and genera. There is a general Mediterranean basis to the flora though more or less strong affinities can be traced in several directions: with Morocco, with Europe, and with America. There is no doubt that many of the endemics are ancient and are survivors of or immediate descendants of Tertiary types. SCHRÖTER (1909, pp. 49–55), discusses the pros and cons of theories of previous land connections and the origin of the Canarian flora. He concludes that the Canaries were probably in land connection with the African continent up to the younger Tertiary and that earlier they perhaps formed a northern extension of "Southern Atlantis". This would explain most simply their Tertiary and rich African but sparse American affinities and the predominant Mediterranean character, while the separation since the Pliocene would account for the conservative and peculiar characters of the flora. Change in conditions for colonization must also have been important.

PITARD and PROUST (1908 ?) note that in North Africa the number of endemics is very markedly reduced from west to east. There is considerable difference in the families with the largest number of species in the floras of the Canary Islands and Morocco. In both the maximum number of species occur in the *Compositae*, *Leguminosae*, and *Gramineae*. Then, in the Canaries the important families are: *Labiatae*, *Crassulaceae*, *Euphorbiaceae*, *Solanaceae*, *Rosaceae*, *Polygonaceae*, *Urticaceae*, *Plantaginaceae*, *Convolvulaceae*, *Boraginaceae*, and *Plumbaginaceae*. While in Morocco the families next largest to the three first mentioned are: *Caryophyllaceae*, *Scrophulariaceae*, *Ranunculaceae*, *Rubiaceae*, *Cistaceae*, *Iridaceae*, *Campanulaceae*, *Resedaceae*, *Orobanchaceae*, *Amaryllidaceae*, *Linaceae*, and *Liliaceae*.

GUPPY in his studies on the flora of the Azores (GUPPY, 1917, pp. 411–4) deals briefly with "The plant-stocking of the Macronesian Islands". He refers to the works of HOOKER and CHRIST and states: "Presumably, therefore, the Canary Islands and Madeira, especially the former, hold the wrecks of many floras. To the exclusion of the Azores, they possess a number of strange genera and peculiar species, that tells us of the ages which preceded the period indicated by the non-European trees and shrubs that are common to the Laurel woods of all three groups. The waves of African, Asiatic, and American plants that have in successive ages passed over this portion of the globe, left their wash on the Canarian and Madeiran groups before the Azorean islands became available for plant-stocking". In a later paper (GUPPY, 1921) the same writer more definitely and clearly distinguishes between two groups of endemics in the Canarian flora: *1*) those common to the Canaries and to other Macronesian islands and regarded by GUPPY as the older group, and *2*) those occurring in the Canaries only and often restricted to a single island of the Archipelago and assumed to be younger in origin than species of the first group. We are not here concerned with the significance of the tables prepared by GUPPY. It is, however, open to question that GUPPY has rightly interpreted HOOKER in saying that the latter clearly distinguished "Canarian" and "Macronesian" endemics as groups of different origin. HOOKER certainly refers to "the very early introduction of the parent forms" of endemic species "belonging for the most part to Mediterranean types" and for *Dracaena, Myrsineae, Sapotaceae* (in Madeira) and *Laurineae* and the Egypto-Arabian types he suggests these were at a remote period "expelled from the continent by altered conditions of climate". There is little evidence in this that HOOKER "recognized a great break in the floral history of the group." GUPPY's references to ENGLER (1879) and to DRUDE (1890) simply support the view that there is an ancient (Tertiary) element in the Canarian flora, and this is not denied.

The point at issue is whether HOOKER claimed that the "endemics" (Canarian and Macronesian respectively) represent two clearly defined groups the former more or less recent and the latter old in origin. On this and similar questions it is easy to become confused by uncritical use of such terms as "species," "types", "elements," and "groups." For example, the species of *Aichryson, Aeonium, Greenowia,* and *Monanthes* as accepted by PRAEGER (1932) may be young as species but may represent an old type or types which entered the Canarian archipelago at least as long ago as *Laurus, Dracaena,* etc. Actually we have few facts on which to judge the age *as species* of either Canarian or Macronesian "*types*".

Tropical Africa:— HOOKER never specialized on the flora of continental Tropical and South Africa, apart from the Flora Nigritiana, with G. BENTHAM (*in* the Niger Flora, ed. W. J. HOOKER, London, 1849) and a few shorter papers. Three such papers, a note, and an introduction to a colleague's paper include some phytogeographical material.

A paper, under the title "On the vegetation of Clarence Peak, Fernando Po; with descriptions of the plants collected by Mr. GUSTAV MANN on the higher parts of that mountain", was published in the Journ. Linn. Soc. Bot. 6: 1–23 (1862). This is mainly a systematic account but an introduction analyzes the floristic relationship on the basis of MANN's collection.

"Of the total 76, 37 are Abyssinian species, and 16 others closely allied to such; and of the 56 temperate, 32 are also natives of the mountains of Abyssinia, most of them being absolutely specifically identical, and others but slightly differing; such differences being in some cases doubtless apparent rather than real, and owing to the want of a larger suite of specimens; 13 others also are very closely allied to Abyssinian species."

"Again, of the Abyssinian mountain plants common to Clarence Peak, no fewer than 17 are absolutely peculiar to these two localities as far as is at present known, including some very remarkable plants; as

Clematis Simensis	*Stachys aculeolata* n. sp.
Thalictrum rhynchocarpum	*Pycnostachys Abyssinica*
Sagina Abyssinica	*Calamintha Simensis*
Trifolium subrotundum	*Cyanotis Abyssinica?*
— *Simense*	*Kyllingia macrocephala*
Helichrysum chrysocoma	*Trisetum lachnanthum*
— *Hochstetteri*	*Festuca Schimperiana*
— *globosum*	*Gymnandropogon,* sp."
Blaeria spicata	

"Besides these are the following, which are not found south of Abyssinia in Africa:—

Galium rotundifolium	*Deschampsia caespitosa*
Parietaria Mauritanica	*Brachypodium sylvaticum*"

"Others are common to Abyssinia, the Mauritius, Madagascar, etc.: as

Viola Abyssinica	*Rubus apetalus*
Hypericum angustifolium	*Carex Boryana*"
Geranium Simense	

"There are, again, other species whose only near affinities are with Abyssinian: as species of

Agrocharis	*Plectranthus*
Gymnosciadium?	*Veronica*
Dichrocephala	*Euphorbia*
Swertia	*Habenaria*"

"Extending the comparison to genera, I find that of the 66 Clarence Peak genera only 7 are not Abyssinian, and of the 45 temperate genera 41 are temperate Abyssinian. Of the 3 remaining, *Luzula* and *Schoenus* may yet be found in Abyssinia, and *Leucothoe* is a Mauritius plant."

"The next affinity is with Mauritius, Bourbon, and Madagascar: of the whole 76 species, 16 inhabit these places, and 8 more are closely allied to plants from there. Three temperate species are peculiar to Clarence Peak and the East African Islands, including *Leucothoë angustifolia, Sebaea brachyphylla,* and *Carex Wahlenbergiae. Ericinella* and *Leucothoë* are the only genera not Abyssinian, which are common to these islands and to Fernando Po."

"Lastly, if compared with the Cape, the contrast is very striking: not only is there a total want of any true Cape types, except such few as are common to Abyssinia or the Eastern African Islands (5 species), but only 12 of the

76 Fernando Po species are known to be South African; and of these all but *Luzula* have been also found in Abyssinia. Only 12 others are nearly related to South African forms. Turning to the genera, *Peddiea* is the only peculiarly South African one; and this is not temperate at Fernando Po, and is subtropical in South Africa."

"Hence the result of comparing the Clarence Peak flora with that of the African continent is—(*1*) The intimate relationship with Abyssinia, of whose flora it is a member, and from which it is separated by 1800 miles of absolutely unexplored country; (*2*) the curious relationship with the East African Islands, which are still further off; (*3*) the almost total dissimilarity from the Cape flora."

"With the West African Islands again, contrary to my expectations, there is no marked relationship whatever, except obscurely with St. Helena through *Wahlenbergia arguta*: the arborescent *Compositae* and *Lobeliaceae*, *Phylicae*, *Melhaniae*, *Frankenia*, *Acalypha*, and frutescent *Hedyotis* of St. Helena, being wholly unrepresented in Fernando Po."

"Taking a still wider range, the temperate flora of Fernando Po belongs to the northern hemisphere. Of the 48 temperate genera, 12 only are not European; whilst the following species are European, and most of them British:—

Oxalis corniculata	*Parietaria Mauritanica*
Sanicula Europaea	*Luzula campestris*
Galium Aparine	*Deschampsia caespitosa*
— *rotundifolium*	*Brachypodium sylvaticum*"
Limosella aquatica	

"The two following are also probably states of European plants:— *Ranunculus pinnatus*, very near *R. philonotis*; *Calamintha Simensis*, near *vulgaris*" (pp. 2–4).

A second paper, "On the plants of the temperate regions of the Cameroons Mountains and islands in the Bight of Benin", appeared in the Journ. Linn. Soc. Bot. 7: 171–240 (1864). Again, the greater part of the paper is a taxonomic account but the introductory essay is largely phytogeographical as the following extracts show.

"From Mr. MANN's descriptions, the Cameroons Mountains present a dense forest-region up to about 7000 ft., when open grassy fields succeed, with bushes of *Hypericum*, *Pittosporum*, *Adenocarpus*, *Pygeum*, *Leucothoë*, *Ericinella*, *Myrica*, and various herbaceous plants. The many peaks which rise above this elevation are either stony and barren (being all formed of lava scoriae or basalt), or are dotted with tufts of grass and a few other herbaceous plants."

"The most interesting plants from the highest summits are, *Umbilicus pendulinus*, *Silene*, *Trifolium*, *Galium Aparine* and *G. rotundifolium*, *Scabiosa succisa*, *Helichrysa*, *Veronicae*, *Bartsia*, *Stachys*, *Trichonema Bulbocodium*, *Deschampsia caespitosa*, *Poa nemoralis*, *Koeleria cristata*, and various other European and even British plants" (p. 175).

"*1*) In the poverty of its flora the Cameroons range, etc. seems to partake of the characteristics of the Abyssinian Alps. We know far too little of the physical geography of either of these districts to hazard many conjectures upon this point, which must to a certain extent be dependent on the arid volcanic nature of the soil and the limited area of the temperate region. Mr. MANN spent many weeks, and at various seasons, in his explorations, and 237 flowering plants were all that rewarded his toil. Geological causes have probably had, in the case of the Cameroons Mountains, much to do with the dearth of species, some parts of the range even now presenting evidence of subterranean heat."

"*2*) The preponderance of Abyssinian forms is proved by almost all of the genera and half the species being natives of Abyssinia, and by many other species being very closely related to, or obvious representatives of, plants of that country. There are, further, several of the genera and many of the species peculiar to Abyssinia and the peaks of Biafra."

"*3*) The number of European genera amounts to 43, and species to 27, by far the greater part of which are British; and a few of them, as *Radiola Millegrana*, have not been found previously anywhere in the African continent. Very few of them extend into South Africa. The greater part are Abyssinian; the remarkable exceptions being *Radiola, Scabiosa succisa, Luzula campestris*, and *Festuca gigantea*, all of which, however, may have been hitherto overlooked in Abyssinia."

"Considering the total isolation of these tropical African mountains from the European regions by hot, low deserts, the existence of these plants in common is most singular, and explicable under two hypotheses: 1st, Mr. DARWIN's theory, which, assumes that during the glacial epoch the plants of the northern zones were driven southwards into the tropics, and on the return of warmth they both retreated northwards and ascended the intertropical mountains; and 2nd, transport by aërial currents and birds—in favour of which is to be urged that, of the whole, six present stuctural adaptations for clinging to the plumage of birds, and all the rest have small or very minute seeds, likely to be transported in mud on the feet of birds. *Solanum nigrum* has rather larger seeds, but with remarkable power of retaining their vitality, and, further, is found in North Africa and many intermediate countries, as are several of the others."

"*4*) The paucity of South-African types was alluded to in discussing the 76 species of the Fernando Po mountain. The great accession of species from the Cameroons has added but few Cape forms, the principal are, *Anthospermum, Anisorhamphus* (perhaps referable to *Hieracium*), a species of *Ilex, Lasiosiphon, Peddiea, Geissorhiza, Hypoxis*, and a few others."

"*5*) Only one new genus has been found, *Ardisiandra* (see Plate I) — a very well marked new form of Primulaceæ, not indicating an affinity with any other flora."

"Of the peculiar genera and species of St. Helena not one has been found; and what genera are common to that island and these mountains are also natives of the Cape region, and far more abundant there" (pp. 180–1).

The last paper on African plants to which reference will be made here was

published under the title „On the subalpine vegetation of Kilima Njaro, E. Africa"
in Journ. Linn. Soc. Bot. 14:141–6 (1874). Though his paper is based on a very
small collection of some 14 specimens made by the Rev. Mr. NEW, HOOKER makes
some rather wide generalizations as shown by the following extract:

"The three isolated mountain districts of tropical Africa form an isosceles
triangle, of which the base, about 1500 miles long, runs nearly north and
south from the Abyssinian Alps to Kilima Njaro; and the two sides stretching across the continent, one in a W.S.W. and the other in a W.N.W.
direction, each being about 2000 miles long, unite these eastern points with
the Cameroons in the Bight of Biafra. So far as is known, no land of equal
height occurs at any intermediate point; and a comparison of their vegetation is therefore a very interesting study. Fortunately, from the Cameroon
Mountains (altitude 13,100 feet) excellent materials were collected by GUS-
TAV MANN, which were published by myself in the seventh volume of our
Journal (p. 171), together with those collected on the adjacent peak of Fernando Po (altitude 9469 feet) by the same enterprising traveller. The botany
of the loftiest Abyssinian peaks has been explored by SCHIMPER and other
travellers in that country; but unfortunately, though contained in our collections, no account of them has hitherto appeared. In the above-mentioned
paper I have compared the vegetation of the Cameroon Mountains with
the Abyssinian, and have shown that out of 64 flowering plants contained
in 56 genera, which occurred chiefly at elevations above 9000 feet, almost
all the genera and no less than half the species are Abyssinian; whilst of the
remainder of the species, many are clearly related to, or are obviously representative of, Abyssinian ones."

"A still more curious fact brought out by that investigation was the number of European forms on both these tropical mountain regions, the
Cameroon possessing 43 European genera and 27 species, 24 of these being
common to Abyssinia and 22 natives of Great Britain."

"These data would have led me to expect many additional temperate and
even alpine plants from the loftier elevation of Kilima Njaro; but in this
I have been disappointed; for Mr. NEW's collection (consisting of about 50
species, of which 20 were collected in the region immediately below the
perpetual snow) contains but few of the European forms found in the Cameroon and Abyssinian mountains, and no additional one whatsoever;
while of the 22 British species found on those mountains, not one appears
upon Kilima Njaro. Of the 20 plants from near the snow, the following are
the most interesting:—

Adenocarpus. Identical with *A. Mannii*, H. f., from the Cameroons, with a
variety.

Helichrysum, 7 species—2 identical with Abyssinian species, one of them a
Cameroon Mountains' plant.

Senecio, 2 herbaceous species.

An *Artemisia,* apparently the South-African *A. afra,* Jacq.

Two Ericaceæ, one a *Blaeria* (probably identical with the Abyssinian species
B. spicata, Hochst. ?), the other an *Ericinella.*

Bartsia, sp. Not in flower; near to, if not identical with, an Abyssinian species.
Protea, sp. Probably the Abyssinian.
Gladiolus abyssinicus, A. Brongn. A Monocotyledon of no significance."

"Now the whole of these, except *Bartsia* and *Adenocarpus* (a Mediterranean and Canary-Islands genus, occurring in Africa only on the Cameroons and Kilima Njaro), are South-African genera and very characteristic of that botanical region. This, taken in conjunction with the fact that the north temperate plants of Abyssinia are almost all absent in South Africa, suggests the view that, whilst South-African plants have migrated far north of the equator, European plants have not advanced correspondingly far to the south. In this there would be little noteworthy if the migration were in either case supposed to take place under existing climatic conditions; for the South-African genera mentioned above are characteristic of a warmer climate than the north temperate forms, and are therefore by so much better suited for advance into a tropical region than the north temperate plants would be. But it is very noteworthy in respect of the hypothesis that such migrations took place only, or chiefly, when the tropical zone was cooler than it now is; and so far as the southward migration of north temperate forms is concerned, this hypothesis is consonant with so many otherwise unexplained phenomena, that it is with many naturalists an accepted truth."

"It may be assumed that the northern forms did once spread over South Africa – the few existing on its Alps being evidence of such an invasion and the survivors of hordes that have been extinguished by the rapid diffusion and differentiation of preexisting South-African forms, which has resulted in a flora of extraordinary variety, richness, and generic and specific peculiarity. It may also be assumed that the migration of the peculiar South-African forms into Abyssinia and the Cameroon Mountains was subsequent to the northern one. I see no valid objection to this; it gains plausibility from a comparison of the temperate Australian flora (where a similar prevalence of endemic types and profusion of species is accompanied by a variety of north temperate genera and species in the subalpine zones) with the temperate American (Chilian and Fuegian) flora, where there is a paucity of endemic types and of species with a great infusion of north temperate genera and species. In short, it would imply simply that, though an area containing few types, and these endemic, may be succesfully invaded by a few imported species (as is now the case in New Zealand, where peculiar species are being supplanted by European weeds), a change of conditions conducive to the rapid differentation of the endemic types will turn the tables on the invaders, and limit their numbers if they do not exterminate them."

"Whether justified or not in assuming the rarity of temperate plants in Kilima Njaro, the fact of the South-African character of its subalpine flora can hardly be questioned; and, taken in conjunction with the fact mentioned by Dr. KIRK of the similarity of the plants brought by Mr. NEW to those of the Dzourba, about 800 miles further south and nearly in the same longitude, it suggests the probability of the South-African flora being represented all along the high lands of western Africa from the Natal district to Abyssinia; and, further, seeing that most of the South-African plants found in the Cameroons are also natives of Abyssinia, it would appear probable that the migration of these to the Cameroons was by and through Abyssinia."

"It is not obvious why none of the temperate European plants found in Abyssinia should in recent times have advanced south to Kilima Njaro and beyond it. Much may, as I have suggested in the case of the Cameroons, be due to the arid soil of these mountains and to their limited area, but more perhaps to recent volcanic disturbances. Mr. MANN observed that some parts of the Cameroons presented evidence of subterranean heat—since which time the Cameroons have been in eruption on the grandest scale; and there can be no doubt that the scene of Mr. MANN's most interesting discoveries is now, in part at any rate, a mass of scoriae, and that some of the endemic species he obtained may exist only in our herbaria, being burnt up in their native locality."

"I will conclude with the hope that these fragmentary observations may prove of some use in calling attention to the great interest of the questions raised by even so small a collection as Mr. NEW's. They will at any rate afford some indication of the debt that science will owe to Mr. NEW when he again visits the scene of his labours, should he succeed in securing a more complete flora of the upper alpine regions of Kilima Njaro" (pp. 143–6).

Attention is called to a note in Nature 30: 635 (1884) which reads as follows:

"A very interesting collection of plants has been brought to Kew by that intrepid African explorer Mr. JOSEPH THOMSON, made during his late journey into the Masai country. They have been examined by Prof. OLIVER, and consist of about thirty-five species from Kilimanjaro at 9000 to 10,000 feet of elevation: a few from a crater near Lake Nairasha at 7000 to 8000 feet elevation; thirty-four from the Kapté plateau at 5000 to 6000 feet; and fifty eight from Lykipia at 6000 to 8000 feet."

"These collections exhibit the mingling of North Temperate types with others characteristic of Southern Africa for which previous discoveries had prepared us. Of these the most interesting are, as new to Tropical Africa, an Anemone, a Delphinium (very different from the Abyssinian *D. dasycaulon*), and a Cerastium of remarkable habit. Of South African forms the most striking is the handsome arborescent Rutaceous plant, *Calodendron capense*, the "wild chestnut" of Natal, to the north of which it had not previously been found. Of northern forms is a Juniper, another genus unknown to Tropical Africa, and which was found forming groves at an elevation of 6000 to 8000 feet, and itself attaining a height of 100 feet! it is the *J. procera* of Abyssinia. A *Podocarpus* gathered along with the Juniper, and also attaining 100 feet in height, is probably the *P. elongata* of Abyssinia, which, or a near ally, also occurs in South Africa. The only other Conifer previously found in the equatorial regions of Africa is the *Podocarpus Mannii* from the peak of St. Thomas in the Gulf of Guinea.   J. D. HOOKER."

7

HOOKER wrote a preface to a paper by D. OLIVER published, under the title "List of the plants collected by Mr. THOMSON F.R.G.S., on the mountains of Eastern Equatorial Africa," *in* Journ. Linn. Soc. Bot. 21:392–406 (1885). The following extract is taken from this preface:

"The localities from which Mr. THOMSON's specimens were brought are, with their elevations:—

	Lat.:	Long.:	Elevation:	Species
Kilimanjaro . . S. 3° 0′		E. 37° 30′	9000–10,000 feet	35
Lykipia . . . . N. 1°–S. 1°		E. 36° 37′	6000–8000 feet	58
Kapté plateau . S. 1°–2°		E. 36° 37′	5000–6000 feet	34
Lake Naivaska. S. 1°		E. 36°.	7000–8000 feet	9"

"The subjects most worthy of comment indicated by a study of these collections may be grouped as follows:—

1) The number and affinities of plants characteristic of the European flora.
2) The number and affinities of plants characteristic of the South-African flora.
3) The comparison of the Eastern Equatorial mountain-flora with that of the western side of the continent.
4) The affinity of the flora with that of the highlands of Abyssinia.
5) Origin of the flora as assumed from these data."

"*1) The Northern or European Element.*— Of the 107 genera and 140 species of flowering-plants, no less than 27 genera, including 37 species, are of a distinctly northern type, and comprise, amongst others, species of *Clematis, Ranunculus, Anemone, Delphinium, Cerastium, Hypericum, Geranium, Trifolium, Lotus, Epilobium, Caucalis, Galium, Scabiosa, Echinops, Artemisia, Sonchus, Erica, Swertia, Bartsia, Leonotis, Rumex, Juniperus,* and *Romulea.* And amongst the species are *Cerastium vulgatum* (two forms), *Caucalis infesta, Galium Aparine,'Scabiosa Columbaria, Sonchus asper, Erica arborea,* and *Rumex obtusifolius.*"

"Of the above, the following genera have not been hitherto detected in South Africa:— *Delphinium*, Caucalis, Echinops*, Artemisia*, Swertia, Bartsia, Leonotis*,* and *Juniperus*.* Those marked with an asterisk have not been found in the mountains of Western Africa; nor have the following:— *Anemone, Lotus, Epilobium,* and *Erica.* Thus no fewer than 9 northern genera are added to the Equatorial African flora by this small herbarium alone. Of all these the Juniper is the most interesting, as indicating the southern limit of that wide-spread northern genus, and the fact of its actually reaching the Equator. The southern limits hitherto ascertained of the genus *Juniperus* are:— In Asia N. lat. 28°, in the Eastern Himalaya, where it is not found under 8000 feet elevation; in America it extends far lower down to Guatemala and the Jamaican mountains, N. lat. about 15°. In Africa the *J. procera* was found by SCHIMPER in the Tigre mountains in N. lat. 14°. Having regard to the comparatively low elevation of the Lykipia forest and its equatorial position, it is evident that a little downward extension of the range of *Juniperus* would constitute it a tropical genus."

"*2) The Southern or Temperate South-African Element.*— There are 35 genera in the above collections which are represented in South Africa, some of which there obtain their maximum, or are even almost peculiar to that region. The most notable of these are all those mentioned above as northern, with the exception of *Delphinium, Artemisia, Echinops, Swertia, Bartsia,* and *Juniperus.* And of other southern types there are the species of *Sparmannia, Calodendron, Psoralea, Alepidea, Felicia, Tripterus, Osteospermum, Berkeleya, Lightfootia, Blaeria, Selago, Struthiola, Podocarpus, Aristea, Gladiolus,* and *Kniphofia.* Of these, *Felicia, Osteospermum,* and *Alepidea* had not been previously found north of the Tropic of Capricorn. One species of *Clematis* is identical with the Cape *C. Thunbergiana,* as is the *Calodendron* with *C. capense,* and the *Alepidea* with *A. amatymbica*; and the *Anemone* is very near *A. capensis.* Of the rest most have representatives in Abyssinia or the mountains of Western Equatorial Africa."

"No less than 15 of these South-African genera appear to be absent on the mountains of Western Equatorial Africa; they are:— *Anemone, Calodendron, Psoralea, Alepidea, Felicia, Tripteris, Berkeleya, Lightfootia, Erica, Selago, Leonotis, Struthiola, Aristea, Gladiolus,* and *Kniphofia.*"

"On the other hand, the mountains of the Gulf of Guinea contain many South-African genera not hitherto found in the Eastern equatorial mountains. Amongst the most notable of these are *Anthospermum, Hieracium, Ilex, Lasiosiphon, Peddiea, Geissorhiza,* and *Hypoxis.*"

"3) *A Comparison of the Eastern with the Western Mountain Vegetation* can only be profitably undertaken when the flora of the former is as diligently gleaned as was that of the latter by Mr. MANN; and we may hope for contributions towards this end from Mr. JOHNSTON's exploration of Kilimanjaro. It is, however, worthy of remark, that of the genera found in the east and not hitherto in the west, the majority are of either Abyssinian or South-African types, whilst the compensating wealth of the western flora is in European types not hitherto detected in the east."

"4) *The Affinity of the Flora with that of the Highlands of Abyssinia* is very marked, as was to have been expected. Most of the genera are in fact Abyssinian, as are all, or nearly all, of the following species:— *Ranunculus oreophytus, Viola abyssinica, Sparmannia abyssinica, Geranium simense?, Trifolium simense, Lotus tigrensis?, Lythrum rotundifolium, Epilobium stenophyllum, Diplolophium abyssinicum, Caucalis melanantha, Coreopsis abyssinica, Lightfootia abyssinica, Erica arborea, Swertia Schimperi, S. pumila,* and *Juniperus procera.* Besides the above, the Abyssinian affinity is shown by the presence of an *Ueberlinia,* a genus hitherto known only as a monotypic Abyssinian one, and by the species of several of the other genera being more nearly allied to plants of that country than of any other."

"5) *On the Origin of the Flora.*— The most striking feature of the flora thus first explored by Mr. THOMSON is the discovery in Lykipia of three such typical forest-trees in close association as the *Juniperus procera* of Abyssinia, the *Calodendron capense* of South Africa, and the noble *Podocarpus,* a close ally both of the Cape *P. elongata* and of the [western] tropical *P. Mannii,* discovered on the top of the peak of the island of St. Thomas by the naturalist whose name it bears. And these three plants no doubt indicate the affinities of the flora being most strong with the countries north and south of it, and less so with that far to the east of it. This is what the configuration of the continent would indicate as most probable, the loftier mountains being on the east side, and being connected by more or less continuity of high land from Abyssinia to the Cape Colony. That the flora of the latter country extended into the former was well known; and this renders the discovery of a locality in the line of continuous migration or distribution, where the most marked type of the northern flora (*Juniperus*) meets the most marked of the southern (*Calodendron*), and this at the respective limits of each, a most interesting one, and only second in importance to the general result of Mr. THOMSON's labours, which is the discovery of so many northern forms in the comparatively isolated equatorial tract which he has been the first to explore."

"Thus I think it may be regarded as most probable that the Equatorial African mountain-flora is in the main an immigrant one from Abyssinia, possessing many genera and species that have advanced even as far as the Cape Colony, besides many others that have not gone so far, and of which latter a few have been collected by Dr. KIRK during Dr. LIVINGSTONE's second expedition, in the mountains of comparatively low elevation near Lake Shirwa in lat. 15° S. In a lesser degree it has been peopled by a return flow of South-African genera and species, of which many have in like manner advanced further and reached Abyssinia, whilst others have been arrested in their northward spread. It would be interesting, but in the present state of our knowledge fruitless, to speculate on the direction in which the wave of migration is now advancing, and whether the later northern preceded the southern, or *vice versa*. Yet when it is considered that the whole area over which Mr. THOMSON's collections were made is volcanic, and probably geologically modern, in its present configuration, it must be evident that the main features of its vegetation are of no great antiquity."

"There is one more point of interest to which a study of MR. THOMSON's collection invites attention, which is that, whereas the lowlands of Eastern Tropical Africa (and indeed of all Tropical Africa) abound in species and representative species of the Deccan peninsula of India, the highlands of these two regions seem to have nothing in common, botanically or zoologically. And what renders this more noteworthy is that, though they have no types in common, there are desiderata common to both, as exemplified by the absence of Cupuliferae, and paucity of Coniferae, Cycadeae, and Palmae, all of which abound in the Eastern Archipelago and in most other tropical countries. Looking still further off, and comparing the African flora with the Australian, a singular difference is observable in this, that whereas the Tropical-Australian flora is in very great measure made up of species belonging to Temperate-Australian genera, the Tropical-African is, except in the highlands, of a totally different type from the South-African. The tropical floras of both have been obtained largely from the Asiatic continent; but whereas in Australia there is a mingling of the Asiatic and endemic southern genera and species, there is no such mingling of the elements in Africa, except at considerable elevations, in its tropical regions." (pp. 393–6).

In his limited phytogeographical studies on African floras HOOKER was forced to utilize exceptionally inadequate materials and data. In the middle of last century Africa was indeed "Dark Africa" so far as concerned the greater part of its tropics. For Fernando Po and for the Cameroons HOOKER had to draw comparisons with Abyssinia and South Africa since nearly all the intervening country was botanically unknown. Nevertheless, in these pioneer studies certain interesting facts emerged which have been confirmed by later investigators with much more material to hand. This is true, for example, of the occurrence of northern genera and species on the higher parts of the tall mountains—a subject that later much interested ENGLER (*see* ENGLER, 1892, 1904*a*, 1904*b*). On the other hand, the very great advances in our knowledge of both the flora and vegetation of Tropical Africa in the past one hundred years has inevitably made HOOKER's papers appear very incomplete and the impression is, perhaps, inevitable that he was insufficiently cautious in some of the conclusions he drew from meagre data.

The best modern account of the vegetation of Fernando Po is by MILBRAED (1922, pp. 164–75, followed by a floristic list). The northeast cape of the island is only 32 km. from the mainland. The highest point, Clarence Peak or O Wassa, is 2850 m. in altitude. The coast is rocky and MILBREAD found no proper mangrove community. Rain forest is naturally the predominant climax vegetation but is now largely replaced in the lower parts by plantations (especially of cocoa) and secondary communities interspersed with forest relicts. At about 500 m. there is an area of transition between the lower and upper tropical forest and here are almost pure secondary communities of elephant grass (*Pennisetum purpureum*). The upper tropical forest is dominated by *Allanblackia monticola* which composes three-quarters of the community (in the parts visited). The undergrowth is typical of the African hylaea, *i.e.* predominantly of brushwood that is woody in all its axes. The most important shrubs are: *Alchornea floribunda, Leptonychia pallida, Strychnos isobelina, Allophylus hirtellus, Clerodendron grandifolium,* and *Uvaria fusca.* The field layer consists especially of species of *Elatostema, Impatiens,* and *Begonia,* with *Coleus decurrens* and *Crossandra guineensis.* Numerous epiphytes are noteworthy and include *Bryophyta, Hymenophyllaceae,* and other ferns, *Acanthaceae,* and *Begonia* species. The epiphytes are particularly abundant in the upper part of the *Allanblackia* zone and at about 800 m. appear *Polyscias fulva, Macaranga occidentalis,* and *Cyathea manniana,* forerunners of a type of vegetation that attains its maximum development about 1400 m. At about 900 m. *Allanblackia* disappears and from this altitude to 1400 m. a temperate rain-forest occurs. Trees include *Ficus clarencensis, Eriocoelum* sp.*, Bakerisideroxylon revolutum,* and *Entandrophragma rederi.* In the undergrowth there is a predominance of *Acanthaceae*. Above 1400 m. the high forest has a broken canopy. The most important trees are *Macaranga occidentalis, Polyscias fulva, Neoboutonia africana* var. *mannii*, and, above 2000 m., *Schefflera mannii* f. *lanceolata*. The forest limit is at about 2400 m. and the last 400 to 500 m. of the peak region is grassland with a large number of species of grasses and forbs including species of *Asplenium, Polystichum, Lycopodium, Andropogon, Avena, Habenaria, Thesium, Trifolium, Peucedanum, Swertia, Calamintha, Pycnostachys, Veronica, Mimulopsis, Galium, Lobelia, Wahlenbergia, Conyza, Helichrysum,* and *Senecio*.

Fernando Po is included in the area covered by The Flora of West Tropical Africa by HUTCHINSON and DALZIEL (1927–36).

EXELL (1944) discusses the plant-life of Fernando Po in his account of the vascular plants of S. Tomé. He notes that its flora is much less insular in character than that of S. Tomé and is closely allied to that of Cameroons Mountain on the mainland. HOOKER's remarks on the floristic element of temperate nature or affinity are confirmed. EXELL suggests that the proportion of supposed endemics to the whole flora, as at present recorded, is too high and will be reduced by further discoveries on the mainland. For Fernando Po, the present figures are: total genera 468; total species 826; species per genus 1.76; largest families: *Rubiaceae* 97, *Leguminosae* 53, *Euphorbiaceae* 40, *Orchidaceae* 38, *Acanthaceae* 37, *Compositae* 35, *Gramineae* 22, *Cyperaceae* 20 species; endemic species 99; percentage of endemism 12.0. Regarding the islands of Fernando Po, Principe, S. Tomé, and Annobon it is interesting that the degree of endemism increases with distance from the mainland and with size. Statistical studies, based on EXELL's 1944 paper, for the last three of these islands are published by WILLIAMS (1947) and accepted, with interesting comments and further comparisons with Fernando Po, by EXELL (1947).

A great deal has been written on the flora and vegetation of the Cameroons since the time of HOOKER. Important references are those of FABER (1908), JENTSCH and BÜSGEN (1909), BÜSGEN (1910), ENGLER (1910 and 1925), JENTSCH (1911), ENGLER (1919), MILBRAED (1922), UNWIN (1920), MIGEOD (1928), DALZIEL (1928), HÉDIN (1930), MILBRAED (1930), DALZIEL (1930), MAITLAND (1932), MILBRAED (1932), GRANDIDIER (1934), FOURY (1934–5), and FOURY (1935). Here it must suffice to outline the results of more modern investigations on the vegetation of Cameroon mountain. The best descriptions of this are those of ENGLER (1910), DALZIEL (1930), and MAITLAND (1932).

The last-named author recognizes the following zones:

The Mangrove Zone.— This is confined to the southern end of the mountain and to the estuaries of the Rio del Rey and Mungu rivers. It is a mere fringe, interrupted in parts where the rugged rocks form the fore-shore.

The Rain-Forest Zone.— This covers the slopes up to 2000 m. (6500 ft.). It contains a sub-zone with its higher limit at 984 m. (3200 ft.), where the coastal components of the rain-forest cease.

The Mountain Forest Zone.— This immediately follows the Rain-Forest zone and is characterized by a formation consisting of *Hypericum lanceolatum, Lasiosiphon glaucus, Rapanea neurophylla, Lachnopylis mannii, Schefflera hookeriana, Myrica arborea, Pittosporum mannii,* etc., with its higher limit at 2616 m. (8500 ft.).

The Mountain Grass-Land Zone, with its higher limit at 3384 m. (11.000 ft.).

The Alpine Desert Zone.— This succeeds the grassland and continues to the summit.

Cameroons Mountain is of volcanic formation and can be regarded as a restricted range. The tropical forest of the lower part below the grassland is one of the finest examples of moist tropical forest in the world and the full reports of the Cambridge 1947–8 expedition of investigations by modern methods are awaited with interest. Etinde, or Little Cameroon Mountain, 1687 m. (5625 ft.) is clothed with evergreen forest to the summit. On the main mountain there are narrow rifts and steep ravines. Above the forest limit there are steep grass slopes or hog-back spurs deeply cut by ravines. Above is a region forming an irregular terrace with grassy craters, mounds, and boulder-strewn hollows, the whole more or less obscured by low vegetation. DALZIEL notes that the wavy boundary between forest and grassland is due to many forces: lava and showers of molten material, fierce winds, and annual grass fires. Here, when the forest is destroyed it never recovers and all agencies, climatic and human, combine to give the victory to the grass. ENGLER (1910) gives the names of many species occurring at various altitudes.

Kilimanjaro (Kilimandscharo, Kilima Njaro, etc.), in Tanganyika, was first explored botanically in a preliminary manner by Baron DER DECKEN in 1861–2. The Rev. CHARLES NEW collected a number of plants on the mountain in 1871 and his specimens were the basis of HOOKER's note. Since HOOKER's investigations on East African botany were very limited indeed no more need be done here than to refer the reader to the admirable historical summary by COTTON (1930) of botanical research on Kilimanjaro which prefaces the account of his own visit, to draw attention to the concise account published in The Times for 2 Dec. 1931, and to record two papers, with remarkable air and ground photographs, on the climate of the mountain in Weather 2: 329–44 (1947).

Apart from the higher mountains, whose flora was unknown to HOOKER when he wrote on tropical African flora and vegetation, the botany of the Kenya highlands has received remarkably little attention in publications. In the Lukipia, Lake Naivaska, and the Kapte plateau areas there has probably been very considerable change in the vegetation since the visits of THOMSON on whose collections HOOKER based his comments. Attention is called to the papers by BATTISCOMBE (1926), EDWARDS (1935), EDWARDS (1940), and ENGLER (1895).

Chapter VI

NORTH AMERICA

In 1877 HOOKER visited the United States and with ASA GRAY travelled extensively in the Rocky Mountains. Three published accounts
must be considered as a direct result of this journey.

In Nature 16: 539–40 (1877) is a short paper (also published in Amer.
Journ. Sci. 14: 505–9, 1877) with the title "Notes on the botany of the
Rocky Mountains". In this, HOOKER refers especially to the section of
the Rocky Mountains in Colorado and Utah and points out that it
contains representatives of very distinct American floras that are respectively characteristic of immense areas of the continent. "There are
two temperate and two cold or mountain floras, *viz.*: (*1*) a prairie flora
derived from the eastward; (*2*) a so-called desert and saline flora
derived from the west; (*3*) a sub-alpine; and (*4*) an alpine flora; the
two latter of widely different origin, and in one sense proper to the
Rocky Mountain ranges" (p. 539).

HOOKER states that "The most singular botanical feature of North
America is unquestionably the marked contrast between its two humid
floras, namely, those of the Atlantic plus Mississippi one, and the
Pacific one ... ...," and hopes that material now collected will enable
the relations of the dry intermediate zone to be discussed.

The route taken by the expedition is outlined.

"The net result of our joint investigation and of Dr. GRAY's previous intimate knowledge of the elements of the American flora is, that the vegetation of the middle latitudes of the continent resolves itself into three principal meridional floras, incomparably more diverse than those presented by
any similar meridians in the old world, being, in fact, as far as the trees,
shrubs, and many genera of herbaceous plants are concerned, absolutely
distinct. These are the two humid and the dry intermediate regions above
indicated."

"Each of these, again, is subdivisible into three, as follows:—

1) The Atlantic slope plus Mississippi region, subdivisible into (α) an Atlantic,
(β) a Mississippi valley, and (γ) an interposed mountain region with a temperate
and sub-alpine flora.

2) The Pacific slope, subdivisible into (α) a very humid cool forest-clad coast
range; (β) the great hot, drier Californian valley formed by the San Juan river
flowing to the north, and the Sacramento river flowing to the south, both into

the Bay of San Francisco; and (γ) the Sierra Nevada flora, temperate, sub-alpine, and alpine.

3) The Rocky Mountain region (in its widest sense extending from the Mississippi beyond its forest region to the Sierra Nevada), subdivisible into (α) a prairie flora; (β) a desert or saline flora; (γ) a Rocky Mountain proper flora, temperate, sub-alpine, and alpine."

"As above stated, the difference between the floras of the first and second of these regions, is specifically, and to a great extent generically absolute; not a pine or oak, maple, elm, plane, or birch of Eastern America extends to Western, and genera of thirty to fifty species are confined to each. The Rocky Mountain region again, though abundantly distinct from both, has a few elements of the eastern region and still more of the western."

"Many interesting facts connected with the origin and distribution of American plants and the introduction of various types into the three regions, presented themselves to our observation or our minds during our wanderings; many of these are suggestive of comparative study with the admirable results of Heer's and Lesquereux's investigations into the pliocene and miocene plants of the north temperate and frigid zones, and which had already engaged Dr. Gray's attention, as may be found in his various publications. No less interesting are the traces of the influence of a glacial and a warmer period in directing the course of migration of Arctic forms southward, and Mexican forms northward in the continent, and of the effects of the great body of water that occupied the whole saline region during (as it would appear) a glacial period."

"Lastly, curious information was obtained respecting the ages of not only the big trees of California, but of equally aged pines and junipers, which are proofs of that duration of existing conditions of climate for which evidence has hitherto been sought rather amongst fossil than amongst living organisms" (p. 540).

The North American flora is briefly referred to by Hooker in his presidential address to the Royal Society in 1877 (Proc. Roy. Soc. 26:444–6:1878). The following extract is from this address:

"*The American Flora.*— Though I have as yet little to say of the results of Dr. Gray's and my own investigations under the Survey, I have every reason to hope that, having been extended through the Sink, Salt, or desert regions west of the Rocky Mountains, and thence across the Sierra Nevada to the Pacific coast, they will, with the materials previously obtained by my fellow traveller and myself, enable us to correlate our several researches into the distribution of North-American plants, and to point out the lines along which the migrations of the existing types were directed, and the countries whence they migrated."

"As regards the components of the United-States flora, these seemed to us to be threefold, and to be intermixed throughout the continent—an endemic American, a European, and an Asiatic; it seemed that the flora was a ternary compound, so to speak, while that of the temperate Old World was in a continental point of view, binary—Europe and Asia having many types in common, but very few representatives of the strictly American flora. The distribution of North-American plants, unlike the European, is mainly in a meridional direction, the difference of the floras of the Eastern, Central, and Western States being wonderfully great—far greater than those of similarly situated regions in the Old World. The European components extend over the whole breadth of the continent, diminishing, however, to the westward. The American components present many localized genera, inhabiting the Eastern, Central and Western States respectively; they increase in numbers and peculiarity, as also in restriction of range, towards the west. The Asiatic components are found both in the Eastern and Western States, but hardly at all in the Central; and some of them are common to both the east and west, while others are peculiar to each. But whereas the European components prevail on the side towards Europe, the maximum of Asiatic representation is on that remote from Asia. This has been conspicuously shown by GRAY's discovery, in the Eastern States, of single representatives of Japanese genera previously supposed to be monotypic; and what is most noteworthy is, that such representatives are in some cases extremely rare and local plants, found in single and very restricted areas, indicating a dying-out of the Asiatic representation of America."

"The evidences of climatic changes in past eras of the existing flora of the continent are seen in the prevalence of arctic and northern species of plants in the alpine zones of the meridional mountain-chains, the Appalachian, Rocky Mountains, and Sierra Nevada, even as far south as the 33rd parallel. These plants had spread southwards during a period of cold, and on its subsequent mitigation had retired to the lofty situations they now inhabit. To the former existence of a warmer climate we may partly look for the extension of Mexican types to the dry regions west of the Rocky Mountains up to the 41st parallel; and to it may be attributed the remarkable northward extension of the Cacti in a very narrow meridional belt, scarcely one hundred miles broad, along the eastern flanks of the same mountains, from their head-quarters in New-Mexico, in the 33rd, almost to the 50th parallel" (pp. 444–5).

To a lecture given to the Royal Institution of Great Britain, delivered on 12 April 1878 and printed in Proc. Roy. Inst. 8: 568–80 (1879), as a separate, and also in Gard. Chron. N.S. 10: 140–2, 216–7 (1878), HOOKER gave the title "The distribution of the North American flora". After commenting upon the large number of plants spread about the world by Anglo-Saxon immigrants and outlining the main physiographical features of North America the principal characters of the flora are given as follows:

"*Polar Area*.— Commencing in the Polar area, the Arctic American flora, though on the whole a uniform one, is distinctly divisible into three; the first extends from Behring's Straits to the mouth of the McKenzie River, and is marked by the presence of certain Asiatic genera and species that advance no farther eastward; the second extends thence onwards to Baffin's Bay, and presents various American genera and species not found either eastward or westward of it; and the third is that of Greenland, which is almost exclusively European, and presents several anomalies, which I shall hereafter discuss. Besides this eastern and western distribution of the Arctic flora, it streams southward along the three meridional mountain chains of the continent."

"*British North-American Flora.*— South of the Arctic flora is that of the British possessions, that is of temperate America north of the 47*th* parallel; it consists of a mixture of North European, North Asiatic, and American genera, in very different proportions, disposed in five meridional belts—(*1*) to the eastward, the Canadian forest region; (*2*) the woodless region, a continuation of the prairie region farther south; (*3*) the Rocky Mountain region, where Mexican genera appear; (*4*) a dry region, a continuation of the Desert or Sink regions to the south of it; and (*5*) the Pacific region, which assimilates very closely in its vegetation to that of Kamtschatka."

"*United States Flora.*— It is on entering the United States that the flora of temperate North America attains its great development of genera and species in all the meridians, and that the boundaries of the meridional belts of vegetation are most strictly defined."

"I. *The Great Eastern Forest region*, extending over half the continent, and consisting of mixed deciduous and evergreen trees, reaches from the Atlantic to beyond the Mississippi, dwindling away as it ascends the western feeders of that river on the prairie. It is noteworthy for the number of kinds especially of deciduous trees and shrubs that are to be found in it, even on a very limited area. Of this I shall select two examples from my Journal. One was a patch of native forest, a few miles from St. Louis, on the Missouri, where in little more than half an hour, and less than a mile's walk, I saw forty kinds of timber trees, including eleven of oak, two of maple, two of elm, three of ash, two of walnut, six of hickory, three of willow, and one each of plane, lime, hornbeam, hop-hornbeam, laurus, diospyros, poplar, birch, mulberry, and horse-chestnut; together with about half that number of shrubs."

"The other example was afforded me by Goat Island, which divides the great cataract of Niagara, and covers less ground than Kew Gardens. Here the vegetation was more boreal and less varied than in Missouri; but with Dr. GRAY's aid I counted thirty kinds of trees, of which three were oaks and three poplars, together with nearly twenty different shrubs."

"I know of no temperate region of the globe in which any approach to this aggregation of different trees and shrubs could be seen in such limited areas, and perhaps no tropical one could afford a parallel."

"No less remarkable is the composition of the flora of the Eastern States. Professor GRAY has shown that most of its genera are common to Europe and Asia, but very many are all but confined to North-eastern Asia and Western America. This generic identity, however, gives but a faint idea of the close relationship between the East American and East Asiatic, especially the Japanese, floras, for there is further specific identity in about two hundred and thirty cases, and very close representation in upwards of three hundred and fifty; and what is most curious is, that there are not a few very singular genera, of which only two species are known, one in East Asia, the other in East America; and in some of these instances the Asiatic species is a widespread plant in East Asia, whilst the American is an extremely scarce and local plant in its country, which and other considerations render it conceivable that the Asiatic element in East America is a dying-out one."

"Leaving out of consideration the purely American genera of this flora, there remain the genera common to Europe, Asia, and America; the genera confined to America and Asia; and the genera confined to America and Europe. I shall give an illustration of the proportions in which these occur by a reference to the principal trees and large shrubs only, their names being familiar to you, though the smaller shrubs and herbs afford infinitely more numerous and striking examples; thus, of those common to the three northern continents, I find in America thirty-eight genera with about one hundred and fifty species, these include maples, ashes, hollies, elms, planes, oaks, chestnuts, nut, hornbeam, birches, alders, willows, beech, poplars, etc. Of those confined to America and East Asia I find in America thirty-three genera and fifty-five species, including magnolias, tulip tree, negundo, wistaria, Virginia creeper, gleditschia, hydrangea, liquidamber, nyssa, tecoma, catalpa, diospyros, sassafras, benzoin, mulberry, walnut, and others, which not being European, are unfamiliar to you. Lastly, of those confined to Europe and America I find only one genus, namely the hop-hornbeam, of which there is but a single representative in each country."

"Here then is conclusive evidence of the close botanical relationship of North-eastern Asia and Eastern North America; a relationship of which there is but little evidence in the vegetation of the Prairies and Rocky Mountains, and still less, perhaps, in the regions farther west."

"II. *The Prairie region* succeeds, a grassy land with many peculiar herbaceous American genera, including Mexican types, of which last the most conspicuous are a yucca and cacti, which latter increase in number as the Rocky Mountains are approached, where they form a noticeable feature in the landscape."

"In the parks and lower valleys of the Rocky Mountains deciduous trees are few and scattered, and the forest is an open one of conifers, amongst which a pine, allied to the Mexican nut-pines, *P. edulis*, first appears. Higher on the mountains the coniferous forests are dense, and almost the only deciduous tree is an aspen, which forms impenetrable brakes on the slopes and in the gullies. Above the forest region are the sub-alpine and alpine regions, presenting a mixture of European, Asiatic, and American types."

"III. Descending to the Sink region the cacti and yucca almost disappear, though they increase to a maximum farther south in this meridian. Deciduous trees are very few, and confined to the gullies of the mountains, and Mexican genera increase in numbers. The hoary sage-bush (*Artemisia*) covers immense tracts of dry soil, and saline plants occupy the more humid districts."

"Another nut pine of Mexican affinity (*P. monophylla*) traverses the centre of this region in a narrow meridional strip, and the proportion of endemic plants, herbaceous especially, is very large."

"IV. *The Sierra Nevada* is clothed with the most gigantic coniferous forest to be found on the globe, amongst which a very few species of deciduous trees are scattered; but none of these are identical with trees of the Eastern forests, though several are representative of them. New Mexican genera occur at all elevations from the crest of the range to its base, and thence extend across the Californian valley and the Coast-ranges to the Pacific, mixed with northern West American genera and species."

"In this slight outline of the botanical features of temperate and Arctic North America, I have alluded to three as most noteworthy, namely, the vegetation of Greenland, the Asiatic character of the vegetation of the eastern half of the continent, and the more southern and even Mexican character of the vegetation of the western half. How are these features to be accounted for?"

"It so happened that Dr. GRAY, Professor of Botany in Harvard College (Boston), and I were contemporaneously, but without concert, engaged in botanical investigations, which have resulted in explanations of the two first features. He was at work on the flora of Japan, I on that of the Polar Zone, and we were both bringing to bear upon our subjects considerations regarding the variation of species which Mr. DARWIN almost simultaneously laid before the public, and which, I need not say, powerfully directed our studies."

"I shall take the vegetation of Greenland first, as being first in order, though second in date of appearance and least in importance. Its chief peculiarities are, (1) that its plants are almost all of them Scandinavian (that is, North-west European), hardly any of the peculiar plants of the American Arctic sea-coast and Polar islands crossing Baffin's Bay and Davis Straits; (2) that of its 300 flowering plants hardly any present even a variation from their Scandinavian prototypes; (3) that it is poorer in species than is any other division of the Arctic flora, and wants many Scandinavian plants that are found in most other Arctic countries; (4) that though Greenland extends 400 miles south of the Arctic circle, its extra-Arctic continuation adds only about 100 species to the flora, and these all cross the Arctic circle in other longitudes; (5) some Greenland species are confined to it and to the mountains of the Atlantic side of America, being found nowhere else in Arctic or sub-Arctic America."

"My explanation of these anomalies was, that at a period previous to the Glacial, a flora common to Scandinavia and Greenland was spread over the whole American Polar area, and that, on the accession of the cold of that period, this flora was driven southwards, and was affected differently in different longitudes. In Greenland many species were exterminated, being as it were driven into the sea at the southern extremity of the penisula, where only the hardiest survived. On the return of warmth the Greenland survivors migrated northward, peopling the peninsula with the hardiest of the species of its former flora, unmixed with American species; and unchanged in aspect, from never having been brought into competition with those of any other flora. On the other hand, the same Scandinavian plants when driven south on the plains of the continent, multiplied there in individuals, and being brought into competition with American species

descending from the continental mountains on to the plains, assumed varietal forms. On the return of warmth, therefore, many Scandinavian species
that had been exterminated in Greenland, would, having survived on the
continent, travel northwards on it, some unchanged, others under varietal
forms, accompanied with American species that had descended from the
mountains during the cooling of the continent. Lastly, as some of the
Scandinavian species were no doubt local, and confined to near the meridian
of Greenland, it is not surprising to find that a few such should survive only
in Greenland and on the eastern alps of North America."

"Thus only could I satisfactorily account for the almost complete identity
of the Greenland flora with the Scandinavian after such changed conditions
of climate; for the paucity of its species; for the absence in it of varieties:
for the rarity in it of peculiarly American species; for the few species which
extra-Arctic Greenland adds to its Arctic flora; and for certain of its plants
being limited in range to Greenland and the eastern American alps" (pp.
5–9, of separate).

The views of AsA GRAY regarding the relationship between the flora
of North-east Asia and Eastern North America are briefly summarized.
In Miocene times many of the existing genera and even species of both
the Japanese and North American floras co-existed in the high latitudes
of America and GRAY assumed that during this period the three northern continents were conjoined, or so contiguous as to allow of a
commingling of their floras.

The glacial period carried "an Arctic climate south to the latitude
of the Ohio, but so gradually, that these plants were not exterminated,
but wholly or in part driven southward, followed in the rear by the
Arctic vegetation. As the temperature rose with the retreating ice,
this flora returned northward, leaving the Arctic and subarctic plants
on the mountains of both East and West America." (p. 9). The retreat
northward of the flora was to a higher latitude than that it now attains
and this resulted in a second commingling of American and Asiatic
plants. "Lastly, DANA's Terrace epoch supervened, when the previously depressed northern region was again raised, cooling the climate,
finally dissociating the Asiatic and American floras, and giving to the
Arctic and sub-Arctic plants of the continent their present limits"
(pp. 9–10).

"It remains now to account for the great rarity of East Asiatic types in
America west of the Prairies, and the presence in those meridians of Mexican
and still more southern ones. Hitherto there have been no other attempts at
a solution of this problem than such unsupported speculations as that the
western half of the continent, though so much the loftier, was submerged
during the southern migration of the northern Miocene plants; or that the
climate of the West was unsuited to the habits of these, which appears to

me to be at variance with the fact that when imported into it they thrive luxuriantly."

"The explanation which I have to offer will be best understood by a reference to our section (p. 4), which shows the western half of the continent to be enormously elevated as compared with the eastern, and to have been singularly adapted for the retention of vast bodies of ice for long after the Glacial period. We find there a valley (the desert region), upwards of 400 miles broad, and upwards of 4000 feet elevation, with many ranges of over 8000 feet in it, bounded by broad and lofty mountains, together occupying at least two-thirds of the breadth of the western half of the continent. We further know that these mountains were clothed with ice during the Glacial epoch, and that the valley was then occupied by a vast lake; for on the uppermost of the many shelves which the retiring waters of this lake cut on the flanks of the Rocky Mountains and Sierra Nevada, the skull of the musk-ox, the most Arctic of land quadrupeds, has been found."

"It is obvious that this whole western region must have retained its glacial mantle for an incalculable period after Eastern America had been sufficiently warmed to admit of the northward return of the plants that had been driven southward in it; and that this glaciated condition must have effectually barred a similar return of the same plants in those western meridians; these must have perished in short on reaching Southern California. Long ages after, when the western ice disappeared, and the climate of the valleys warmed, the Mexican and more southern plants would, as a matter of course, take possession of the unoccupied soil, and advance northward till they encountered the boreal vegetation of North-western America, with which they now comingle" (p. 10).

As examples of East Asiatic types in Western America that escaped extinction HOOKER refers to the species of *Sequoia*. He discusses their existing and fossil ranges, the ages they now attain when growing, and the destruction by man of the remaining forests.

In the Bulletin of the United States Geological and Geographical Survey of the Territories 6: 1–62 (1880), ASA GRAY and J. D. HOOKER published a joint paper under the title "The vegetation of the Rocky Mountain Region and a comparison with that of other parts of the world". Though it is impossible to know how much in most parts of this paper to attribute to HOOKER separately, it is of sufficient importance to summarize and illustrate by extracts.

The Rocky Mountain region is divided into three altitudinal zones ("botanical districts"):

1) An arid and woodless district, which occupies far the greater part of the area.

2) A wooded district, in some places covering, in others locally adorning, the mountain slopes.

3) An alpine unwooded district above the belt where trees exist. But in some places, slopes woodless from dryness merge into tracts woodless from cold, no proper forest belt intervening (p. 3).

These three zones (in the later used headings and text, they are termed "regions") are considered separately at some length.

The "Alpine Region."— It is noted that "Botanically the alpine regions of the temperate zone in the northern hemisphere are southward prolongations of arctic vegetation, almost pure in the boreal parts, but more mixed with special types in lower latitudes, these special types being a part of the flora which is characteristic of each continent in those latitudes" (p. 3).

In the Eastern United States, the "peculiar element in the scanty alpine flora" consists of only five species, whilst the Pacific "alpine" flora has a higher proportional number of non-arctic species, but the number pertaining to non-arctic genera is small and all of their genera are peculiar to America. Of the whole 111 species (of which a list is given) about 50 are not known in Europe and Asia in identical species. In the list there are 184 "alpine" species for the Rocky Mountains and 52 for the mountains in the northeastern part of the Atlantic states. "Notwithstanding the geographical extent of the country over which it is spread, the North American alpine flora is meagre in species compared with that of Europe."

The *"Forest Region."*— In stating that "The most conspicuous portion of the vegetation of a country, and the most important under more than one point of view, is its trees" our authors would appear to anticipate some of the conclusions reached by applying the concept of climatic climaxes. A list is given of about 50 species of trees recorded over the whole of the Rocky Mountain region "in the widest extent". While forests are much better developed towards the north they are richer in species in the south.

"This is not the place to institute a comparison between the Rocky Mountain forest and the eastern; but it may be remarked that, while angiospermous, round headed, and deciduous-leaved trees prevail in the latter, largely in the number of species and genera and conspicuously in the extent of surface occupied, the Rocky Mountain sylva, in its characteristic features, is gymnospermous, spiry, and evergreen. In the importance of its useful products, such as lumber, the difference between the two sorts, as a whole, in the Altantic forest cannot be great. But with perhaps only one exception, that of the so-called Mountain Mahogany, *Cercocarpus ledifolius* (a small tree or more commonly a shrub), the economical value of the Rocky Mountain forest is almost wholly in its coniferous trees, and in the mountains these alone strike the eye" (p. 11).

An account is given of the constituent species of the forests arranged in the order of their conspicuousness and importance.

"Without entering here into a comparison of the Rocky Mountain forest with any other, it may be noted that the species are peculiar to the

8

region or the vicinity of it, with a few exceptions. *Prunus Pennsylvanica, Populus balsamifera, monilifera,* and *tremuloides,* may be said to come in from the northeast and only the last extends far into the district. The *Negundo* and *Juniperus Virginiana,* with *Fraxinus viridis,* belong to the Atlantic forest region, and do not penetrate far, unless we count the Californian *Negundo* as a derivative form. The connection with Pacific forest species is closer; and for the rest they are mainly Mexican plateau types, of which the botanical district in question may be regarded as a northern extension" (pp. 16–7).

The characteristics of the shrub and herbaceous vegetation of the Rocky Mountain forest region is dealt with in outline.

"The peculiar shrubs of the Rocky Mountains (including the Wahsatch Range and corresponding ranges farther north) are only *Jamesia Americana,* a Hydrangeous genus of no near affinity to any other, except *Fendlera,* which (equally unique) belongs to a lower region in New Mexico and Western Texas, *Robinia Neo-Mexicana,* which is an outlying species on the south-eastern border, *Quercus undulata, Rubus deliciosus, Philadelphus microphyllus, Ceanothus Fendleri,* and *Berberis Fendleri,* the latter a species of the *Vulgaris* type. They are all southern; the Northern Rocky mountains have no characteristic shrub, as they have no characteristic tree. The principal shrubs which they share with the Pacific forest region are *Acer glabrum, Prunus demissa, Rubus Nutkanus, Spiraea discolor, Ribes,* 3 or 4 species, *Symphoricarpus oreophilus* and *rotundifolius, Ledum glandulosum, Salix Geyeriana,* and, if we come down to such low frutescent growth, *Pachystima Myrsinites,* and *Berberis repens*" (p. 18).

The more characteristic and restricted genera of herbs are mentioned with some notes on their distributions.

"*Woodless Regions below Forest*".— Our authors distinguish three districts under this heading.

1) The Lower Rocky Mountain Slopes, including the "parks" of Colorado and non-saline valleys. The prevalent characteristic shrubs are largely of the family *Rosaceae* (*Cercocarpus parvifolius, C. ledifolius, Cowania mexicana, Purshia tridentata, Spiraea discolor, S. millifolium, S. caespitosum, Prunus andersonii*). Other shrubs are: *Ceanothus velutinus, Ribes cereum,* and *Ephedra* sp. Amongst herbs important genera are: *Gilia, Pentstemon, Phacelia, Eriogonum, Astragalus,* and *Oenothera.* There are also many *Compositae.*

2) The arid or desert interior district. This includes the Great Basin. "The region, in a general botanical view, is one of *undershrubbiness;* and the prevalent growth is composed of *Artemisia, Chenopods,* and lignescent small-flowered *Compositae*" (p. 21). A list of genera more or less peculiar to the Great Basin proper and its borders is given.

3) The eastern woodless plains. "If the arid district of the interior of the United States west of the eastern Rocky Mountains is denominated the region of "Sage Brush" (*i.e.,* of shrubby *Artemisia* and *Chenopods*), the mostly less arid, less saline, equally homogeneous, and even more extensive plains between the Rocky Mountains and the eastern forest region may be characterized as the region of

Buffalo Grasses. Its full development is between latitude 35° and 45°, where it occupies an average of ten degrees of longitude. North of this it is narrowed or interrupted, and then merges into a district which is woodless from cold or from the nature of the soil, and at length arctic. Southward it is equally broad, and it trends westward and loses itself in the New Mexican plateau region, which has a certain character of its own, but in which the eastern forms of vegetation mingle first with those of the Rocky Mountains, with those of the Mexican plateau, and at length with those which prevail in the Great Basin." (p. 24).

It is suggested that annual burning in the autumn by Indians is a cause of extending the treeless prairie region. Another cause of prairie conditions over areas that can now carry trees, if not fired, may be that the climate was formerly drier than it is now.

"The western portion of these plains is not only drier, but in some parts alkaline, or with other characters of soil uncongenial to forage grasses, especially at the north, where there are only two inches of rain in the three summer and no more in the three winter months. A good deal of the southern part gets about four inches of summer rain, but only half as much in winter. In some parts, accordingly, the characteristic vegetation of the ultramontane plateau intrudes. The Pulpy Thorn, *Sarcobatus*, and its Chenopodeous associates are largely developed on the Upper Missouri waters, accompanied by a peculiar Sage-Brush, *Artemesia cana*, while the *A. tridentata* is rather rarely established on this side of the mountains."

"We have termed this district the region of Buffalo Grass. The grasses form such an inconspicuous and unimportant a feature in the interior arid region that it has not been worth while to mention them, and even on the mountains, except in the alpine region, they are of small account. On the eastern plains they are the characteristic feature. When we get beyond the eastern prairie border, the grasses of which are prevailingly eastern in character, we come upon plains which are generally covered with the very low and tufted grasses peculiar to the drier plains, which form, if not a sward, yet something which serves as a substitute for it, not green, except in early spring, but of dull grayish hue, and the characteristic species usually rising only a hand-breadth above the surface. These are the Buffalo Grasses or Bunch Grasses, which have nourished hordes of bison and flocks of antelopes down to a few years ago, and which are now the capital of the herdsmen or ranchmen, and the nutritious food of increasing numbers of domestic cattle."

"The Buffalo Grass, *par excellence*, and by its abundance, is *Buchloë dactyloides* of ENGELMANN. This is a dioecious Chlorideous grass, the male and the comparatively scarce female plants of which were very naturally thought to be of quite different genera until their relationship was suspected and determined by Dr. ENGELMANN, and this apt name was applied to it" (pp. 25–6).

Other genera of grasses mentioned are: *Munroa*, *Bouteloua*, *Pleuraphis*, *Vaseya*, *Eriocoma*, *Sporobolus*, *Beckmannia*, *Distichlis*, *Atropis*, *Stipa*, *Aristida*, *Hordeum*, and *Elymus*.

"Of other dominant and more or less peculiar forms of vegetation—having chiefly in view the central tract—we should mention a great white-flowered *Argemone* (*A. hispida*, Gray); *Stanleya*, and the greater part of the known species of *Vesicaria*; *Cleome integrifolia*; the whole genus *Callirrhoë*; a *Krameria*; a *Glycyrrhiza*; the herbaceous *Sophora sericea*; the principal development of the peculiar genus *Petalostemon*, and southward numerous species of *Dalea* (which go on increasing into Mexico); also of *Psoralea*; most of the species of *Gaura*, several of *Œnothera*, and the peculiar genus *Stenosiphon*, allied to *Gaura*; a good number of *Cactaceae* (chiefly *Opuntiae* and *Mammillariae*), increasing southward; a thick-rooted perennial *Cucurbita* (*perennis*), with some relatives southwestward; the species of *Machaeranthera* · or biennial Asters, *Aplopappus spinulosus* and some other species; *Bigelovia* and *Gutierrezia* in characteristic forms which are shared with the ultramontane arid district, and a great development of Senecionoid Compositae, perhaps not exceeding the other parts of the United States, yet more conspicuous to the eye; the two species of *Solanum* with prickly calyx closed over the fruit; *Pentstemon* in species equalled only by California; *Hedeoma* and *Monarda*; *Leucocrinum*, which, however, extends westward."

"Besides those variously mentioned, a goodly number of genera are peculiar to this and the more western districts, which we need not here enumerate. Of absolutely peculiar genera, there is *Selenia*, in Cruciferae; *Cristatella*, in Capparidaceae; *Musenium*, *Polytaenia*, and *Trepocarpus*, in Umbelliferae; *Thelesperma* (except for a Buenos Ayrean species), *Engelmannia*, *Bradburia*, *Diaperia*, etc., among Compositae; *Stephanomeria*, *Lygodesma* and *Troximon* are very characteristic Cichoraceous genera, which also abound far westward."

"The characteristics of the Rocky Mountain flora—whether taken as a broad whole or in its constituent geographical parts—are in no small degree negative. What this flora lacks is perhaps more remarkable than what it possesses. This will appear on a comparison of the vegetation of the three great regions: the Atlantic naturally wooded region; the Central region, woodless except on mountains; the Pacific region, largely but not wholly wooded" (pp. 26–7).

A comparison is made, group by group, of the Atlantic, Pacific, and Rocky Mountain Region floras. The groups are nearly all what are now generally accepted as families. From internal evidence it appears that this section of the paper was written by GRAY.

It is pointed out that many North American species or "types" have extended southwards into South America. The migration has been mainly on the western side of the American continent on which the mountains abut on the coast. A list of 90 such species or genera is given.

"The natural and obvious line of communication between the botany of the northern and southern temperate zones has been along the central part of North America and Mexico, and along the western part of South America. When our cool temperate flora flourished only along or near the southern borders of the United States, the warm-temperate (to which most of the above-enumerated forms belong) were still further south. When the climate became again warmer, a portion of these plants were as well placed for southward as for northward retreat" (p. 60).

Conclusions as to the origin of the North American flora are reached as follows:—

"Our knowledge—fragmentary, yet real—of the flora around us begins with a period when it or its direct ancestors occupied the zone between the arctic circle and the pole, and doubtless several lower degrees of latitude. There it must have flourished until the coming on of that change of climate which culminated in the glacial period. It must at that time have encircled that portion of the earth much as the arctic flora now does. During the period of maximum refrigeration, its northern limits, abutting upon an arctic flora then in low latitude, must have been so far south in the Atlantic States that the vegetation of the northern shore of the Gulf of Mexico probably resembled that of the southern shore of the Gulf of Saint Lawrence now. Of this northern limit there cannot be much doubt; yet we could not hazard an opinion as to where the warm-temperate vegetation of that day merged into the subtropical, as it now does in Southern Texas."

"The change between that period and the present, in the opposite direction, has been an amelioration of climate which has carried the arctic flora back to the arctic circle, with which we now associate it, excepting the portions which, in the retreat, have ascended the mountains and persisted there, forming the arctic-alpine vegetation. This, as we have seen, is very scanty in the Atlantic district, where it has abided only on the most northern mountains; while the more elevated ranges of the western part of the continent have afforded ampler refuge."

"A similar advance and ensuing retrogression, consequent upon the coming and going out of the Glacial epoch, must have taken place in other parts of the northern hemisphere. Under these great and protracted movements of transference, we suppose that a common flora, which was comparatively homogeneous round the new arctic zone, has been differentiated into the several existing north-temperate floras, and that their common features, and the occasional very unexpected identities or similarities (such as those between Japanese and North American botany) are thus explained. Their respective peculiarities are thought to have resulted from the different vicissitudes and the different climatic conditions to which the primeval stock has been exposed in Asia, Europe, and America, and upon the opposite sides and great interiors of continents, the climates of which—greatly different now—have probably been so from very early times. The plants which were most adapted or adaptable to the one could not be expected to survive in another, or in any other than one of similar or analogous climate. But this is not the place for considering the application of these principles to the botany of the northern hemisphere generally. When they come to be applied to the theoretical elucidation of the great difference between the

Atlantic and the Pacific floras it will need to be noted that the two sides of the continent, at the time when they received the progenitors of the present vegetation, were more completely separated than now; that they seem to have been, as it were, two long peninsulas stretching southward from a mainland at the north, the great plains between our eastern district and the Rocky Mountains being then under water."

"It may be inferred that the Atlantic side of the continent was more open than the western to the reception of the ancestral flora from the north, and so received it in larger measure and variety, or that it has been since that time more free from disturbance and catastrophe. Probably the two causes may have conspired in the production of the result. There is, moreover, reason to suspect that the recession of the glaciation was earlier on the Atlantic side of the continent than in the more elevated central and Pacific regions; and that, from all these causes, its preglacial flora was more completely restored to it than to that of the Pacific side."

"And, finally, we infer that the Pacific region, while preserving through all vicissitudes a moderate number of boreal types, and receiving a few Eastern Asiatic ones probably at a later date, has been mainly replenished from the Mexican plateau, and at a comparatively late period. A large part of the botany of California, still more of Nevada, Utah, and Western Texas, and, yet more, that of Arizona and New Mexico, may be regarded as a northward extension of the botany of the Mexican plateau."

"This may, at least, be said: that two types have left their impress upon the North American flora, and that its peculiarities are divided between these two elements. One we may call the *boreal-oriental element*; this prevails at the north, and is especially well represented in the Atlantic flora and in that of Japan and Manchuria; the other is the *Mexican-plateau element*, and this gives its peculiar character to the flora of the whole southwestern part of North America, that of the higher mountains excepted" (pp. 61–2).

* * *

J. D. HOOKER's special interest in the flora and vegetation of North America was concerned with his own experiences on his one expedition to the Rocky Mountains and to his friendly and intimate relationships with ASA GRAY. That his father, Sir WILLIAM JACKSON HOOKER, was the author of "Flora boreali-americana," London, 1829–40; in two quarto volumes, leads one to assume that he had for long been familiar with the earlier known floristics of North America. There is a good account of HOOKER's visit to the United States in L. HUXLEY's "Life and letters of Sir JOSEPH DALTON HOOKER," vol. 2, chapter XXXVIII, pp. 205–27.

Plate 17 shows Sir JOSEPH and ASA GRAY in their camp in the Sierra Sangre de Cristo of Colorado (summer, 1877). Dr. JOSEPH EWAN (Tulane Univ.) kindly sent us the following details about this interesting photograph:—

Apes' hill (Africa).

Plate 16. — APES' HILL, N. AFRICA, in background; from Gibraltar.
From HOOKER'S Sketches, at Kew. Photo of original.

Plate 17. — Encampment of Hooker and Gray in the Sierra Sangre de Cristo of Colorado (*see* p. 119).

"In the summer of 1877, when Sir JOSEPH DALTON HOOKER visited the United States to study the distribution of forests and the broad relationships of our vegetation, he crossed the continent in the company of Professor ASA GRAY and Mrs. GRAY, pausing at St. Louis to visit Dr. GEORGE ENGELMANN, and botanized at two principal localities in Colorado before continuing on to the Great Salt Lake and San Francisco. HOOKER wrote in *Nature* (16:539. 1877): "I availed myself of an oft-repeated invitation to us both from Dr. HAYDEN... to join the Topographical and Geological, Survey in Colorado and Utah." The first stop of three days was near the summit of La Veta Pass in southern Colorado, where this photograph of their campsite was taken. Dr. GRAY is seated on the ground, press in hand, with specimens of *Abies concolor* spread on the ground before him. On his right is HOOKER. It is interesting to recall that Mrs. GRAY, who is seated at the table and is wearing the veil, considered these photographs of HOOKER and of her husband "perhaps the best ever taken". Sir RICHARD STRACHEY, in gray topper, map in hand, and his wife across the table, dressed in black, accompanied HOOKER to America primarily to inspect the system of narrow gauge railroads and, incidentally, to indulge his taste for natural history. They were en route to India where General STRACHEY had travelled and served for many years with the Royal Bengal Engineers. Dr. F. V. HAYDEN is at the extreme right, coffee cup in hand. The geologist and geographer HAYDEN had spent a quarter century in the Rockies and was then chief of the "HAYDEN Survey" as it came to be called. His assistant, Capt. JAMES STEVENSON (1840–1888), standing behind General STRACHEY, joined the HAYDEN Survey in the "Northwest" before he was sixteen years old. Seated at the table to the right of the General is Dr. ROBERT HENRY LAMBORN (1835–95), metallurgist, vice-president of the Denver, Rio Grande and Western RR., friend of Dr. ENGELMANN, who had joined the party from his interest in railroading and his wish to assist General STRACHEY. The guide, standing at the left, and the unseen photographer are unknown. WILLIAM HENRY JACKSON, who was a member of the HAYDEN Survey during those years, was absent in New Mexico at this time. Dr. HAYDEN probably hired the guide for the party, as he arranged other particulars of the journey, and accompanied the scientists to San Francisco. The other unidentified persons in the photograph may have been personnel from Forts Lyon and Garland, which the party visited on this trip to the Sierra Sangre de Cristo."

As already remarked, the paper published in 1882, the most important of HOOKER's publications on the botany of North America, was prepared jointly by ASA GRAY and HOOKER. Internal evidence suggests that much of it was the work of ASA GRAY though L. HUXLEY (*l.c.*, p. 218) states that "The full botanical results [of the 1877 journey] were worked over when ASA GRAY, on his visit to Europe in 1881–2, spent a couple of months at Kew."

So much research has been carried out on the flora and vegetation of the United States since 1882 that any attempt to summarize it would be presumptuous and even to approach completion and success would require a volume to itself. HARSHBERGER (1911) produced a volume on the phytogeography of North America which has been rather severely criticized by some American botanists with regard to various, often minor, details. It contains a valuable bibliography (that for the Rocky Mountains occupying pp. 71–4) and a summary of the botanical features of the Rocky Mountain Region (pp. 546–67). HARSHBERGER adds little to the wider phytogeographical problems that so greatly interested GRAY and HOOKER.

The following are some of the more important and general papers and books bearing on North American phytogeography since 1911. Many of them contain useful lists of published works by consulting which the student can obtain a fairly complete bibliography on this subject: BERRY (1927), CAIN (1944), CHANEY (1947), DAUBENMIRE (1943), DICE (1943), LIVINGSTONE and SHREVE (1921), OOSTING (1948), RAUP (1941), and WEAVER and CLEMENTS (1929, 1938).

Chapter VII

THE GALAPAGOS ISLANDS

It may be some comfort to botanists of the post-Hitler period to know that HOOKER's paper "On the vegetation of the Galapagos Archipelago, as compared with that of some other tropical islands and of the continent of America", in Trans. Linn. Soc. Lond. 20: 235–62 (1851) which was read to the Linnean Society on 1 and 15 December 1846, did not appear until five years later! It is based on a preceding paper by HOOKER: "An enumeration of the plants of the Galapagos Archipelago; with descriptions of those which are new" read on 4 March, 6 May, and 16 December 1845 and printed in the Trans. Linn. Soc. 20: 163–234 (1851). This latter paper is a systematic list of the specimens collected in the Galapagos Islands by CHARLES DARWIN, JAMES MACRAE, and a few other naturalists. DARWIN visited the islands during the voyage of H.M. Ship 'Beagle' and devotes a chapter to them in his "Journal of Researches" (London, ed. 1, 1839; ed. 2, 1852; chapter XVII).

That HOOKER devoted much time and thought to his two papers on the Galapagos Islands is evident from the facts published in "Life and Letters". His studies led to wider considerations than it was at first anticipated they would.

"The results of my examination have been, that the relationship of the Flora to that of the adjacent continent is a double one, the peculiar or new species being for the most part allied to plants of the cooler parts of America, or the uplands of the tropical latitudes, whilst the non-peculiar are the same as abound chiefly in the hot and damper regions, as the West Indian islands and the shores of the Gulf of Mexico; also that, as is the case with the Fauna, many of the species, and these the most remarkable, are confined to one islet of the group, and often represented in others by similar, but specifically very distinct congeners."

"This examination has led me to take a survey of the vegetation of several other tropical islands, whose plants present much peculiarity, and to trace the effects of isolation in geographical position upon vegetation; as well as certain characters in some orders, their distribution and proportions, which seem to distinguish insular floras from the continental" (pp. 235–6).

He summarizes the vegetation as follows;

"The genera *Avicennia* and *Rhizophora*, species of both of which bear the name of Mangrove in different parts of the world, prove that in some of the islands at least (Charles and Chatham) there is a phaenogamic vegetation below high-water mark. On the other hand, from the steepness of the coasts and dryness of the soil near the ocean, there appear to be few maritime plants. Those which I presume to be more strictly such are *Cissampelos Pareira*, *Tephrosia littoralis*, *Scaevola Plumieri*, *Convolvulus maritimus*, *Calystegia Soldanella*, *Verbena littoralis* and *Heliotropium Curassavicum*, all natives of the South American coast, and to which may probably be added some of the peculiar *Amaranthaceae*."

"The lower parts of the island are very arid and rocky, presenting thickets of starved shrubs and leafless trees, and to these situations are assigned the weeds of the Flora, such as herbaceous or suffrutescent *Malvaceae* and *Euphorbiaceae*, many species of *Borreria*, some *Compositae*, various *Lycopersica*, *Verbenae*, *Galapagoa*, *Boerhaavia* and some grasses; to which may be added some larger shrubs, as small trees of *Acacia*, *Castela*, *Cactus* and *Opuntia*. Where marshy land occurs, and this is not uncommon on the summits, several species of *Cyperus* and *Mariscus* appear; and to a salt lake, which is beautifully fringed with succulent plants, belong *Portulaca*, some *Amaranthaceae*, *Pleuropetalum*, and probably *Sesuvium*."

"On ascending the hills the climate and vegetation both suddenly change, the sea-vapours are condensed on the higher parts of the islands, and a comparatively luxuriant flora is the consequence. From these more favoured localities are brought the greater number of the very peculiar vegetable forms of the island; curious arborescent *Compositae*, which have no near allies in other parts of the globe, and of which there are eight species in this group, all closely related to one another. Associated with these are trees of *Phytolacca*, *Leguminosae*, *Psidium*, *Psychotria*, *Chiococca* and *Clerodendron*, all tropical in appearance, accompanied by others no less characteristic either of a warm and equable temperature, humid atmosphere or wooded region; such are the genera *Passiflora*, *Viscum*, *Ipomaea*, *Epidendrum* and *Peperomia*, with the great majority of the Ferns, and all the *Jungermanniae* and *Musci* that have been collected on the group" (pp. 237–8).

HOOKER considers the flora is "an exceedingly poor one when compared with that of other tropical islands of their own or even less extent" (p. 238). He notes that one peculiarity is the paucity of monocotyledonous plants, "which hardly equal 1/9 of the Dicotyledons", though "the tropical islands in general possess proportionally more *Monocotyledones* than do the continents" (p. 240). He reviews the proportion of monocotyledons in the floras of various islands and then says:

"From the above facts it may be assumed that equable, temperate, and rather humid climates are most favourable to a Monocotyledonous vegetation, for it diminishes both under the extreme cold of the arctic zone and the great heat of the tropics; on the other hand increasing towards the southern temperate and antarctic zones, where such conditions are best fulfilled, proportionally with the latitude, to as far south as a Phaenogamic vegetation extends" (p. 241).

"The prevailing natural orders in the Galapagos are the Ferns, containing 28 species; *Compositae* 28; *Leguminosae* 24; *Euphorbiaceae* 18; *Rubiaceae* 15; *Solaneae* 13; *Gramineae* 12; *Amaranthaceae* 10; *Verbenaceae* 9; *Cyperaceae* and *Boragineae* each 7: of the other 43 orders none are so extensive, or are otherwise worthy of particular mention, except *Cordiaceae*, of which there are six species, only one or perhaps two of which inhabit the adjacent continent. All of these orders will be recognised as forming a great part of the vegetation of every tropical country, except the *Amaranthaceae*, which however find their maximum on the west coast of South America. Hence it is not to the prevalence of any particular natural order, or the undue number of species contained in any one, that the Galapagos owe their extraordinary amount of novelty. All the general features of a tropical vegetation are retained, and even the genera to a great extent, but the change is in the species, of which one half are confined to that archipelago; and this peculiarity in species not only relates to the difference existing between the Galapagos and the mainland of America, of which it is a botanical province, but to the separate islets of the archipelago, which, as Mr. DARWIN aptly remarks, should be called "'a group of satellites, physically similar, organically distinct, yet intimately related to each other, and all related in a marked though much less degree to the great American continent'" (pp. 241–2).

The paucity of monocotyledons is due to the scarcity of species in the petaloid families and grasses. There is a relative abundance of *Cyperaceae*. His account of the *Compositae* is not only illuminating in itself but is an excellent example of what may be termed his "phytogeographical style." It is, therefore, given here in full.

"The *Compositae* are in every respect the most remarkable family in th Galapagos, both as regards number of new species and new genera, and from their forming much of the wood of the islands. They also are the most instructive, as the species are very clearly defined; the peculiar genera have representatives in the different islets; and whilst the new species are almost wholly allied to plants from the Andes or extra-tropical parts of America, the old are almost universally the weeds of the low coast of the same continent. It is not therefore with this family as with some others, that the new species are, though permanently, only partially distinct from the continental ones, and possibly varieties due to climatic causes; but they are the representatives of species which are only found beyond the reach of direct migration or are to a great extent entirely new genera."

"In respect of the peculiarity of their *Compositae*, the Galapagos may be compared with some other tropical islands, as the Sandwich group and St. Helena; also with two extra-tropical islands, Juan Fernandez and New Zealand. All of these have a larger amount of peculiarity in their floras than any other tracts of land of the same size. It has been noticed that the four last-named islands or groups are remarkable for possessing a great proportion of arborescent *Compositae*, and in this too the Galapagos share, though the comparison can be carried no further between any of them; for whilst the order is here represented by *Melampodineae* and *Heliantheae*, in Juan Fernandez it is by *Senecioneae* and *Cichoraceae*, in St. Helena chiefly by *Asteroideae*, in the Sandwich group by *Verbesineae* and *Bidentineae*, and in New Zealand by *Helichryseae* and *Astereae*. In all these cases, the further the islands are from the mainland, the less evidence do the *Compositae* they contain afford of the botanical province to which each may belong. Thus the Galapagos contain, in the peculiar plants of this order alone, internal evidence of a strong botanical relation between that Archipelago and Mexico, which a further examination of other orders confirms. Juan Fernandez in like manner abounds in a tribe peculiarly copious in Chili, and the New Zealand arborescent *Compositae* are allied to, though generically and specifically very different from, those of New Holland: but on the other hand, the peculiar genera of the Sandwich group are scattered through many tribes, belonging some to the old world and others to the new; whilst in St. Helena (the whole of whose *Compositae* are shrubby or arborescent and all belonging to peculiar genera), the order seems made up of the fragments of groups charcteristic of very remote parts of the world: the majority belong to a genus of *Astereae* related to what occurs in New Zealand; others to such *Labiatiflorae* as Juan Fernandez possesses; a third genus to the *Melampodinous* family of the Galapagos, and the fourth belongs to the same tribe of American *Compositae*."

"This order here equals $^1/_8$ of the whole Phaenogamic plants, or is nearly the same as its proportion is for the flora of the whole world, and the same as that of the Sandwich group, but smaller than that of Juan Fernandez, and especially of St. Helena, where it equals one third of the flowering plants remaining there. On the other hand, the Society group, in possessing only $^1/_{35}$ of *Compositae*, the smallest number relatively to the whole flora of any tropical country, betray their relationship to the flora of the torrid zone in the old world, which in this respect is strikingly contrasted with that of the new; for it is not improbable that there are more species of this order contained in the comparatively narrow belt of land comprised between the tropics of America, than the same latitudes produce from the west coast of Africa eastwards ro the remotest of the Pacific islands."

"Except St. Helena, there is no part of the globe whose *Compositae* are so nearly unexceptionably different from those of any other country as the Galapagos. Of the 17 genera in which they are included, 5 are widely different from any previously known; and of the species, 28 in number, 23 are peculiar and 5 are tropical weeds, readily introduced by man, and found in the colonized islets alone; whence their origin is suspicious. Of the 12 remaining genera, 9 are almost exclusively American, and the remainder of more general distribution. The last circumstance connected with this order to which I shall allude, is the gummy exudation for which the shrubby *Scalesiae* are conspicuous, and which is equally a characteristic of some of the St. Helena *Compositae*. The species in both instances are inhabitants of arid spots, fully exposed to the sun of the torrid zone, which together seem favourable to the copious secretion of gums and gum-resins in various parts of the world" (pp. 243–5).

In the second part of the essay, HOOKER divides the flora of the Galapagos Islands into "two types", the West Indian (including Panama), to which almost all the wider ranging plants belong, and the Mexican temperate American type, in which are classed the great majority of the peculiar species.

"The species which I have referred to the Mexican type (from the affinities of the remarkable *Compositae*) include those whose nearest allies belong to Mexico or the higher levels in Columbia, or to the lower latitudes of the Southern United States, California or Chili; unlike those of the West Indian type, they are all specifically entirely distinct from their continental congeners, and are about 45 in number, belonging to such genera as *Discaria*, *Dalea*, *Phaca*, *Galactia*, *Opuntia*, *Cereus*, *Viscum*, all the new genera of *Compositae*, besides *Aplopappus* and *Hemizonia*, species of *Ipomaea*, *Psidium*, *Cordia*, *Tournefortia*, *Croton*, *Peperomia*, *Epidendrum*, *Eutriana* and *Aristida*. Those belonging to the Savannah lands of the United States, or dry parts of the tropics rather than the damp, hot, low grounds, are 24 (out of the 45) of *Polygala*, *Galapagoa*, *Elaterium*, *Sicyos*, 7 species of *Borreria*, 6 of *Acalypha*, and 5 of *Euphorbia*, besides a *Brandesia* and *Alternanthera*. Thirteen from the following genera, though very distinct, are exceptional, as being allied to plants of the same range as those included in the West Indian type; they belong to *Desmodium*, *Phaseolus*, *Acrolasia*, *Pleuropetalum*, *Pisonia*, *Froelichia*, 3 species of *Bucholtzia*, *Mariscus*, *Cyperus*, and *Paspalum*" (pp. 250–1).

"The position of the group between the Pacific Islands and America, points to these as the only mother-countries form which plants could have migrated. We have seen that many are common to the latter country; but as at least 15 species are also found in the South Sea Islands, it may be supposed that there has been migration from that quarter, especially as many plants are dispersed in a very remarkable manner over every group in the Pacific, establishing themselves very soon after the formation of any new land, and whose further extension to the Galapagos might have been deemed possible" (p. 252).

HOOKER then discusses the means of transport by which the plants could have been introduced into the islands. Oceanic currents are held to account for the majority of the littoral species and for some non-littoral species belonging to the *Leguminosae*, *Boraginaceae*, *Verbenaceae*, and *Solanaceae*. These have "seeds too large for probable transport by winds; they possess no means of attaching themselves to birds, etc. whilst the indurated seed-coats of some, and the exalbuminous embryos of many, probably aid them in resisting for some time the effects of salt water" (p. 253).

A few species with wings or other appendages or with small seeds may have been transported by winds.

"That birds are active agents in transporting species may be presumed from the very considerable number of widely diffused plants which are admirably adapted for availing themselves of this means of transport; though, on the other hand, the exquisite care with which sea-fowl plume themselves must not be overlooked, nor the slender chance there is of a seed remaining attached to a body subjected to such violent motion and constant immersion as these birds undergo. The plants which may have been thus introduced are species of *Tribulus, Siegesbeckia, Nicotiana, Dicliptera, Plumbago, Pisonia, Boerhaavia, Poa ciliaris* and *Setaria Rottleri*: all belonging to this section are ubiquitous plants throughout the tropics."

"As no land-bird is common to the Galapagos and mainland of America, this group is deprived of one very frequent means of transport,—the stomachs of birds, which often receive seeds as the food, especially of the migratory species; these pass undigested from them in a locality far removed from that where they were collected, not only with unimpaired vitality, but with the process of germination accelerated."

"Man is the last agent to which I alluded: that he has been already active is very perceptible from the fact, that Charles Island, the only colonized island, contains the smallest proportion of peculiar plants, and numerically far the most of these common to and probably introduced from the coast with cultivation."

"If the non-peculiar plants of the Galapagos then have been introduced from the continent of America, it is the currents and winds that we must regard as the agents; of these, the winds are steady south-east trades, blowing from the coast of Peru, by which the West Indian species cannot have been carried. The currents are more variable; and to these I would direct attention, and have brought together all the information on this subject I could command, from the voyages of the English and French in the seas between the Galapagos and American shores" (pp. 253–4).

HOOKER obviously lays very considerable stress on oceanic currents as agents of fruit and seed transport to the Galapagos Islands. About 90 species are presumed to have been introduced into the Galapagos Islands through various agencies.

"The last feature in the Galapageian Flora to which I alluded is, that the several islets are tenanted for the most part by different plants; this difference between the Florulae is as decided as that which exists between the botany of the whole coast and that of America, or even more so in proportion, if it be remembered how very similar the islets are in climate and geological structure, and how close to one another in geographical position."

"Were this peculiarity effected only by those species which may have come from the continent, it would have admitted of some explanation, so capricious are the elements which regulate the interchange of species, and so uncertain in their effects even when apparently most uniform in their action. But in this case, the difference is most marked in the distribution of the species that are Galapageian only, the individuals of which are not common to every part of the archipelago, but for the greater part confined each to one solitary islet; only 13 of the 128 peculiar flowering plants and ferns having been found hitherto on two of the four whose Flora we know, two upon three of the islets, and but one upon all four. On the other hand, the amount of difference, though great numerically, is as regards its nature restricted within very narrow limits, the plants of one island being represented in others by similar though not identical species, producing a similarity in all general features combined with a difference in details."

"Such well-marked and at the same time very narrow limits to the dispersion of nearly 130 species, is probably nowhere to be met with but amongst the Galapagos, and, wonderful though it must appear, it is still very much the accident of their birth-place; it is in a great measure due to the want of means of intercourse, especially atmospheric storms, between the several islets, and argues no physical peculiarity or want of vigour in the species themselves. Supposing all the species now inhabiting the Galapagos to be collected on a continuous surface, equalling in area the aggregate of the islets forming that archipelago, then would the Flora lose much of its characters; the strife with its neighbours for position, which marks all stages of the life of any two or more contiguous plants, would terminate in a few replacing the many, and the introduced species bearing a greater proportion to the indigenous, whilst the individuality of the Flora would thus be lessened in degree or wholly destroyed. It must be admitted, that the first steps towards ensuring the continuance of many species in a given area, are to isolate them, and to cut off the means of migration; exactly as in a garden the plants are protected from encroachment mechanically, and the seeds of the more volatile collected betimes, to prevent a like effect being naturally brought about."

"Though, however, this in some degree explains why the florulae of the islets should be distinct in character, it can give no clue to the representation of species amongst them; which representation, whether it be regarded in the light of the whole group bearing the imprint of America, with but few of the productions of that continent, or of the several islets each individually distinct combining to form an harmonious whole, is a mystery which it is my object to portray, but not to explain; and I shall proceed to show the amount of this difference, and its relation to the physical features of the islets" (pp. 258–9).

James Island is the richest in species,

"As might be expected from its central position in the archipelago, and from its containing very elevated land. Albemarle, though the largest, is on the other hand singularly deficient in individuals and kinds, and, as well as Chatham Island, is described as peculiarly sterile and arid. Charles Island, the smallest of all, is almost the richest in species; and though it does not follow that it is hence peculiarly productive for man, we cannot but couple its varied flora with the fact, that it is the only one hitherto colonized."

"With regard to the relative amount of peculiar species possessed by each islet, it would seem to be affected by its climate, and may be thus expressed:

Charles Island   has 22 species common to other islets, which are as 1:4.4 of its
                  whole flora.

James Island   has 23 species common to other islets, which are as 1:4.3 of its
                  whole flora.

Albemarle Island has 18 species common to other islets, which are as 1:2.6 of its
                  whole flora.

Chatham Island   has 17 species common to other islets, which are as 1:2.4 of its
                  whole flora."

"This accordance of the proportions obtained for the two fertile islets and of those for the two sterile is very striking, and especially as they are obtained from collections made by six different and wholly independent voyagers, and indicate that sterility of soil has proved an important agent in preventing the confusion of the floras, and also shows how few are the agents of migration; for Albemarle being the westernmost, and Chatham the easternmost of the whole archipelago, they would otherwise have shown very different proportions."

"If we analyse the florulae still further, and seek to know how far each has profited by immigrants from America, a similar difference will be found between the fertile and the sterile islets.

Charles Island   contains 49 American plants, which are to whole flora as 1:1.9
James Island   contains 52 American plants, which are to whole flora as 1:1.9
Albemarle Island contains 20 American plants, which are to whole flora as 1:2.3
Chatham Island   contains 19 American plants, which are to whole flora as 1:2.1"

"Whence it appears that the fertile islets, though in position not more favourably placed for receiving the plants of the American coast, still show, not only numerically but proportionally, their aptitude for supporting a richer flora than that which is peculiar to the group."

"The nature of the collections is hardly such as to warrant the drawing any further conclusions, the numbers representing the relationship of the peculiar and non-peculiar plants of each islet to one another being small. There is, however, one point which demands a notice, and that is, the obvious relation between the distribution of the peculiar species over the four islets in question and the direction of the easterly current. Chatham Island being situated east of the group, it follows that the current can never transport insular species to it: on the other hand, Albemarle, on the west, lies directly in its course. Now, excluding the American plants altogether, we have the following evidence of the western islands being peopled by colonists from the eastern; shown by the proportion each islet contains of the Galapageian species found on others.

Chatham Island;   its Galapageian species found on other islets are to whole
                   florula as 1:5.0.

James Island;   its Galapageian species found on other islets are to whole
                   florula as 1:4.8.

Albemarle Island; its Galapageian species found on other islets are to whole
                   florula as 1:3.9.

Charles Island;   its Galapageian species found on other islets are to whole
                   florula as 1:3.1."

"The amount of difference between the islets is, as I have stated above, mainly specific, and is apparent in no less than fifty-eight of the peculiar species of the archipelago, which thus represent one another, and for whose names I would refer to the catalogue of the species already before the Society" (pp. 260–1).

*   *   *

Since the time of HOOKER's essay a considerable number of important papers dealing with the flora and vegetation of the Galapagos Islands have appeared. As is well known, the fauna is also of special interest and LOTSY, "Vorlesungen über Deszendenztheorien" 1: 367–73, Jena, 1906, summarizes the influence the biological peculiarities of the archipelago had on DARWIN. The Galapagos Islands are situated about 500 miles west of the coast of Ecuador and, as might be expected, the flora is almost entirely of American affinities. The main points of phytogeographical interest are the influence of the environments as limiting and selecting agents and the debatable question of the origin of the flora: whether it arrived from the American mainland to colonize the virgin surface of newly formed volcanic islands, ecologically initiating a primary succession, or whether it is a survival with subsequent modifications of the flora of a piece of land formerly joined to Central or South America. BAUR published two papers (1891, 1897) in which he supported the view of the origin of the Galapagos Islands by subsidence. He based his conclusions particularly on acceptance of the view that the flora and fauna are more or less *harmonic*, or like that of the continent of which the islands are literally a detached part. The premises and logic of BAUR's arguments are open to criticism and, as we shall see, most subsequent authors do not accept his conclusions. ROBINSON (1902) published a bibliography, compiled by M. A. DAY, of the botany of the islands, and here further discussion will be limited to ROBINSON's paper and subsequent publications in their bearings on the earlier work of HOOKER.

ROBINSON (1902) gave a systematic list of the known flora and summarized the general features of the vegetation. In his account of the affinities of the flora he emphasized that the links are almost entirely with America, only the genus *Lipochaeta* connecting the flora westwards with Hawaii. He said, however, that it is not possible to trace closely a relationship with any one section of Pacific America. In a general way, nearly all the plants of the archipelago are identical with, or obviously related to, species of the Sierras and Andes or of the Pacific slope between Lower California and northern Chile. The xerophytes in the flora show a considerable resemblance to the desert flora of southern

Peru and the drier parts of the Andes, while the mesophytes correspond most nearly to plants of Ecuador, Colombia, Central America,
and southern Mexico.

ROBINSON pointed out that when HOOKER refered to the "West
Indian" relationship of the flora of the Galapagos Islands he used the
phrase very definitely to include Panama and the adjacent lowlands
of the continent, a qualification not always sufficiently regarded by
later authors. Since HOOKER wrote new discoveries have shown that
there are much greater differences between the floras of the West
Indies and Central America than were known to him. We know now
that there is no special affinity between the flora of the Galapagos Islands and that of the West Indies proper.

ROBINSON has, perhaps, over-emphasized HOOKER's conclusion of
a double relationship in the flora of the Galapagos Islands. There
seems nothing in HOOKER's account that is opposed to ROBINSON's
own conclusion that the diverse floral elements have reached the
archipelago, probably, at different times from widely different habitats.
HOOKER was interested in the flora rather than in the vegetation.
ROBINSON, like HOOKER, commented upon the paucity of Monocotyledons in the flora and noted that, as with most insular floras, the absence of certain great groups is more striking than the number and
diversity of genera and families present. He considered in some detail,
and very fairly, the opposed views of (1) the pelagic origin and (2) separation from the continent by subsidence of the archipelago. He concluded that the evidence was in favour of the emergence theory. On
this view the present plant population reached the islands from the
mainland and there developed into more and more highly differentiated
forms, varieties, and species characteristic of different islands. The differences in the florulae must find their explanation in peculiarities of
climate and soil, together with an element of chance.

Two papers by STEWART (1911, 1916) may be considered together.
The first paper is extremely valuable, amongst other reasons for the
account of the "botanical regions" based on the ecology of the vegetation. Four main regions, above the strand, are recognized: dry,
transition, moist, and grassy. All of these regions only occur on the
more elevated islands and there is often a great difference in the elevations at which a region begins and ends on the same sides of different islands as well as on different sides of the same islands. Moisture
is the most important controlling ecological factor. The moist region
owes its origin and position mainly to fog banks which strike the
windward sides of the mountains at various elevations. These important ecological descriptions and explanations supplement HOOKER's
account for in spite of its title, "On the Vegetation of the Galapagos

Archipelago", this deals mainly with the flora and not with the vegetation as these two words are now most often used.

With regard to the origin of the Galapagos Islands, STEWART considered that the weight of evidence favours the view that the islands were formerly more or less united amongst themselves but not with the mainland. Assuming that the archipelago is of oceanic and not of continental origin, STEWART stated that seeds and spores could, apart from man, have been brought only by winds, oceanic currents, and migratory birds. He concluded that while migratory birds must not be considered as the only factor in distribution, they seem to be the most important cause, as the presence of many of the plants found on the islands, especially those of a mesophytic character, can be explained in no other way.

In his second (1916) paper STEWART dealt with the vegetation and flora of the individual islands. A great many details are given but as such an analysis needed data far beyond those available to HOOKER and the treatment is so very different from that followed in HOOKER's essay it must suffice here merely to record that HOOKER's main conclusions are not affected by it.

KROEBER (1916) analyzed anew the figures of ROBINSON's and STEWART's accounts by considering the percentages of common occurrence of species in any two islands and by the mean of percentages for every island paired with every other island. He concluded that so far as the number of joint species is concerned the florulae of the Galapagos Islands show exactly such connections as might be expected. The number of species found in each island is the most important factor determining the number of species which two islands have in common. A second, but much less, important factor is geographical position. In general, islands in proximity have more species in common than those far apart. Thirdly, the southeastern islands have a slightly greater influence on the smaller islands than have the western and central groups. KROEBER thought that the origin of the Galapagos Islands—by emergence or subsidence—is scarely soluble by botanical evidence. As regards the internal floral relationships there is little that is not explainable on the basis of mathematical chance operating evenly as if all the islands formed a unit; this factor is disturbed in some measure by ordinary geographical influences.

A very readable, first-hand, and general account of the archipelago will be found in BEEBE's book (1924). As this has only incidental references to the flora, this brief reference to it must suffice.

CHUBB (1933), dealing with the geological history of the Galapagos Islands, recognized three phases in their building: those of great volcanoes, of minor eruptions, and emission of lava from fissures. An uplift of a few hundred feet probably occurred between the last two phases but since then the islands have been stationary except for slight local subsidence.

SVENSON (1935) gave an outline account of the vegetation of the archipelago with more details for some of the islands and a systematic list of species collected on them by the Astor Expedition. He quotes TOWNSEND (Zoölogica 4: 71, 1925) as holding that a former land conection between the Galapagos Islands and Central or South America is not very probable. A concise review of the Galapagos flora is given by SVENSON *in* Plants and Plant Science in Latin America (*ed.* F. VERDOORN, 1945). In a slightly later and longer paper (1946) he published an annotated catalogue of 328 species with details of their distribution. He considered that STEWART's estimate of 40 per cent. endemism in the plants of the Galapagos Islands is too large and concluded that the flora seems to be of comparatively recent introduction.

SWARTH (1934) summarizes the zoological evidence on the history of the islands, particularly that derived from a study of the avifauna. He shows that this is clearly not derived from the South American mainland directly opposite but from various sources. There is even an important element definitely recognizable as of West Indian derivation. He concludes that, "The bird population of the Galapagos, abundant as regards individuals, is, as regards representation of different groups, of the sparse and miscellaneous character to be expected of chance-controlled wanderers to distant islands." The evidence thus favours an oceanic and not a continental origin for the islands.

Reference may be made finally to LACK's book (1947) which, though it deals mainly with the fauna, and especially the ornithology, of the Galapagos Islands, contains references to the vegetation. He discussed the "emergence" and "subsidence" theories from the zoological point of view and concluded in favour of the former.

We thus see that, apart from BAUR and a number of other zoologists, biologists who have studied the flora and fauna of the Galapagos Islands either favour the view of HOOKER that the archipelago is of oceanic origin and has been stocked, with very few exceptions, by transport of disseminules from continental Central and South America or reserve their opinion. The evidence presented by the authors quoted, or deduced from their systematic and distributional accounts, is, on the whole, clearly in favour of the origin of the main mass of the flora by transport from different parts of the American mainland. With islands far from mainland we can generally only observe the results of happenings long since completed. Natural introduction, if it occurs, is at long and irregular intervals and can, therefore, be very rarely observed by botanists. Hence, an element of doubt remains in any deductions from the facts of present distribution. Where the geological evidence is clear, unequivocal, and accepted by all qualified geologists who have specialized on it, the botanist is on safer ground in weighing his own evidence for emergence and transport of the flora or subsidence and the relict nature of the flora.

Chapter VIII

ANTARCTICA

HOOKER's "Antarctic Voyage", as it is often and somewhat mislead-
ingly called, was the first of his important botanical expeditions. A
full outline account of it is given in L. HUXLEY: „Life and Letters of
Sir JOSEPH DALTON HOOKER", vol. 1, chapters II to VI inclusive
(1918). The "voyage" consisted of three expeditions to the south with
the two ships EREBUS and TERROR with breaks between in Tasmania,
Sydney, and New Zealand and in the Falklands (with an excursion to
Hermite Islands in Tierra del Fuego and west of Cape Horn). General
details of the whole voyage are given in Captain Sir JAMES CLARK
Ross's "A voyage of discovery and research in the Southern and Antarctic Regions", two volumes, London, 1847. In this interesting and
readable work there are various, acknowledged, botanical contributions by HOOKER, who is introduced (p. 82) as "Dr. HOOKER, the Assistant Surgeon of the Erebus". The following are the references:

> Kerguelen Island, 1: 83–7.
> Auckland Islands, 1: 144–8.
> Campbell Island, 1: 158–63.
> Fossil wood in Tasmania (Van Diemen's Land), 2: 5–11.
> Falkland Islands, 2: 261–77.
> Cockburn Island, 2: 335–42.

These contributions should not be lost to the knowledge of botanists for they are first-hand accounts of unusual floras not entirely
included in the volumes of the "Flora Antarctica". Two extracts are,
therefore, given here, those for Campbell Island and Cockburn Island.

"Although Campbell's Island is situated 120 miles to the southward of
Lord Auckland's group, and is of much smaller extent, it probably contains
fully as many native plants. This arises from its more varied outline, and
from its steep precipices and contracted ravines, affording situations more
congenial to the growth of grasses, mosses, and lichens. Its iron-bound
coast and rocky mountains, whose summits appear to the eye bare of vegetation, give it the aspect of a very desolate and unproductive rock, and it is
not until the quiet harbours are opened, that any green hue save a few grassy spots is seen. In these narrow bays the scene suddenly changes; a belt
of brushwood, composed of some of the trees mentioned as inhabitants of
the last-visited island, but in a very stunted state, form a verdant line close

to the beach. This is succeeded by bright green slopes, so studded with the *Chrysobactron* as to give them a yellow tinge, visible a full mile from the shore. Most of the beautiful plants of Lord Auckland's group, including the elegant caulescent ferns, are equally abundant here, and from many of them growing in this higher latitude at a proportionally lower elevation, their beauty strikes every one on first landing."

"The stay of the expedition here was necessarily very short, and though two days sufficed to collect between 200 and 300 species, the island cannot be considered as sufficiently explored to justify any rigid numerical comparison between its Flora and that of the Aucklands; still some few relative observations may be offered. Sixty-six flowering plants were detected, of which fourteen were not seen in the neighbouring group. Thus, in two degrees of latitude, thirty-four species had disappeared from the Flora of this longitude, and been replaced by at least twenty other plants, producing as great a concomitant change in the proportions of the two groups of flowering plants as was to be expected from the higher latitude. The new species are almost all typical of an antarctic climate, and consist both of species confined to the island, and of others hitherto considered peculiar to Antarctic America. The proportion of monocotyledonous plants is increased from being 1 : 2.2, to 1 : 1.4. The grasses, instead of bearing the small ratio of 1 : 14, which they do in Lord Auckland's group, here appear as 1: 4.5. *Cyperaceae* and *Orchideae* have proportionally decreased, and the *Compositae*, which were to all *Dicotyledones* as 1 : 10.4, are here as 1: 5.6. These are not the signs of the vegetation of a more rigorous latitude alone, but of one differing more widely from that of New Zealand than Lord Auckland's group did, where only one-seventh of the plants were common to other antarctic regions, whilst in Campbell's Islands fully one-fourth are natives of other longitudes in the Southern Ocean."

"Considering the aggregate of the plants in the islands to the southward of New Zealand as composing one Flora, a comparison of it with those of other countries is not out of place here. The flowering plants amount to one hundred species, or about the same number as have been collected in the whole group of arctic islands to the northward of the American coast. Of these one fourth have been found in New Zealand, whilst many of the others belong to genera whose abundance is characteristic of that country. Only one-thirteenth of the whole are known to be Tasmanian, and one-sixth are common to Tierra del Fuego. Since there is no other country with which these islands possess any marked botanical features in common, their Flora may be considered a continuation of that of New Zealand, differing only in that it is more typical of the antarctic regions."

"The remarkable points of resemblance to the last-named group with which we have compared this Flora, are the preponderance of *Rubiaceae*, *Araliaceae*, *Epacrideae*, *Orchideae*, and *Myrsineae*; the small amount of surface occupied by *Compositae*, *Caryophylleae*, *Cruciferae*, and *Ericeae*; and the entire want of *Saxifrageae*, *Leguminoseae*, *Labiatae*, and *Amentaceae*, all scantily represented in New Zealand. The more striking points of difference are the increased proportion of *Monocotyledones*, which are there as 1:3.2, and in these two islands as 1 : 1.8; of grasses, which bear a proportion there to other flowering plants of 1 : 13, and here of 1 : 6.8; and of *Compositae*, which there appear as 1 : 8, and as 1 : 4.4 here. This Flora further departs from that of New Zealand in possessing none of its numerous species of pine or beech, of which latter genus five are now known to grow there, and this is the more remarkable because all the beeches and several of the pines are alpine, both in New Zealand and in Van Diemen's Land, only reaching the level of the sea in the southern parts of those islands. The pines of the southern hemisphere are, however, exceedingly local, nor are they so ant-arctic as some of those in the northern hemisphere are arctic. Of the ten New Zealand species it is not certain that more than two or three are natives of the middle island, or that any of them are peculiar to a latitude south of 40°. Not only do Lord Auckland's group and Campbell's Island exhibit no in-considerable number of Fuegian plants, considering the immense inter-vening tract of ocean (upwards of 4,000 miles), but in all the particulars in which their Flora differs from that of New Zealand, it more closely approx-imates to that of Antarctic America. Strong though the resemblance is in the numerical proportions of the orders, and in the similarity of many of the smaller plants, the trees and shrubs of the one differ in every respect from those of the other locality; for beeches extend from a latitude in the Ameri-can continent which corresponds to their principal parallel in New Zealand beyond the latitude of Lord Aucklands' group, as far south as Cape Horn itself, in the 57*th* degree."

"The relation between the Flora now under consideration and that of the northern regions is but slight; and the same may be said, though not to an equal extent, of any two countries in the higher latitudes of the opposite hemispheres. This group lies in the latitude of England, yet we recognise in it only three indigenous plants of our own island,—the *Cardamine hirsuta*, *Montia*, and *Callitriche*. Of the sixty genera twenty-two are English, and twenty eight natives of a more northern latitude than England. Hardly any of these belong to the divisions *Calyciflorae*, *Compositae*, or to the higher orders of the *Monocotyledones*; while, on the other hand, they include the whole of the *Thalamiflorae*, *Monochlamydeae* and grasses, and most of the *Cyperaceae*. Such genera as *Sieversia*, *Trisetum*, and *Hierochloe* have their analogues chiefly in the arctic regions; whilst *Myosotis*, *Ranunculus*, *Carda-mine*, *Stellaria*, *Veronica*, *Luzula*, *Juncus*, and all the grasses, are predominant in the arctic Flora. There are, however, slight points of resemblance, ren-dering the want of a larger amount of their congeners more remarkable, and also of others which in the north generally accompany them, as saxifrages, heaths, and *Vaccinia*, *Leguminosae*, pines, beech, and especially oak, birch, and willow; for most of which no representative has hitherto been found in the high southern latitudes" (1, pp. 158–63).

"As regards its botany, this island may be considered one of a group, lying immediately south of Cape Horn, beyond the sixtieth degree of lati-tude. The number of plants ascertained to inhabit them hardly exceeds twenty-six; and one of these, a grass, the only flowering plant, does not pass the sixty-second degree; nor, consequently, reach that island, to whose vegetation the following observations more immediately refer. Previous to the voyage of the "Erebus and Terror," almost nothing was known of the vegetation which approaches nearest to the Antarctic Pole. We had yet to learn whether a flora, so situated, would be found to consist of plants which inhabit the elevated and comparatively rigorous regions of a milder clime; or of those growing in a similar latitude of the opposite hemisphere; or finally, if Nature had not there produced new and isolated species, adapted to the peculiarities of the locality."

"The Flora of Cockburn Island contains nineteen species, all belonging to the orders, *Mosses, Algae,* and *Lichens.* Twelve are terrestrial; three inhabit either fresh water or very moist ground; and four are confined to the surrounding Ocean. Of these nineteen plants, seven are restricted to the island in question, having been hitherto found nowhere else (besides an eighth, which is a variety of a well known species); the others grow in various parts of the globe, some being widely diffused."

"The greatest amount of novelty is found here, as in other cryptogamic floras, among the most highly organized class: for example, of the *Mosses,* two out of five are new. There are seven *Algae,* and two of them, or less than a third, are new. Of six species of *Lichen,* four are already described, (perhaps five), so that only one, or at most two, can be considered peculiar."

"The twelve plants of Cockburn Island that are common to other parts of the world, may be arranged according to their greater or less diffusion; for while some may be seen in all latitudes, others are sporadic, appearing in certain remote spots; and a few are confined to the regions in the vicinity of Cockburn Island."

"The four following plants are the most generally dispersed:— *Bryum argenteum, Ulva crispa, Lecanora miniata,* and *Lecidea atro-alba.* The first is a very frequent British moss, found likewise in Arctic latitudes, in many parts of the tropics, and at the Falkland Islands. The second is an arctic *Alga,* also abounding in the temperate parts of the northern hemisphere, in the tropics, and the Falklands. *Lecanora miniata* is an arctic lichen, and seen in all intervening countries down to Cockburn Island; while the other lichen (*Lecidea atro-alba*) inhabits Britain, sub-arctic Europe, and New Zealand."

"Of the sporadic plants which follow, it is probable that some may yet be discovered in intermediate stations, having either escaped observation from their minuteness, or been described as different species; they are two mosses, *viz. Tortula gracilis,* indigenous to Europe and Cockburn Island; and *Tortula laevipila,* found in Europe and the Falklands; two sea-weeds, *viz. Desmarestia aculeata,* var. *media,* originally detected in Unalaschka (lat. 55° N.); and *Oscillatoria aerugescens?* if this latter be identical with the Irish species of that name, it had hitherto been found in one loch in Ireland only: and a *lichen* (*Collema crispum*), which is a native of Britain and other parts of Europe, where it generally grows on walls, though occasionally as in Cockburn Island, on the ground. To this list should be added another *lichen,* recognised as a Falkland Island and European *Parmelia,* the specimens of which were unfortunately lost. The remaining two plants are well known sea-weeds, natives of several parts of the southern temperate, and antarctic ocean; *viz.*, *Iridaea micans* and *Adenocystis Lessoni.*"

"The two most striking vegetable productions of this island are a noble sea-weed, called *Sargassum Jacquinottii*, and a *Lichen*. The first of these was not found attached, but floating in the ocean among the ice, by which it was sometimes much mutilated. Though belonging to a highly variable order, it is a perfectly distinct as well as conspicuous species, first discovered at Deception Island, one of the South Shetlands, by the surgeon of H.M.S. *Chanticleer*, and afterwards by Admiral D'URVILLE, who collected his specimens nearly in the same latitude. It attains a length of three feet, is flat, and the margin runs out into longish lobes with a solitary bladder at the base of each; the colour is a dirty chocolate brown."

"On approaching Cockburn Island, the cliffs above are seen to be belted with yellow, which, as it were, streams down to the ocean, among the rocky débris. The colour was too pale to be caused by iron ochre, which it otherwise resembles; and this appearance was found to be entirely owing to the abundance of a species of *lichen* (*Lecanora miniata*) that prevails in the vicinity of the sea throughout the Antarctic Islands, and in other parts of the globe. It grows nowhere else in such profusion: a circumstance which may arise from its preference for animal matter: the penguin rookery of Cockburn Island, which taints the air by its effluvium, being, perhaps, peculiarly congenial to this lichen."

"Immediately on landing, one plant, and only one, is easily discernible, the *Ulva crispa*. Like the *Lecanora*, it abounds in the south, and vegetates upon or near decomposing organised substances. It consists of pale green membranous fronds, barely one fourth of an inch high, and crowded together in great numbers."

"The *Mosses* grow in the soil which is harboured in the fissures of rocks: they are excessively minute, the closest scrutiny being requisite to detect them. There were, as above mentioned, only five species: two of them bore unripe capsules, and all were confined to spots having a northern exposure, and even there they were so hard frozen into the ground that they could not be removed without a hammer."

"One of the *Algae* was collected in a pool of fresh water, hardly two spans across, and sheltered by a projecting rock that faced the north. The surface of the water was slightly coated with a steel-blue scum: the earth at bottom, perhaps half an inch below, was hard frozen; and the water itself just thawing, for it was an unusually warm day, the thermometer standing at 40°. *Collema crispum*, a British plant, grew on the borders of this pool, and with it a green microscopic *Conferva.*"

"A small and beautiful undescribed *lichen* (*Lecanora Daltoni*) occurred very sparingly on the rocks: it is allied to *L. chrysoleuca* of the Swiss Alps. The other plants of this order were exceedingly inconspicuous, and only discoverable by carefully examining the surface of the rocks."

"The sea-weeds gathered on the shores of Cockburn Island were all floating, and carried along by a strong current, loaded with masses of ice."

"Vegetation could not be traced above the conspicuous ledge of rocks, with which the whole island is girt, at fourteen hundred feet elevation. The *lichens* ascended the highest. The singular nature of this flora must be viewed in connexion with the soil and climate; than which perhaps none can be more unfriendly to vegetable life. The form of the island admits of no shelter: its rocks are volcanic, and very hard, sometimes compact, but more frequently vesicular. A steep stony bank descends from the above-mentioned ledge to the beach; and to it the plants are almost limited. The slope itself is covered with loose fragments of rock, the débris of the cliff above, further broken up by frost, and ice-bound to a depth which there was no opportunity of ascertaining; for on the day the island was visited, the superficial masses alone were slightly loosened by the sun's rays. Thus the plants are confined to an almost incessantly frozen locality, and a particularly barren soil, liable to shift at every partial thaw. During nearly the entire year, even during the summer weeks which the Expedition spent in sight of Cockburn Island, it was constantly covered with snow. Fortunately the ships occupied a position that permitted of landing, on almost the only day when it was practicable to form a collection. The vegetation of so low a degree of latitude might be supposed to remain torpid, except for a few days in the year; when if the warmth were genial, and a short period of growing weather took place, the plants would receive an extraordinary stimulus. But far from such being the case, the effect of the sun's rays, when they momentarily appear, is only prejudicial to vegetation. The black and porous stones quickly part with their moisture; and the *Lecanora* and *Ulva* consequently become so crisp and parched, that they crumble into fragments when an attempt is made to remove them."

"The conducting power of the minerals in Cockburn Island is too feeble to melt the ice immediately beneath them; and the air was so dry during our visit, that Daniell's Hygrometer, placed hardly six inches above the ice and on the stones, indicated twenty-two degrees of difference on one occasion; and upon another, it fell from 40° to 13°, without producing any condensation. Such dryness is eminently injurious to all vegetables but *lichens*, which, in many cases, seem to thrive best under excessive atmospheric changes. The preponderance of the *Lecanora* in Cockburn Island cannot arise from this exsiccation stimulating its growth; but may be caused by the reaction that takes place afterwards, on the rapid condensation of vapour previously heated by the temperature of the rocks upon which it grows" (2, pp. 335–42).

"The Botany of the Antarctic Voyage of H.M. discovery ships *Erebus* and *Terror* in the years 1839–43, under the command of Captain Sir JAMES CLARK ROSS" was published in three parts:

Part. I. Flora Antarctica. London, 1844–7, vols or parts I. and II. *See* B. DAYDON JACKSON, *in* Journ. Bot. 50; 284–5 (1912) and F. G. WILTSHEAR, *op. cit.* 51: 355–58 (1913), for dates of publication of the constituent parts.

Part. II. Flora Novae-Zelandiae. Vol. 1, Flowering Plants. London, (1853–5). Vol. 2, London, 1855. *See* B. DAYDON JACKSON, *in* Bull. Herb. Bois. 1: 299 (1893), and WILTSHEAR *l.c.* for dates of publication of the constituent parts.

Part III.   Flora Tasmaniae. Vol. 1, Dicotyledones. London, 1855–60. Vol. 2,
Monocotyledones and Acotyledones. London, 1860. *See* B. DAYDON JACKSON, *in*
Journ. Bot. 47: 106–7 and WILTSHEAR, l.c.

These works are mainly systematic accounts of the collections made
by HOOKER illustrated by beautiful coloured plates by FITCH or FITCH
and HOOKER. They have, however, renowned introductions which
are essentially phytogeographical and from which a selected series of
quotations must be made.

Flora Antarctica:—

"Lord Auckland's Group.—   A view of this small and very limited group,
of about twenty miles long and eleven in its greatest breadth, as it appears
on approaching from the sea, presents an almost equal distribution of wood,
shrubs, and pasture-land. The mountains are low and undulating, nowhere
exceeding 1400 or 1500 feet, clothed for their greater part, but scarcely to the
very summits, with long grass, and frequently covered during November
and December, though not generally, with snow. The climate is rainy and
very stormy, so that on the windward sides the plants are stunted and
checked, and resemble those of a higher southern latitude, or of an elevation
several hundred feet above that which the same species inhabit on the sheltered parts. The whole group of islands appears formed of volcanic rocks,
mostly of black trap, whose decomposition, especially among the ranker
vegetation of the lower grounds, produces a deep rich soil. A *Myrtaceous
tree* (*Metrosideros umbellata*) forms the larger proportion of the wood near the
sea, and intermixed with it grow an arborescent species of *Dracophyllum*,
several *Coprosmas*, *Veronicas* (frutescent), and a *Panax*. Under these, and
particularly close to the sea-beach, many *Ferns* abound; conspicuous among
them is a species with caulescent or subarborescent stems half a foot and upwards in diameter, crowned with handsome spreading tufts of fronds.
Beyond the wooded region, some of the same plants, in a dwarf state, mingled with others, compose a shrubby broad belt, which ascends the hill to an
elevation of 800 or 900 feet, gradually opening out into grassy slopes, and
succeeded by the alpine vegetation. It is especially towards the summits of
these hills that the most striking plants are found, vying in brightness of
colour with the Arctic Flora, and unrivalled in beauty by those of any other
Antarctic country. Such are the species of *Gentian*, and a *Veronica* with
flowers of the intensest blue, several magnificent *Compositae*, a *Ranunculus*,
a *Phyllachne*, and a *Liliaceous* plant whose dense spikes of golden flowers are
often so abundant as to attract the eye from a considerable distance. Here
too the vegetable types of other Antarctic lands may be seen in the greatest
number, and even such as are analogous to the Arctic productions, none of
which can be more decided than a species of *Hierochloe*, *Potentilla*, *Cardamine*,
Iuncus, *Drosera*, *Plantago*, *Epilobium*, several *Grasses*, and *Mosses* belonging
to the genera *Andraea*, *Conostomum* and *Bartramia*. Many of the plants in the
lower grounds are no less striking and beautiful, as an arborescent *Veronica*
bearing a profusion of white blossoms, a maritime *Gentian*, a handsome

large-flowered *Myosotis*, the magnificent *Aralia polaris* (Hombr. and Jacq.), two fine kinds of *Anisotome*, and several beautiful *Ferns*" (1, p. 2).

"*Tierra del Fuego.*— The botanical features exhibited by this country are not circumscribed by its geographical limits; along the north-east shores the very distinct Flora of East Patagonia accompanies the geological formation prolonged there from the Patagonian plains. On the south-west and south sides again, the vegetation is a continuation of that of West Patagonia, and is characteristic of the western flank of the Cordillera, from South Chile to Cape Horn. Thus it is that we find the Andes dividing two botanical regions from the North Polar almost to the Antarctic circle. The greater part of Fuegia is formed by the Andes alone; but the plants of the north-east portion, where the granitic formation of Patagonia introduces a change in the vegetation foreign to that of Tierra del Fuego, will be necessarily included in the present Flora."

"The Deciduous Beech (*Fagus antarctica*), is the most distinguishing botanical production of this country. In company with the Evergreen Beech (*F. Forsteri*), it covers the land, especially on the west coasts, as far north as the Chonos Archipelago, in latitude 45° south. It is hardly seen in the north-east portions of Fuegia proper, northward of Staten Land, and though abundant on the west flanks of the Andes, through fourteen degrees of latitude, is unknown on the Atlantic side of Patagonia. I have assumed therefore the shores of the strait of Magalhaens to be the northern limit of the Fuegian Flora eastward of Port Famine, and have included in, or rather added to that Flora, all the known plants of the Pacific side of the Andes, reaching north to the Chonos Archipelago. The latter position is peculiar, in the *Beech* being there replaced, at the level of the sea, with other trees; by the sudden change in the aspect of the coast vegetation that the flora of Chilöe, immediately to the northward, presents; and by its being only a few miles beyond the "glacier-bound Gulf of Penas," where perennial ice descends to the level of the ocean in a latitude nearly midway between the Equator and the Antarctic Pole" (2, pp. 212–3).

"The Falkland Islands rank next in botanical importance to Fuegia. Though lying to the northward of the main body of that country, their vegetation is so influenced by climate and by some other peculiarities common to these islands and the Patagonian plains, that they produce no tree whatever. They are situated between the parallels of 51° and 53°, and the meridians of 57½° and 61½° west, and consist of an eastern and western island, nearly equal in size, and together forming an oval, whose axis lies east and west and extends about 160 miles. The general outline is jagged, like that of Fuegia, and similarly indented by deep inlets and ramifying bays; but their level or undulating surface, never rising above 2000 feet, and the geological formation, bear no resemblance to an archipelago formed by a submerged chain of mountains. Altogether, the Botanical and other characters of the Falklands are allied to the Atlantic coast of Patagonia, opposite to the strait of Magalhaens, from whence they are only 300 miles distant."

"The most evident causes for the absence of trees in the Falkland Islands are the dislocation or removal of that group from the main land; their

comparatively plane surface, everywhere exposed to the violence of the westerly gales, and more especially to the rapid evaporation and sudden changes
in temperature and in other meteorological phenomena. The southerly and
westerly winds are violent, cold, and often accompanied by heavy snow-
storms; the easterly and northerly arrive saturated with warmer sea vapours,
which, quickly condensing over the already chilled surface of the soil, form
fogs and mists that intercept the sun's rays; whilst the north-westerly winds
are singularly dry and parching, from the influence of the Patagonian plains
over which they blow. Such sudden alternations from heat to cold, and
from damp to dry, are particularly inimical to luxuriant vegetation, and no
foliage but perhaps the coriaceous growth of Australia could endure them.
The characteristics both of Fuegia and Patagonia may be seen mingled in the
Falklands, and except *Veronica elliptica* (Part I, p. 58), which is chiefly confined to the western coasts of the western island, and plants of both these
countries appear together, overspreading the whole surface of the islands.
Few species are peculiar, and no genus or order predominates to any remarkable extent, unless it be the *Gramineae*: the species themselves are well
marked and do not run much into varieties. Though the want of shade is
unfavourable to the fruiting of *Mosses* and *Hepaticae*, there are a considerable
number of species of those orders, and some are identical with those of the
American mountains and of Europe" (2, pp. 213–4).

"Considering the distance of the Falkland Islands from the continent,
their size, the extent of surface covered with vegetation, and above all, their
geological formation and the nature of their climate, the number of peculiar species is very insignificant; such circumstances generally accompanying
or being indicative of a concomitant change in botanical features, specific
difference itself being by some attributed to the operation of these causes,
and the immutability of species thence called in question. The Falkland
Islands appear ill adapted to the more striking vegetation of Fuegia or of
Patagonia, if we may judge from the absence of trees and even of such bushes
as *Berberis, Escallonia, Fuchsia, Ribes*, etc., which grow in the former country
and to all of which the changeable nature of the climate is injurious; while,
on the other hand, the mean temperature is too low for the *Leguminosae,
Malvaceae*, and other predominant Orders of Patagonia. It is more remarkable that some of the plants of each are seen, composing together the
whole vegetation, yet appearing unchanged by a climate that is certainly
unfavourable to the general flora of those distant regions where these very
species most abound. To conclude by an example, *Sisyrinchium* and *Oxalis
enneaphylla* will not associate themselves with the *Tussac* and *Empetrum* in
Cape Horn, nor are *Astelia* and *Caltha appendiculata* to be found in company
with *Nassauvia* and *Calceolaria Fothergillii* on the coast of Patagonia, though
all these may be seen growing side by side in the Falklands in the greatest
profusion."

"Immediately to the south of Cape Horn are groups of islands, and possibly a larger body of land. Vegetation in the Southern Hemisphere reaches
the northern shores of these inhospitable spots, where, at a distance of no
less than thirty-six degrees from the actual Pole and three degrees to the

northward of the Antarctic circle, the flora of the south finds its extreme limit" (2, p. 215).

"Proceeding westward from Antarctic America, the next island that requires notice, as exhibiting an Antarctic vegetation, is Tristan d'Acunha. Though only 1000 miles distant from the Cape of Good Hope, and 3000 from the Strait of Magalhaens, the Botany of this island is far more intimately allied to that of Fuegia than Africa. Captain CARMICHAEL's list (Linn. Trans., vol. xii, p. 483), contains twenty-eight flowering plants (I exclude *Sonchus oleraceus*); only one species of *Phylica*, and one *Pelargonium*, amounting to one-fourteenth of the whole, are Cape forms; whilst seven others, or one-fourth of the flora, are either natives of Fuegia or typical of South American Botany, and the *Ferns* and *Lycopodia* exhibit a still stronger affinity. There are some points in which the vegetation of Tristan d'Acunha resembles that of St. Helena and Ascension. Though these islands are separated from one another by nearly thirty degrees of latitude, they lie within eight degrees of longitude, and all are the exposed summits of ancient volcanoes, such as the highest peaks of the Andes might present, if that mighty chain were partially submerged. The relation between the floras of Ascension and St. Helena is evident, though to enumerate them would be out of place here; those between the latter islands and Tristan d'Acunha are indicated by the genera *Phylica* and *Geranium*, and also by some of the Ferns and *Lycopodia*: as, however, it is also through those genera that the botany of Tristan d'Acunha resembles that of the Cape, it may fairly be doubted whether the apparent affinity with St. Helena is not imaginary. It is a very remarkable circumstance that while these three islands all posess some of the features of the African Flora, the predominant ones are absent; thus, whilst the St. Helena Flora is allied, and exclusively so, to that of the Cape in *Geranium, Melhania*, and *Phylica*, it has no representatives of entire Orders, namely *Proteaceae, Rutaceae, Oxalideae, Crassulaceae, Ericeae, Restiaceae*, and many others, far more characteristic of the African vegetation than are any of the plants inhabiting St. Helena."

"The other islands whose plants will find a place in this division of the 'Antarctic Flora' are situated south of the Indian continent, widely apart from the American, and so far as geographical position is concerned, belong to Africa or India; these are, Prince Edward's and Marion Islands, the Crozets, Kerguelen's Land, and the islands of Amsterdam and St. Paul" (2, pp. 216-7).

Flora Novae-Zelandiae:— The Introductory Essay to the Flora Novae-Zelandiae, published on 6 Dec. 1853, consists essentially of three chapters concerned respectively with the history of botanical discovery in New Zealand, the limits, ranges, and variation of species, and the composition and relationships of the New Zealand flora. While the third chapter is our main concern here, the second chapter is of considerable interest and importance as throwing light on HOOKER's theoretical views and general conclusions on "species prob-

lems" in the decade immediately preceding the publication of DAR-
WIN's "Origin of Species". He calls these problems "obscure subjects"
and it is very evident that in discussing them he shows that he has
thought deeply about them and at the same time reflects the wide general
interest amongst botanists at this time regarding the nature and origin
of species, subjects that were soon so much illuminated by DARWIN's
great work. At least amongst professional botanists, DARWIN's theories were possibly less "bolts from the blue" than we are apt to think.
At any rate many of the problems were to the fore and ripe for solutions.

HOOKER assumes "certain positions" and adopts them as principles or axioms. These are:

"§ 1. That all the individuals of a species (as I attempt to confine the term) have
proceeded from one parent (or pair), and that they retain their distinctive (specific) characters.
§ 2. That species vary more than is generally admitted to be the case.
§ 3. That they are also much more widely distributed than is usually supposed.
§ 4. That their distribution has been effected by natural causes; but that these
are not necessarily the same as those to which they are now exposed" (p. viii).

In the next paragraph our author tones down the first "principle"
to a pragmatic application.

"Although in this Flora I have proceeded on the assumption that species,
however they originated or were created, have been handed down to us as
such, and that all the individuals of a unisexual plant have proceeded from
one individual, and all of a bisexual from a single pair, I wish it to be distinctly understood that I do not put this forward intending it to be interpreted into an avowal of the adoption of a fixed or unalterable opinion on
my part. Whether or not such a theory be consonant with that great mystery,
the origin of organic beings, animate and inanimate, is not the point I would
here dwell upon; but the fact that it appears to me essential that the systematist should keep some such definite idea constantly before him, to give
unity to his design, and to guide him in the more or less arbitrary restriction
of the species of a variable genus, to which he is unfortunately often obliged
to resort. Except he act upon the idea that for practical purposes at any rate
species are constant, he can never hope to give that precision to his characters of organs and functions which is necessary to render his descriptions useful to others; for in groups where the limits of species cannot
be traced (or, what amounts to the same thing in the opinion of many,
where they do not exist), the object of the systematist is the same as in
groups where they are obvious, – to throw their forms into a natural arrangement, and to indicate them by tangible characters, whose value is approximately relative to what prevails in genera where the limitation of species is
more apparent" (p. viii).

He carefully states views on "species" held by different naturalists and sets forth "the arguments in favour of the permanence of specific characters in plants." Reading these "arguments" one is struck by many of them having the same underlying principle—the absence of any known satisfactory "cause" for evolution. HOOKER essentially rejects changes directly induced by environments as sufficient to account for specific evolution. Nevertheless, he was not fully satisfied with his tentative conclusions and says "I would again remind the student that the hasty adoption of any of these theories [of permanence of species] is not advisable."

HOOKER's feelings of doubt and uncertainty are further evident in his remarks on variation within species. "The views entertained as to the limitation of species appear to be quite arbitrary: no general principles have been discovered for the guidance of the systematist; and those that are adapted vary in kind and in value with every natural group." These remarks are illustrated by reference to genera and species of the New Zealand flora. He argues for a broad concept of species "in working up incomplete floras." The importance of hybridization as "an element in confusing and masking species" he considers has been exaggerated. Modern research, however, may demand some modification of the statement: "The most satisfactory proof we can adduce, of hybridization being powerless as an agent in producing species (however much it may combine them), are the facts that no hybrid has ever afforded a character foreign to that of its parents, and that hybrids are generally constitutionally weak, and almost invariably barren" (p. xv).

Theory and criticisms of a practical nature are combined in the next quotation:

"These considerations lead us to others still more elusive of the naturalist's grasp. The reference of all varieties to a species, and of its individuals to a single parent, argues the existence at some epoch of a type or form around which all varieties may be grouped. It has been observed that two or more created or induced types or species may resemble one another so closely, that, amid the multitude of varieties of each, the naturalist shall seek in vain for that which best demonstrates the species. No one can deny the possibility of such creations, nor perhaps their probability, when he considers the infinite varieties of climates, how insensibly they pass into one another, and how nicely the functions of some plants appear to be adapted to certain modifications of these, and to no others. Had, moreover, every climate its own species, and were there any difficulty in propagating the majority of the plants of one climate in a very different one, such creations would appear to be indispensable: but the facts of botanical geography assure us, that it is by far the smaller half of the vegetable kingdom that is confined to narrow geographical or climatic areas, and that very few plants indeed are absolutely

10

local, whilst the operations of the gardener and agriculturist prove, that a vast proportion of the plants of the two temperate zones are capable of growing in any moderate climate. I do not think that those who argue for narrow limits to the distribution and variation of species, can have considered a garden in a philosophical spirit, or have weighed such facts as that there have been cultivated, within the last seventy years, in the open air of England (at Kew) upwards of twenty thousand species of plants from all quarters of the globe, and this within a space that, had it been left to nature, would not have contained two hundred indigenous species! The fact that an overwhelming proportion of these have come up true to their parent, and have continued so under every possible disadvantage of transportation and transplantation, of altered seasons, and amount and distribution of temperature and humidity, of unsuitable soil exposure, and of the multitude of errors in management which unavoidable ignorance of their natural locality and habit engenders. Such appears to me the most forcible argument in favour of the power of plants to retain their original characters under altered circumstances."

"To return however to the idea of a type, I must remind the New Zealand reader that the word is often used in a vague and unphilosophical manner: in the too frequent sense of the term it denotes that individual of a species which was first cultivated, described, figured, or collected, or that form which is most abundant in the neighbourhood of the writer; whereas all the individuals thus referred to may represent anomalous or exceptional states of the true type. The fact is, that we have no clue whatever to the originally created typical form of any plant, consistent with the view of its origin in a single parent, and its powers of varying. If we take a species of universal distribution, a careful examination of all its variations, and a contrast between these and those of its allies, may lead to the detection of a form, which for various reasons may be assumed as the real or ideal standard; for we have no reason to suppose that the whole globe is so altered that the circumstances under which the assumed type originally appeared do not now exist anywhere. But with local plants the case is different; they may have originated where they are now found, but it is more consistent with geological truths to assume that many did not, and that, however slight the induced changes have been, and however powerless to obliterate specific character, they may still mask the original form."

"Practically, then, the type is a phantom; what was once the typical state may no longer be the common one, or that which now fulfils the office the species did at an earlier epoch. For practical purposes we must assume the most common form to be the most typical, for it is that which is best known. In doing this, however, there is extreme difficulty in combating local prejudices; the general botanist cannot give a higher place in the great scheme of Nature to a natural object on account of its beauty, rarity or local associations, any more than he can call a doubtful plant a native because it looks well in his flora or herbarium; but there are local observers who cannot be brought to see things in such a light, and who take the exclusion of plants accidentally introduced into the flora of their neighbourhood, and the reduction of supposed local types to varieties of better known and wider spread plants, as little short of an insult to their understandings, and a slight upon the natural history of their village or island, and suppose that because the systematist cannot see with their eyes he therefore takes a less true interest in what he observes" (pp. xvi–xvii).

He concluded that species are more widely distributed than is usually thought and that the number of species supposed to be known to botanists is "a greatly exaggerated one." His own estimate (presumably for vascular plants, though this is not clearly stated and he may mean for all plants) is nearer 50,000 than 100,000.

The modern phytogeographer will appreciate the statement that:

"Of all the branches of Botany there is none whose elucidation demands so much preparatory study, or so extensive an acquaintance with plants and their affinities, as that of their geographical distribution. Nothing is easier than to explain away all obscure phenomena of dispersion by several speculations on the origin of species, so plausible that the superficial naturalist may accept any of them; and to test their soundness demands a comprehensive knowledge of facts, which moreover run great risk of distortion in the hands of those who do not know the value of the evidence they afford" (p. xix).

With regard to plurality of centres of origin HOOKER notes that the flora of New Zealand shows many phenomena which bear upon the problem and enumerates these as follows:

"*1*) Seventy-seven plants are common to the three great south temperate masses of land, Tasmania, New Zealand, and South America."

"*2*) Comparatively few of these are universally distributed species, the greater part being peculiar to the south temperate zone."

"*3*) There are upwards of 100 genera, subgenera, or other well-marked groups of plants entirely or nearly confined to New Zealand, Australia, and extratropical South America. These are represented by one or more species in two or more of these countries, and they thus effect a botanical relationship or affinity between them all, which every botanist appreciates."

"*4*) These three peculiarities are shared by all the islands in the south temperate zone (including even Tristan d'Acunha, though placed so close to Africa), between which islands the transportation of seeds is even more unlikely than between the larger masses of land."

"*5*) The plants of the Antarctic islands, which are equally natives of New Zealand, Tasmania, and Australia, are almost invariably found only on the lofty mountains of these countries" (pp. xix–xx).

HOOKER considers the possibility of transport of disseminules between "Antarctic" continents and islands and rejects its feasibility as an explanation of the known facts of distribution. He then says: "It was with these conclusions before me, that I was led to speculate on the possibility of the plants of the Southern Ocean being the remains of a flora that had once spread over a larger and more continuous tract of land than now exists in that ocean; and that the peculiar Antarctic genera and species may be the vestiges of a flora characterized by the predominance of plants which are now scattered throughout the southern islands" (p. xxi). He discusses the views of LYELL, DARWIN, WATSON, and FORBES and the following extract gives his main conclusions:

"To extend a theoretical application of these views to the New Zealand Flora, it is necessary to assume that there was at one time a land communication by which the Chilian plants were interchanged; that at the same or another epoch the Australian, at a third the Antarctic, and at a fourth the Pacific floras were added to the assemblage. It is not necessary to suppose that for this interchange there was a continuous connection between any two of these localities, for an intermediate land, peopled with some or all of the plants common to both, may have existed between New Zealand and Chili when neither of these countries was as yet above water. To account, however, for the Antarctic plants on the lofty mountains, a new set of influences is demanded; no land connection between these islands and New Zealand could have effected this, for the climate of the intermediate area must necessarily have prevented it. But changes of relation between sea and land induce changes of climate, and the presence of a large continent connecting the Antarctic islands, would, under certain circumstances, render New Zealand as cold as Britain was during the glacial epoch. Sir C. LYELL first demonstrated this, and showed what such conditions should be; and by consulting the 'Principles of Geology,' my reader will understand how such a climate would reign in the latitude of New Zealand, as that its flora should consist of what are Antarctic forms of vegetation. The retirement of the plants to the summit of New Zealand mountains, would be the necessary consequence of the amelioration of climate that followed the isolation of New Zealand, and the replacement of the Antarctic continent by the present ocean."

"The climate throughout the south temperate zone is so equable, and the isothermal lines are so parallel to those of latitude, that it is not easy for the New Zealand naturalist to realize the altered circumstances that would render the plains of his island suitable for the growth of plants that now inhabit its mountains only; but if he glance at the map of the isothermal lines of the northern hemisphere, he will see how varied are the climates of regions in the same latitude; that London, with a mean temperature of 51°, is in the same latitude as Hudson's Bay, where the mean temperature is 30°, and the soil ever frozen: and we will further be able to understand by a little reflection, how a change in the relative positions of sea and land would, by isolating Labrador, raise its temperature 10°– 15°, causing the destruction of all the native plants that did not retire to its mountain-tops, and favouring the immigration of the species of a more genial climate."

"The first inference from such an hypothesis is that the Alpine plants of New Zealand, having survived the greatest changes, are its most ancient colonists; and it is a most important one in many respects, but especially when considered with reference to the mountain floras of the Pacific and southern hemisphere generally. These may be classed under three heads:

1) Those that contain identical or representative species of the Antarctic Flora, and none that are peculiarly Arctic; as the Tasmanian and New Zealand Alps.

2) Those that contain, besides these, peculiarities of the Northern and Arctic Floras; as the South American Alps.

3) Those that contain the peculiarities of neither; as the mountains of South Africa and the Pacific Islands."

"We thus observe that the want of an Arctic or Antarctic Flora at all in the Pacific islands, and the presence of an Arctic one in the American Alps, are the prominent features; and I shall confine my remarks upon these to the fact that, with regard to the isolated islands of the Pacific, they are situated in too warm a latitude to have had their temperature cooled by changes in the relative position of land and ocean, so as to have harboured an Antarctic vegetation. With regard to the South American Alps, there is direct land communication along the Andes from Arctic to Antarctic regions; by which not only may the strictly Arctic genera and species have migrated to Cape Horn, but by which many Antarctic ones may have advanced northward to the equator."

"There is still another point in connection with the subject of the relative antiquity of plants, and in adducing it I must again refer to the 'Principles of Geology', where it is said, "As a general rule, species common to many distant provinces, or those now found to inhabit many distant parts of the globe, are to be regarded as the most ancient ...... their wide diffusion shows that they have had a long time to spread themselves, and have been able to survive many important changes in Physical Geography." If this be true, it follows that, consistently with the theory of the antiquity of the Alpine flora of New Zealand, we should find amongst the plants common to New Zealand and the Antarctic islands, some of the most cosmopolitan; and we do so in *Monita fontana, Callitriche verna, Cardamine hirsuta, Epilobium tetragonum*, and many others" (pp. xxiii–xxv).

The last chapter of the Introductory Essay deals with the composition and affinities of the New Zealand flora itself.

"The traveller from whatever country, on arriving in New Zealand, finds himself surrounded by a vegetation that is almost wholly new to him; with little that is at first sight striking, except the Tree-fern and *Cordyline* of the northern parts, and nothing familiar, except possibly the Mangrove; and as he extends his investigations into the Flora, with the exception of *Pomaderris* and *Leptospermum*, he finds few forms that remind him of other countries. Of the numerous Pines, very few recall by habit and appearance the idea attached either to trees of this family in the northern hemisphere, or to the *Callitris* of New Holland, or to the *Araucariae* of that country and Norfolk

Island; while of the families that on examination indicate the only close affinity between the New Zealand Flora and that of any other country (the *Myrtaceae*, *Epacrideae*, and *Proteaceae*,) few resemble in general aspect their allies in Australia. A paucity of Grasses, an absence of *Leguminosae*, an abundance of bushes and Ferns, and a want of annual plants, are the prevalent features in the open country, whilst the forests abound in *Cryptogamia* and in phaenogamic plants with obscure green flowers, and very often of obscure and little-known Natural Orders."

"Considerably more than two hundred of the New Zealand species have either unisexual or polygamous flowers, or are otherwise incomplete in their reproductive organs, even when their floral envelopes are more or less developed. The number of Natural Orders is large in proportion to the genera; being as 92 to 282, that is, about one to three: while the genera are to the species as 282 to 730, each genus having on the average only two and a half species; whence it follows that there are, on the average, but eight species to each Natural Order."

"Considering these circumstances, and the additional one, that very many of the Natural Orders cannot be recognized by the flower alone, by fruit alone, or by habit or foliage, it may, I think, safely be said that the New Zealand Flora is, for its extent, much the most difficult on the globe to a beginner. Indeed, the mere fact that the student must know a Natural Order for every eight species he has to investigate, offers as direct a means of proving this by comparison as any datum could do, for the probable proportion of species of plants on the globe to the known Natural Orders, exceeds three hundred and fifty to one; in Tasmania the proportions are eleven to one, and in Great Britain they average fourteen to one."

"It is, therefore, not surprising that the vegetation of New Zealand should be wanting in any conspicuous or prevailing feature, which is the case to so great a degree that, excluding Ferns, I do not think any two botanists would, without investigation, characterize any part of the islands as the region of any particular order, genus, or species. The *Coniferae*, when known, prove to be perhaps the most universally prevalent natural family; but the majority of their species, not being social, but growing intermixed with other trees, give no character to the landscape. The vast number of trees, the paucity of herbaceous plants, and the almost total absence of annuals, are the most remarkable features of the Flora; for of flowering trees, including shrubs above twenty feet high, there are upwards of 113, or nearly one-sixth of the Flora, besides 156 shrubs and plants with woody stems. Of the largest Natural Orders, so far as regards the number of species, the individuals are often so few, that the botanist would form a very erroneous estimate of the numerical force of such in the whole island from an examination of some of its parts only: thus the Orders most numerous in species are: *Compositae*, 90; *Cyperaceae*, 66; *Gramineae*, 53; *Scrophularineae*, 40; *Orchideae*, 39; *Rubiaceae*, 26; and *Epacrideae* and *Umbelliferae*, each 23; none of which can be said to form prevalent features in the landscape, though none are rare."

"In the neighbouring island of Tasmania, where the same Orders predominate to a great extent, the case is widely different: there the Grasses everywhere form a prominent feature; the *Cyperaceae*, from their size, strength, and cutting foliage, arrest the traveller's progress through the forest; *Orchideae* of many kinds carpet the ground in spring with beautiful blossoms; the heaths are gay with *Epacrideae*; herbs, trees, and shrubs of *Compositae* meet the eye in every direction; whilst the *Myrtaceae* and *Leguminosae* are characteristics both of the arboreous and shrubby vegetation. The difference is so marked, that I retain the most vivid recollection of the physiognomy of the Tasmanian mountains and valleys, but a very indifferent one of the New Zealand forest, where all is, comparatively speaking, blended into one green mass, relieved at the Bay of Islands by the symmetrical crown of the Tree-fern, the pale green fountain of foliage of the *Dacrydium cupressinum*, and the poplar-like *Knightia* overtopping all. It is true that there is more variety in the latter country than is expressed by this selection of a few individuals, and a little reflection recalls a vast number of noble, and some beautiful botanical objects, but with the exception of groves of the Kaikatea Pine (*Podocarpus dacrydioides*) on the swampy river banks, the *Pomaderris* and *Leptospermum* on the open hill-sides, and *Dammara* on their crests, there is little to arrest the botanist's first glance; and nothing in the massing or grouping of the species of any Natural Order renders that Order an important element in the general landscape, or gives individuality to any of its parts, by flowers and gaiety, or by foliage and gloom. The same features prevail even so far south as Lord Auckland's Group, where *Dracophyllum*, *Coprosma*, *Metrosideros*, *Panax*, and a shrubby *Veronica* unite to form an evergreen mantle: and I suspect, from the accounts I have heard and read, that they are repeated on the damp cool coasts of Chili, to the north of the region of the sombre Beech-forests which clothe the Fuegian islands" (pp. xxvii–xxix).

A very marked feature of the flora is the large number of endemics, which HOOKER gives as 26 genera and 507 species, or more than two-thirds of the whole. Yet a close relationship to other countries may be traced in most of the endemic genera and species, though there are some 7 or 8 genera that are exceptionally isolated.

"The remaining third of the New Zealand Flora may be divided into five groups, for illustrating the relations of the plants to those of other countries,— *viz.*,

1) 193 species, or nearly one-fourth of the whole, are Australian.
2) 89 species, or nearly one-eighth of the whole, are South American.
3) 77 species, or nearly one-tenth of the whole, are common to both the above.
4) 60 species, or nearly one-twelfth of the whole, are European.
5) 50 species, or nearly one-sixteenth of the whole, are Antarctic Islands', Fuegian, etc."

"1) *Those of Australian affinity.*— The decided preponderance of Australian forms is not confined to this large number of absolutely identical species; I have shown it to prevail in the genera containing peculiar species also. There are no Natural Orders in New Zealand which are not also found in Australia and Tasmania, except *Coriariae, Escalloniae, Brexiaceae*, and *Chloranthaceae*. Upwards of 240 of the 282 New Zealand genera are Australian, and of these more than fifty are all but confined to these two countries. New Zealand, however, does not appear wholly as a satellite of Australia in all the genera common to both, for of several there are but few species in Australia, which hence shares the peculiarities of New Zealand, rather than New Zealand those of Australia: this is the case with *Pittosporum, Coprosma, Olearia, Celmisia, Forstera, Gaultheria, Dracophyllum, Veronica, Fagus, Dacrydium*, and *Uncinia*; of which there are comparatively few species in Australia and Tasmania: on the other hand, *Stackhousieae, Pomaderris, Leptospermum, Exocarpus, Persoonia, Epacris, Leucopogon, Goodenia*, and a few other large Australian genera, are very scantily represented in New Zealand."

"If the number of plants common to Australia and New Zealand is great, and quite unaccountable for by transport, the absence of certain very extensive groups of the former country is still more incompatible with the theory of extensive migration by oceanic or aerial currents. This absence is most conspicuous in the case of *Eucalypti*, and almost every other genus of *Myrtaceae*, of the whole immense genus of *Acacia*, and of its numerous Australian congeners, with the single exception of *Clianthus*, of which there are but two known species, one in Australia, and the other in New Zealand and Norfolk Island."

"The rarity of *Proteaceae, Rutaceae*, and *Stylideae*, and the absence of *Casuarina* and *Callitris*, of any *Goodeniae* but *G. littoralis* (equally found in South America), of *Tremandreae, Dilleniaceae*, and of various genera of *Monocotyledones*, admit of no explanation, consistent with migration over water having introduced more than a very few of the plants common to these tracts of land. Considering that *Eucalypti* form the most prevalent forest feature over the greater part of South and East Australia, rivalled by the *Leguminosae* alone, and that both these Orders (the latter especially) are admirably adapted constitutionally for transport, and that the species are not particularly local or scarce, and grow well wherever sown, the fact of their absence from New Zealand cannot be too strongly pressed on the attention of the botanical geographer, for it is the main cause of the difference between the floras of these two great masses of land being much greater than that between any two equally large contiguous ones on the face of the globe. If no theory of transport will account for these facts, still less will any of variation; for of the three genera of *Leguminosae* which do inhabit New Zealand, none favour such a theory; one, *Clianthus*, I have just mentioned; the second, *Edwardsia*, consists of one tree, identical with a Juan Fernandez and Chilian one, and unknown in New Holland; and the third genus (*Carmichaelia*) is quite peculiar, and consists of a few species feebly allied to some New Holland plants, but exceedingly different in structure from any of that extensive Natural Order."

"2) *Species of South American affinity.*— The South American species in New Zealand amount to 89, or one-eighth: of these some are absolutely peculiar to the two countries, as *Myosurus aristatus*, two species of *Coriaria*, *Edwardsia grandiflora*, *Haloragis alata*, *Hydrocotyle Americana*, and *Veronica elliptica*. Of these the *Edwardsia* is by far the most striking case, from the size of the tree: it appears to have a much wider range in New Zealand than in Chili, and supposing it to have been transported between these countries, it is difficult to say which was the parent one; its affinities would, however, incline us to consider it amongst the aborigines of the former. It is by representative genera and species that the affinity of the New Zealand and South American floras is best shown, and this most conspicuously by *Fuchsia* and *Calceolaria*, two most remarkable genera, confined to these two countries, but by far the most abundant to the west of the Andes. Here again the amount of affinity is differently displayed by each; of the Calceolarias one is so closely allied to an American species, that I doubt the propriety of keeping them separate, while the other appears a very distinct species; the Fuchsias are both extremely peculiar, one of them being the only species that has no petals. Altogether there are 76 genera common to New Zealand and South America, and 17 of these are not found in Australia, or elsewhere in the Old World. It is curious that none of the latter belong to those peculiarly Arctic and north temperate genera mentioned in the note to p. xxiv, except *Caltha*, to a southern form of which, however, the New Zealand species belongs."

"3) *Plants common to New Zealand, Australia, and South America.*— Of the 77 plants common to these three countries, which include one-tenth of the flora of New Zealand, the majority are Grasses, 10; *Cyperaceae*, 7; moisture-loving Monocotyledons, 9; *Monochlamydeae*, 8; *Umbelliferae* and *Compositae* each 4; and fully 50 of the whole number are also found in Europe, and do not indicate any peculiar affinity between these three southern masses of land: of those that are not European, some are Antarctic plants found in mountainous districts of Australia and Tasmania, as *Oxalis Magellanica*. Of genera and species which, from their near affinity with one another, and marked distinction from any others, may be said to be represented in all three countries, the majority are Antarctic, and will be noticed under the fifth head."

"4) *European plants in New Zealand.*— These, amounting to 60, or about one-twelfth of the whole flora, are in many respects the most interesting, and to their identification (which I consider approximate only) I have given a great deal of care. Many I consider still open to inquiry, which may reduce their supposed numbers; but on the other hand I am sure that future discoveries will add to them. To some extent these are distributed according to well defined laws, which do accord with facilities for migration by transport, thus:— *a*) 17 are sea-shore plants, or inhabitants of salt marshes, as *Ruppia*, *Zanichellia*, *Atriplex*, and their allies; *Dodonaea*, *Arenaria rubra*, and *Calystegia Soldanella*, also affect coasts;— *b*) 16 are fresh-water plants, or natives of very marshy spots, for whose transport, however, it appears to me as difficult to account as if they were land-plants;— *c*) 5 are *Compositae*,

of which four have pappus; a facility for aerial transport, which loses its
significance and weight from the fact that the species of *Compositae* (which
of all Orders is the largest and most universal) are the most local. The fact
of these five being found in so very many parts of the globe, and being the
only ones that are so, is extremely remarkable, for it points to oceanic transport as the means of their diffusion: though the probabilities are against
their all having thus accidentally met in that most isolated area which they
all inhabit;— *d*) 19 of the species are *Glumaceae*, including seven Grasses
and three aquatic *Cyperaceae* (which latter have also been included under *b*)."

"This large proportion of the lower Orders of Phaenogamic plants is in
accordance with a general law of geographic distribution, but not the more
intelligible on that account, for I cannot recognise in their structure or
physiology any peculiarities that render them fitted for such diffusion. And
I may add, that after a most careful microscopic study of the structure of
the seeds of all the plants common to Europe and New Zealand, I have come
to the conclusion that, as a body, they present no such facility for transoceanic or aerial transport, as would account for their having migrated further
than the majority of other plants. To this may be added the fact that the Orders
to which they belong, are not those whose seeds after transport are found
to vegetate most surely or freely in gardens."

"Many of the European species occurring in New Zealand are also Australian, Tasmanian, and Antarctic; some of the more remarkable exceptions
are,—of plants not hitherto found in South America, *Hierochloe borealis*,
Alopecurus geniculatus, some *Carices*, and other Monocotyledons. Of plants
not found in Australia, *Agrostis canina* and *Taraxacum officinale*. Of those not
found either in Australia or South America, *Carex stellulata* and *Pyrenaica*,
and *Sparganium natans*."

"It should also be mentioned here, that some very widely diffused European and Australian plants are absent from New Zealand, as *Lythrum Salicaria, Alchemilla arvensis, Portulaca oleracea, Hydrocotyle vulgaris, Zapania nodiflora, Verbena officinalis, Prunella vulgaris, Samolus Valerandi, Vallisneria spi-
ralis, Potamogeton perfoliatus* and *crispus, Alisma Plantago, Caulinia oceanica,
Juncus maritimus* and *effusus, Carex caespitosa, Cladium Mariscus, Isolepis fluitans, Cyperus rotundus, Glyceria fluitans*, and *Arundo Phragmites*."

"5) *Antarctic plants in New Zealand.*— Of these Antarctic plants, about
50 inhabit the mountains and southern extreme of New Zealand; a number
which (as I have stated at p. 15) will probably be greatly increased by future
discoveries. They may be geographically grouped as follows:— *a*) Those
of general distribution, being common also to Europe, as *Callitriche, Montia,
Cardamine hirsuta, Potentilla anserina, Epilobium tetragonum, Myriophyllum,
Calystegia Soldanella* and *C. Sepium, Limosella*, many *Monochlamydeae*, and
more *Monocotyledones.*— *b*) Those found also in Tasmania, and chiefly on
its mountains, but not elsewhere; as *Oxalis Magellanica, Acaena*, some *Epilo-
bia, Colobanthus, Scleranthus, Tillaea, Apium, Coprosma, Leptinella, Hierochloe
antarctica*, etc."

"The botanical affinity between extra-tropical South America, the Antarctic islands, New Zealand, and Tasmania, is, however, much better indicated by the peculiar genera, by groups of those, or by individual species which, as it were, represent one another in two or more of these localities, and which give a peculiar botanical character to the flora of southern latitudes beyond latitude 35°."

"Of these genera, there are 50 which afford botanical characters in common, and give as decided a proof of close affinity in vegetation, as do the 50 identical species above mentioned. The most conspicuous of these genera common to all the above-named localities are, *Colobanthus, Drosera, Acaena, Gunnera, Oreomyrrhis, Leptinella, Lagenophora, Forstera, Pratia, Gaultheria, Gentiana, Euphrasia, Plantago, Drapetes, Fagus, Astelia, Juncus, Carpha, Chaetospora, Oreobolus, Uncinia, Carex,* and many Grasses, especially *Hierochloe, Alopecurus, Trisetum, Deyeuxia,* etc." (pp. xxx–xxxiii).

HOOKER then gives a list of 228 species showing "parallelism" in distribution between New Zealand, Australia and Tasmania, and Temperate and Cold S. America. As a result he says: "Enough is here given to show that many of the peculiarities of each of the three great areas of land in the southern latitudes are representative ones, effecting a botanical relationship as strong as that which prevails throughout the lands within the Arctic and Northern Temperate zones, and which is not to be accounted for by any theory of transport or variation, but which is agreeable to the hypothesis of all being members of a once more extensive flora, which has been broken up by geological and climatic causes.

"I have alluded to Pacific Island peculiarities in the New Zealand Flora; these are few, but very well marked by some otherwise local genera, as *Coprosma, Astelia, Exocarpus, Dammara, Geniostoma, Cyathodes, Santalum, Elatostemma, Ascarina, Cordyline,* and others, of which *Ascarina* is the most remarkable, as the genus has hitherto been found nowhere but in New Zealand and the Sandwich Islands. Until the New Caledonian and Hebridean vegetation especially is known, however, we cannot follow out this affinity, as I do not doubt that their rich floras will connect the Botany of the Pacific, Australian, New Zealand, and Malay Islands in a very remarkable manner, and exhibit affinities of the utmost importance."

"There has lately indeed been discovered a most remarkable and unique instance of representation by close botanical affinity between very distant spots, *viz.* the existence of three of the most peculiar Antarctic, New Zealand, and Tasmanian genera on the lofty mountain of Kini-Balu, in Borneo, situated under the equator, *viz. Drapetes, Phyllocladus,* and *Drimys*" (p. xxxvi).

Finally, HOOKER gives lists of New Zealand genera whose species are extremely variable. Of these genera 45 are "of very general distribution" and 34 "endemic, or of confined geographical distribution". He also draws special attention to 21 genera "whose species alter in form or habit." Very cautious conclusions are reached as to the extent of "variability" in the New Zealand flora compared with that in other floras.

"On the whole I am inclined to think that the New Zealand Flora is remarkable for the number of plants which vary thus, but that this peculiarity is rendered conspicuous by the prevalence of *Coniferae* and *Araliaceae*, which are variable in all parts of the world" (p. xxxix).

* * *

Fortunately for the student of the plant-life of New Zealand, COCK-AYNE (1928) has brought together in a very readable book most if not all of the important data available up to 1927, with a very full bibliography. In this work, part 1 deals with the history of botanical investigations in New Zealand and with the environmental factors, parts 2 and 3 with the vegetation, part 4 with the distribution and composition of the flora, and part 5 with the history of the flora. The last two parts are of particular interest as showing the advance on HOOKER's pioneer researches. The whole area, which includes the groups of the Kermadec, Chatham, and Subantarctic Islands, is divided into six provinces and sixteen major and six other districts.

COCKAYNE gives the total number of species of vascular plants, "together with such varieties as are of equal rank to many admitted species" as 1843, "of which 166 are pteridophytes (*Filices* 147), 20 gymnosperms and 1657 angiosperms (monocotyledons 428 and dicotyledons 1229), and they belong to 109 families and 383 genera. The average number of species to each family is 16.9 and 4.8 to each genus." One great advance has been the recognition of widespread hybridization and "no less than 290 groups of such [hybrids] have been noted."

The elements of the flora are classified as follows:

The endemic element.— "The endemic species of *Pteridophyta, Gymnospermae*, and *Angiospermae* (monocotyledons 294, dicotyledons 1077) number 1451 and constitute 78.6 per cent of the vascular flora. Taking each group separately, 36 per cent of the pteridophytes are endemic, 67 per cent of the monocotyledons, and 88 per cent of the gymnosperms (20) and dicotyledons taken together."

The Australian element.— 38 genera, or subdivisions of such, are confined to New Zealand and Australia. Of species, excluding the cosmopolitan and subcosmopolitan, the number supposed to be common to New Zealand and Australia (probably few are truly identical) is 236. The importance of the Australian element is easily exaggerated, as also the floristic relation between the two floras. *"It is hardly going too far to say that it would be possible for one to have an excellent acquaintance with the botany of Eastern Australia and yet to be acquainted with very few indeed of the species which extend to New Zealand"* (p. 411).

The Subantarctic element.— "It is the presence of a well-defined element common to New Zealand, Eastern Australia and Tasmania, subantarctic South America, and the extra-New Zealand subantarctic Islands that has given rise to endless speculations as to its origin."

There are many genera common to New Zealand and Subantarctic South America. As to species, there are at least 58 identical, or almost so, in the New Zealand and Subantarctic flora which have remained unchanged, intense isolation notwithstanding. Moreover there are 44 vicarious species.

The palaeotropic element.— These are the Malayan, Australian-Malayan, Melanesian, and Polynesian elements. Most of the species are endemic but they form "a distinct element of the flora which is in large measure restricted to the frostless parts of the Region comparatively few species gaining the subalpine belt or in the south leaving the coast-line" (p. 414).

The cosmopolitan element of species with a considerable range in either temperate or warm climates.

The Lord Howe-Norfolk element.— Two identical and about 6 vicaious pairs of species link Lord Howe Island and New Zealand and 3 identical and about 8 vicarious pairs show Norfolk Island and New Zealand floristic relationships.

The general conclusions are:

"The flora of New Zealand, notwithstanding its strong endemism, possesses two very distinct elements not floristic only but ecological. The first, and, as I believe , the more primitive, is not one simple floristic entity, but consists of a combination of the palaeozelandic and subantarctic elements of the flora, now difficult to disentangle. They have this one property in common, the power, for the most part, to endure a fair amount of cold. In other words, the element is a temperate one."

"The second element, also largely endemic, consists of descendents of an ancient palaeotropic stock, so ancient indeed that endemic genera have been developed (*Rhabdothamnus, Ixerba, Alectryon* etc.) as well as many distinct endemic species" (pp. 416–7).

"Yet notwithstanding this great age of the members, and their long isolation far from the tropics, but *few have become really fitted to the present average climate of New Zealand, in fact the majority can tolerate very little frost.* For the most part, the species of this class are confined to the lowlands, and in the south some are only found near the coast. This element, in fact, is eminently subtropical; so that the present-day climate is one to which it is not perfectly attuned ......"

"It has been shown that while there is a considerable Australian element, it is made up largely of Subantarctic and Paleaotropic species, while the true Australian element does not play a conspicuous part in the vegetation. Especially is the absence of characteristic Australian genera noteworthy, e.g. *Eucalyptus*, *Acacia* etc., although virtually all the Tasmanian species are not quite hardy in the warmer parts of New Zealand, but some can spread spontaneously. Bearing these facts in mind, the possibility of direct land-connection with Eastern Australia except at a very remote period cannot be entertained" (pp. 416–7).

Regarding the history of the flora, COCKAYNE says it is "in great measure, a matter of speculation merely." A "burning question" in New Zealand biogeography is that of ancient land-connections. Answers are varied. There is doubt as to the efficacy of wind dispersal. "Were spores as readily carried by the wind as is supposed, there should be no special fern-floras, which is not the case." Further, it is *not* the *species* which move but the *associations* to which they belong. "Ecesis, rather than the possibility of bird-carriage etc. during long periods of time, is the great stumbling-block ...... *between the arrival of seed or spore and its becoming a mature plant in a situation favourable, not only for its well-being, but for its increase, is altogether another matter.*"

Just as the botanical and zoological evidence make out a strong case for a Subantarctic or Antarctic "land-bridge", so does the great depth of the ocean to the south and east of the New Zealand continental shelf shake one's belief in the possibility of such connection. Yet COCKAYNE says he must declare for the problematical "bridge". It seems fairly certain that New Zealand has never been completely submerged since early Mesozoic times. In the Cretaceous there was a "Great New Zealand" which was linked with what are now island groups and even with New Caledonia, New Guinea, and northern Queensland. During the "Great New Zealand period" the palaeotropic element would people the north while from fairly warm Antarctica would come the subantarctic element. The palaeozelandic element which in part had its beginnings in the Jurassic, would advance north and south.

"The great elevation and extension of the land was succeeded by an equally great depression which is *the most critical occurrence in regard to any conclusions concerning the origin of the flora, or its distribution in New Zealand itself*" (p. 423). Maximum submergence occurred by the Early Middle Tertiary and was followed by re-elevation. The great Oligocene-Miocene reduction of the land must have brought about the extinction of many species especially among the high mountain plants whose time of origin or arrival is still problematical. Elevation occurred during the Pleistocene towards the end of which came the great extension of glaciation to be followed by minor submergences and elevations.

It is not intended here to attempt to summarize the detailed accounts of New Zealand vegetation given in COCKAYNE's book. HOOKER dealt with the floristics and phytogeography of New Zealand, not with its ecology in the modern sense. Since COCKAYNE's book was published in 1928 ecological research has continued and readers are referred to the following papers: ALLAN (1937), COCKAYNE (1929), COCKAYNE and CHALDER (1932), COCKAYNE, SIMPSON and THOMSON (1932), COCKAYNE and TURNER (1928), COCKAYNE and TEICHELMANN (1930), FLINT (1938), LAING and OLIVER (1929), LOGAN (1934), McINDOE (1932), OLIVER (1930), SIMPSON and THOMSON (1938), THOMSON (1935), THOMSON and SIMPSON (1939), ZOTOV (1938), ZOTOV (1939), and ZOTOV, ELDER, BEDDIE, SAINSBURY, and HODGSON (1938).

The destruction of the natural vegetation and its replacement by modified and often quite alien communities and species has much concerned New Zealand botanists. Recent (post–1927) works on this subject are: ALLAN (1936), ALLAN (1937), ALLAN (1940), COCKAYNE, SIMPSON, and THOMSON (1932), COCKAYNE and SLEDGE (1932), and CUMBERLAND (1941).

On more general phytogeographical problems are the following: ANDERSON (1931), and COCKAYNE (1930). MARTIN (1946) deals with the distribution of the New Zealand moss flora. The fossil flora has not been neglected and papers on this subject include those by OLIVER (1928), OLIVER (1936), and PENSELER (1930). CRANWELL (1938) dealt with fossil pollens in New Zealand. Finally, it is important to record that the cytology of New Zealand plants is beginning to receive attention as shown by the papers of FRANKEL (1940), FRANKEL and HAIR (1937), and HAIR (1942).

Since 1928, little new and important research has been published on the general phytogeography of New Zealand. Advances can be expected when researches along the modern lines of synthetic systematics have been carried out on a considerable number of native species. These must include their cytology and genetics as well as their autecology. CHEESEMAN's Manual (1925) provides an adequate position from which to plan such new researches since it brings up-to-date the earlier taxonomic work of HOOKER on the New Zealand flora which is outside our present terms of reference.

Flora Tasmaniae:— The Introductory Essay, published on 6 Febr. 1860, is of double interest. It has value for its own factual con-

tents and as material for botanical history it provides evidence for the
change in theoretical views of a great botanist as a result of the publication of DARWIN and WALLACE's account of natural selection in Journ.
Linn. Soc. Zool. 3, No. 9: 45–62 (1858). The theoretical views are
not easy to summarize and their importance is such that a long quotation is given of the section entitled "On the General Phenomena of
Variation in the Vegetable Kingdom."

"*1*) All vegetable forms are more or less prone to vary as to their sensible
properties, or (as it has been happily expressed in regard to all organisms),
"they are in a state of unstable equilibrium." No organ is exactly symmetrical, no two are exact counterparts, no two individuals are exactly alike, no
two parts of the same individual exactly correspond, no two species have
equal differences, and no two countries present all the varieties of a species
common to both, nor are the species of any two countries alike in number
and kind."

"*2*) The rate at which plants vary is always slow, and the extent or degree
of variation is graduated. Sports even in colour are comparatively rare phenomena, and, as a general rule, the best-marked varieties occur on the confines of the geographical area which a species inhabits. Thus the scarlet
Rhododendron (*R. arboreum*) of India inhabits all the Himalaya, the Khasia
Mountains, the Peninsular Mountains, and Ceylon; and it is in the centre of
its range (Sikkim and the Khasia) that those mean forms occur which by a
graduated series unite into one variable species the rough, rusty-leaved
form of Ceylon, and the smooth, silvery-leaved form of the North-western
Himalaya. A white and a rose coloured sport of each variety is found growing with the scarlet in all these localities, but everywhere these sports are few
in individuals. Also certain individuals flower earlier than others, and some
occasionally twice a year, I believe in all localities."

"*3*) I find that in every Flora all groups of species may be roughly classified into three large divisions; one in which most species are apparently
unvarying; another in which most are conspicuously varying; and a third
which consists of a mixture of both in more equal proportions. Of these the
unvarying species appear so distinct from one another that most botanists
agree as to their limits, and their offspring are at once referable by inspection
to their parents; each presents several special characters, and it would require many intermediate forms to effect a graduated change from any one to
another. The most varying species, on the contrary, so run into one another,
that botanists are not agreed as to their limits, and often fail to refer the
offspring with certainty to their parents, each being distinguished from one
or more others by one or a few such trifling characters, that each group may
be regarded as a continuous series of varieties, between the terms of which
no hiatus exists suggesting the intercalation of any intermediate variety. The
genera *Rubus*, *Rosa*, *Salix*, and *Saxifraga*, afford conspicuous examples of
these unstable species; *Veronica*, *Campanula*, and *Lobelia*, of comparatively
stable ones."

"*4*) Of these natural groups of varying and unvarying species, some are

large and some small; they are also very variously distributed through the
classes, orders, and genera of the Vegetable Kingdom; but as a general rule,
the varying species are relatively most numerous in those classes, orders,
and genera which are the simplest in structure. Complexity of structure is
generally accompanied with a greater tendency to permanence in form:
thus Acotyledons, Monocotyledons, and Dicotyledons are an ascending
series in complexity and in constancy of form. In Dicotyledons, *Salices*,
Urticeae, *Chenopodiaceae*, and other Orders with incomplete or absent floral
envelopes, vary on the whole more than *Leguminosae*, *Lythraceae*, *Myrtaceae*,
or *Rosaceae*, yet members of these present, in all countries, groups of notor-
iously varying species, as *Eucalyptus* in Australia, *Rosa* in Europe, and
Lotus, *Epilobium*, and *Rubus* in both Europe and Australia. Again, even
genera are divided: of the last named, most or all of the species are variable;
of others, as *Epacris*, *Acacia*, and the majority of such as contain upwards
of six or eight species, a larger or smaller proportion only are variable. But
the prominent fact is, that this element of mutability pervades the whole
Vegetable Kingdom; no class nor order nor genus or more than a few
species claims absolute exemption, whilst the grand total of unstable forms
generally assumed to be species probably exceeds that of the stable."

"5) The above remarks are equally applicable to all the higher divisions
of plants. Some genera and orders are as natural, and as limitable by characters, as are some species; others again, though they contain many very
well-marked subordinate plans of construction, yet are so connected by
intermediate forms with otherwise very different genera or orders, that it is
impossible to limit them naturally. And as some of the best marked and
limited species consist of a series of badly marked and illimitable varieties,
so some of the most natural and limitable orders and genera may respectively consist of only undefinable groups of genera or of species. For instance,
both *Gramineae* and *Compositae* are, in the present state of our knowledge,
absolutely limited Orders, and extremely natural ones also; but their genera
are to a very eminent degree arbitrarily limited, and their species extremely
variable. *Orchideae* and *Leguminosae* are also well-limited Orders (though not
so absolutely as the former), but they, on the contrary, consist of comparatively exceedingly well-marked genera and species. *Melanthaceae* and
Scrophularineae, on the other hand, are not limitable as Orders, and contain
very many differently constructed groups, but their genera, and to a great
extent their species also, are well-marked and limitable. The circumstance
of a group being either isolated or having complex relations, is hence no
indication of its members having the same characters."

"Again, as with species, so with genera and orders, we find that upon the
whole those are the best limited which consist of plants of complex floral
structure; the Orders of Dicotyledons are better limited than those of Monocotyledons, and the genera of *Dichlamydeae* than those of *Achlamydeae*".

"Now my object in dwelling on this parallelism between the characteristics of individuals in relation to species, of species in relation to genera,
and of genera in relation to Orders, is because I consider (Introd. Essay to
Fl. N.Z.) that it is to the extinction of species and genera that we are indebted

for our means of resolving plants into limitable genera and orders. This view is now, I believe, generally admitted, even by those who still regard species as the immutable units of the Vegetable Creation; and it therefore now remains to be seen how far we are warranted in extending it to the limitation of species by the elimination of their varieties through natural causes."

"*6*) The evidence of variability thus deduced from a rapid general survey of the prominent facts elicited from a study of the principles of classification, are to a certain extent tested by the behaviour of plants under cultivation, which operates either by hastening the processes of Nature (in rapidly inducing variation), or by effecting a prolepsis or anticipation of those processes (in producing sports, *i.e.* better marked varieties, without graduated stages), or by placing the plant in conditions to which it would never have been exposed in the ordinary course of natural events, and which eventually either kill it or give origin to a series of varieties which might otherwise have never existed."

"*7*) Now the prominent phenomena presented by species under cultivation are analogous in kind and extent to those which we have derived from a survey of the affinities of plants in a state of nature: a large number remain apparently permanent and unalterable, and a large number vary indefinitely. Of the permanent there is little to remark, except that they belong to very many orders of plants, nor are they always those which are permanent in a state of nature. Many plants, acknowledged by all to be varieties, may be propagated by seed or otherwise, when their offspring retains for many successive generations the characters of the variety. On the other hand, species which have remained immutable for many generations under cultivation, do at length commence to vary, and having once begun, are thereafter peculiarly prone to vary further."

"*8*) The variable cultivated species present us with the most important phenomena for investigating the laws of mutability and permanence; but these phenomena are so infinitely varied, complex, and apparently contradictory, as to defeat all attemps to elucidate the history of any individual case of variation by a study of its phases alone. It would often appear doubtful whether the natural operations of a plant tend most to induce or to oppose variation, and we hence find the advocates of original permanent creations, and those of mutable variable species, taking exactly opposite views in this respect, the truth, I believe, being that both are right. Nature has provided for the possibility of indefinite variation, but she regulates it as to extent and duration; she will neither allow her offspring to be weakened or exhausted by promiscuous hybridization and incessant variation, nor will she suffer a new combination to external conditions to destroy one of these varieties without providing a substitute when necessary; hence some species remain so long hereditarily immutable as to give rise to the doctrine that all are so normally, while others are so mutable as to induce a belief in the very opposite doctrine, which demands incessant lawless change."

"*9*) It would take far too long a time were I to attempt any analysis of the phenomena of cultivation, as illustrative of those of variability in a state of nature. There are however some broad facts which should be borne in mind in treating of variation by cross impregnation and hybridity."

"*10*) Variation is effected by graduated changes; and the tendency of varieties, both in nature and under cultivation, when further varying, is rather to depart more and more widely from the original type, than to revert to it: the best marked varieties of a wild species occurring on the confines of the area the species inhabits, and the best marked varieties of the cultivated species being those last produced by the gardener. I am aware that the prevalent opinion is that there is a strong tendency in cultivated, and indeed in all varieties, to revert to the type from which they departed; and I have myself quoted this opinion, without questioning its accuracy, as tending to support the views of those who regard species as permanent. A further acquaintance with the results of gardening operations leads me now to doubt the existence of this centripetal force in varieties, or at least to believe that in the phrase "reversion to the wild type", many very different phenomena are included. In the first place, the majority of cultivated vegetables and cerealia, such as the Cabbage and its numerous progeny, and the varieties of wall-fruit, show when neglected no disposition to assume the characters of the wild states of these plants; they certainly degenerate, and even die if Nature does not supply the conditions which man (by anticipation of her operations, or otherwise) has provided; they become stunted, hard, and woody, and resemble their wild progenitors in so far as all stunted plants resemble wild plants of similar habit; but this is not a reversion to the original type, for most of these cultivated races are not merely luxuriant forms of the wild parent. In neglected fields and gardens we see plants of Scotch Kale, Brussels Sprouts, or Kohl-rabi, to be all as unlike their common parent, the wild *Brassica oleracea*, as they are unlike one another; so, too, most of our finer kinds of apples, if grown from seed, degenerate and become crabs, but in so doing they become crab states of the varieties to which they belong, and do not revert to the original wild Crab-apple. And the same is true to a great extent of cultivated Roses, of many varieties of trees, of the Raspberry, Strawberry, and indeed of most garden plants. It has also been held, that by imitating the conditions under which the wild state of a cultivated variety grows, we may induce that variety to revert to its original state; but, except in the false sense of reversion above explained, I doubt if this is supported by evidence. Cabbages grown by the seaside are not more like wild Cabbages than those grown elsewhere, and if cultivated states disseminate themselves along the coast, they there retain their cultivated form. This is however a subject which would fill a volume with most instructive matter for reflection, and which receives a hundred-fold more illustration from the Animal than from the Vegetable Kingdom. I can here only indicate its bearing on the doctrine of variation, as evidence that Nature operates upon mutable forms by allowing great variation, and displaying little tendency to reversion. With this law the suggestive observation of M. VILMORIN well accords, that when once the constitution of a plant is so broken that variation is induced, it is easy to multiply the varieties in succeeding generations."

"It may be objected to this line of argument that our cultivated plants are, as regards their constitution, in an artificial condition, and are, if unaided, incapable of self-perpetuation; but an artificially induced condition of constitution is not necessarily a diseased or unnatural one, and, so far as our cultivated plants are concerned, all we do is to place them under conditions which Nature does not provide *at the same particular place and time.* That Nature might supply the conditions at other places and times may be inferred from the fact that the plant is found to be provided with the means of availing itself of them when provided, while at the same time it retains all its functions, not only unimpaired, but in many cases in a more highly developed state. We have no reason to suppose that we have violated Nature's laws in producing a new variety of wheat,—we may have only anticipated them; nor is its constitution impaired because it cannot, unaided, perpetuate its race; it is in as sound and unbroken health and vigour during its life as any wild variety is, but its offspring has so many enemies that they do not perpetuate its race. In the case of annual plants, those only can secure the succession of their species which produce more seeds annually than can be eaten by animals or destroyed by the elements. Cultivated wheat will grow and ripen its seed in almost all soils and climates, and as its seeds are produced in great abundance, and can be preserved alive in any quantity, in the same climate, and for many years, it follows that it is not to the artificial or peculiar condition of the plant itself, and still less to any change effected by man upon it, that its annual extinction is due, but to causes that have no effect whatever upon its own constitution, and over which its constitutional peculiarities can exercise no control."

"*77*) Again, the phenomena of cross impregnation amongst individuals of all species appear, according to Mr. DARWIN's accurate observations, to have been hitherto much underrated, both as to extent and importance. The prominent fact that the stamens and pistil are so often placed in the same flower, and come to maturity at the same epoch, has led to the doctrine that flowers are usually self-impregnated, and that the effect is a conservative one as regards the permanence of specific forms. The observations of CARL SPRENGEL and others have, however, proved that this is not always the case, and that while Nature has apparently provided for self-fertilization, she has often insidiously counteracted its operation, not only by placing in flowers lures for insects which cross-fertilize them, but often by interposing insuperable obstacles of self-fertilization, in the shape of structural impediments to the access of the pollen to the stigma of its own flower. In all these instances the double object of Nature may be traced; for self-impregnation (or "breeding in"), while securing identity of form in the offspring, and hence hereditary permanence, at the same time tends to weakness of constitution, and hence to degeneracy and extinction: on the other hand, cross-impregnation, while tending to produce diversity of form in the offspring, and hence variation and apparent mutability, yet by strengthening the offspring favours longevity and apparent permanence of specific type. The ultimate effect of all these operations is of course favourable to the hypothesis that variability is the rule, and permanence the exception, or at any rate only a transitory phenomenon."

"*12*) Hybridization, or cross-impregnation between species or very well
marked varieties, again, is a phenomenon of a very different kind, however
similar it may appear in operation and analogous in design. Hybridizable
genera are rarer than is generally supposed, even in gardens, where they are
so often operated upon, under circumstances the most favourable to the
production of a hybrid, and unfavourable to self-impregnation. Hybrids
are almost invariably barren, and their characters are not those of new varieties. The obvious tendency of hybridization between varieties or other very
closely allied forms (in which case the offspring may be fertile) is not to en-
large the bounds of variation, but to contract them; and if between very
different forms, it will only tend to confound these. That some supposed species may have their origin in hybridization cannot be denied, but we are
now dealing with phenomena on a large scale, and balancing the tendencies
of causes uniformly acting, whose effects are unmistakable, and which can
be traced throughout the Vegetable Kingdom. In gardening operations the
number of hybridized genera is small, their offspring doomed, and since they
are more readily impregnated by the pollen of either parent than by their own,
or by that of any other plant, they eventually revert to one of their parents:
on the other hand, the number of varieties is incalculable, the power to vary
further is unimpaired in their progeny, and these tend to depart further and
further in sensible properties from the original parent."

"In conformity with my plan of starting from the variable and not the
fixed aspect of Nature, I have now set down the prominent features of the
Vegetable Kingdom, as surveyed from this point of view. From the preceding paragraphs the evidence appears to be certainly in favour of prone-
ness to change in individuals, and of the power to change ceasing only
with the life of the individual; and we have still to account for the fact that
there are limits to these mutations, and laws that control the changes both
as to degree and kind; that species are neither visionary nor even arbitrary
creations of the naturalist; that they are, in short, realities, whether only
temporarily so or not."

"*13*) Granting then that the tendency of Nature is first to multiply forms
of existing plants by graduated changes, and next by destroying some to
isolate the rest in area and in character, we are now in a condition to seek
some theory of the *modus operandi* of Nature that will give temporary permanence of character to these changelings. And here we must appeal to
theory or speculation; for our knowledge of the history of species in relation
to one another, and to the incessant mutations of their environing physical
conditions, is far too limited and incomplete to afford data for demonstra-
ting the effects of these in the production of any one species in a native state."

"Of these speculations by far the most important and philosophical is
that of the delimitation of species by natural selection for which we are indebted to two wholly independent and original thinkers, Mr. DARWIN and
Mr. WALLACE. These authors assume that all animal and vegetable forms are
variable, that the average amount of space and annual supply of food for
each species (or other group of individuals) is limited and constant, but that
the increase of all organisms tends to proceed annually in a geometrical

ratio; and that, as the sum of organic life on the surface of the globe does not increase, the individuals annually destroyed must be incalculably great; also that each species is ever warring against many enemies, and only holding its own by a slender tenure. In the ordinary course of nature this annual destruction falls upon the eggs or seeds and young of the organisms, and as it is effected by a multitude of antagonistic, ever-changing natural causes, each more destructive of one organism than of any other, it operates with different effect on each group of individuals, in every locality, and at every returning season. Here then we have an infinite number of varying conditions, and a superabundant supply of variable organisms, to accommodate themselves to these conditions. Now the organisms can have no power of surviving any change in these conditions, except they are endowed with the means of accommodating themselves to it. The exercise of this power may be accompanied by a visible (morphological) change in the form or structure of the individual, or it may not, in which case there is still a change, but a physiological one, not outwardly manifested; but there is always a morphological change if the change of conditions be sudden or when, through lapse of time, it becomes extreme. The new form is necessarily that best suited to the changed condition, and as its progeny are henceforth additional enemies to the old, they will eventually tend to replace their parent form in the same locality. Further, a greater proportion of the seeds and young of the old will annually be destroyed than of the new, and the survivors of the old, being less well adapted to the locality, will yield less seed, and hence have fewer descendants."

"In the above operations Nature acts slowly on all organisms, but man does so rapidly on the few he cultivates or domesticates; he selects an organism suited to his own locality, and by so modifying its surrounding conditions that the food and space that were the share of others falls to it, he ensures a perpetuation of his variety, and a multiplication of its individuals, by means of the destruction of the previous inhabitants of the same locality; and in every instance, where he has worked long enough, he finds that changes of form have resulted far greater than would suffice to constitute conventional species amongst organisms in a state of nature, and he keeps them distinct by maintaining these conditions."

"Mr. DARWIN adduces another principle in action amongst living organisms as playing an important part in the origin of species, viz. that the same spot will support most life when peopled with very diverse forms, as is exemplified by the fact that in all isolated areas the number of Classes, Orders, and Genera is very large in proportion to that of Species" (pp. v–xii).

Next, HOOKER, examines with considerable detail general principles of distribution. The circumscription of the area of species "forcibly suggests the hypothesis that all individuals of each species have sprung from a common parent, and have spread in various directions from it". The same is true for varieties, genera, and other groups.

"The universality of this feature (of groups having defined areas) affords to my mind all but conclusive evidence in favour of the hypothesis of similar forms having had but one parent, or pair of parents. And further, this circumscription of species and other groups in area, harmonizes well with that principle of divergence of form, which is opposed to the view that the same variety or species may have originated at different spots. It also follows that, as a general rule, the same species will not give rise to a series of similar varieties (and hence species) at different epochs; whence the geological evidence of contemporaneity derived from identity of fossil forms may be relied upon."

"The most obvious cause of this limitation in area no doubt exists in the well-known fact that plants do not necessarily inhabit those areas in which they are constitutionally best fitted to thrive and to propagate; that they do not grow where they would most like to, but where they can find space and fewest enemies. We have seen (13) that most plants are at warfare with one or more competitors for the area they occupy, and that both the number of individuals of any one species and the area it covers are contingent on the conditions which determine these remaining so nicely balanced that each shall be able at least to hold its own, and not succumb to the enervating or etiolating or smothering influences of its neighbours. The effects of this warfare are to extinguish some species, to spare only the hardier races of others, and especially to limit the remainder both as to area and characters. Exceptions occur in plants suited to very limited or abnormal conditions, such as desert plants, the chief obstacles of whose multiplication are such inorganic and principally atmospheric causes as other plants cannot overcome at all; such plants have no competitors, are generally widely distributed, and not very variable" (p. xiii).

As a general rule, "those tracts of land present the greatest variety in their vegetation that have the most varied combinations of conditions of heat, light, moisture, and mineral characters" (p. xiv). Some areas of the globe have a very uniform phanerogamic vegetation, poor in species. It is in those with most varied conditions that most species are found.

"The Floras of islands present many points of interest. The total number of species they contain seems to be invariably less than an equal continental area possesses, and the relative numbers of species to genera (or other higher groups) is also much less than in similar continental areas."

"The further an island is from a continent, the smaller is its Flora numerically, the more peculiar is its vegetation, and the smaller its proportion of species to genera. In the case of very isolated islands, moreover, the generic types are often those of very distant countries, and not of the nearest land. Thus the St. Helena and Ascension forms are not so characteristic of tropical Africa as of the Cape of Good Hope. Those of Kerguelen's Land are Antarctic American, not African nor Indian. The Sandwich Islands contain many North-west American and some New Zealand forms. Japan presents us with many genera and species unknown except to the *eastward* of the Rocky Mountains, in North America. So too American, Abyssinian, and even South African genera and species are found in Madeira and the Canary Islands, and Fuegian ones in Tristan d'Acunha."

"22) There is a strict analogy in this respect between the Floras of islands and those of lofty mountain-ranges, no doubt in both cases owing to the same causes. Thus, as Japan contains various peculiar N.E. American species which are not found in N.W. America nor elsewhere on the globe, and the Canaries and Azores possess American genera not found in Europe nor Africa, so the lofty mountains of Borneo contain Tasmanian and Himalayan representatives; the Himalayas contain Andean, Rocky Mountain, and Japanese genera and species; and the alps of Victoria and Tasmania contain assemblages of New Zealand, Fuegian, Andean, and European genera and species. We cannot account for any of these cases of distribution between islands and mountains except by assuming that the species and genera common to these distant localities have found their way across the intervening spaces under conditions which no longer exist" (pp. xiv–xv).

The importance of the study of the floras of islands is stressed and it is concluded that many of the:

"Facts in the general distribution of species cannot be wholly accounted for by the supposition that natural causes have dispersed them over such existing obstacles as seas, deserts, and mountain-chains; moreover, some of these facts are opposed to the theory that the creation of existing species has taken place subsequent to the present distribution of climates, and of land and water, and to that of their dispersion having been effected by the now prevailing aquatic, atmospheric, and animal means of transport" (p. xvi).

Changes of climates and changes in the relative positions and elevations of land have had an important effect in determining the distribution of plants.

The then known evidence of fossil plants in relation "to the antiquity of vegetable forms and types on the globe" is briefly outlined. Much of this is now very much "out-of-date," but the following paragraph must be quoted:

"Such facts, standing at the threshold of our knowledge of vegetable palaeontology, should lead us to expect that the problem of distribution is an infinitely complicated one, and suggest the idea that the mutations of the surface of our planet, which replace continents by oceans, and plains by mountains may be insignificant measures of time when compared with the duration of some existing genera and perhaps species of plants, for some of these appear to have outlived the slow submersion of continents" (p. xxii).

HOOKER raises a pertinent question with regard to progressive evolution in the plant kingdom.

"What is the standard of progression? Is it physiological or morphological? Is it evidenced by the power of overcoming physical obstacles to dispersion or propagation, or by a nice adaptation of structure or constitution to very restricted or complex conditions? Are cosmopolites to be regarded as superior to plants of restricted range, hermaphrodite plants to unisexual, parasites to self-sustainers, albuminous-seeded to exalbuminous, gymnosperms to angiosperms, water plants to land, trees to herbs, perennials to annuals, insular plants to continental? and, in fine, what is the significance of the multitudinous differences in point of structure and complexity, and powers of endurance, presented by the members of the Vegetable Kingdom, and which have no recognized physiological end and interpretation, nor importance in a classificatory point of view? It is extremely easy to answer any of these questions, and to support the opinion by a host of arguments, morphological, physiological, and teleological; but any one gifted with a quick perception of relations, and whose mind is stored with a sufficiency of facts, will turn every argument to equal advantage for both sides of the question."

"To my mind, however, the doctrine of progression, if considered in connection with the hypothesis of the origin of species being by variation, is by far the most profound of all that have ever agitated the schools of Natural History, and I do not think that it has yet been treated in the unprejudiced spirit it demands. The elements for its study are the vastest and most complicated which the naturalist can contemplate, and reside in the comprehension of the reciprocal action of the so-called inorganic on the organic world. Granting that multiplication and specialization of organs is the evidence and measure of progression, that variation explains the *rationale* of the operation which results in this progression, the question arises, What are the limits to the combinations of physical causes which determine this progression, and how can the specializing power of Nature stop short of causing every race or family ultimately to represent a species? While the psychological philosophers persuade us that we see the tendency to specialize pervading every attribute of organic life, mental and physical; and the physicists teach that there are limits to the amount and duration of heat, light, and every other manifestation of physical force which our senses present or our intellects perceive, and which are all in process of consumption; the reflecting botanist, knowing that his ultimate results must accord with these facts, is perplexed at feeling that he has failed to establish on independent evidence the doctrines of variation and progressive specialization, or to co-ordinate his attemps to do so with the successive discoveries in physical science" (p. xxiv).

At this date, the view that species were immutable creations was not yet unorthodox. HOOKER comments that "we have no direct knowledge of the origin of any wild species: that many are separated by numerous structural peculiarities from all other plants; that some of them invariably propagate their like; and that a few have retained their characters unchanged under very different conditions and through geological epochs" (p. xxv).

The fact of genetic resemblance is the most important.

"To the tyro in Natural History all similar plants may have had one parent, but all dissimilar plants must have had dissimilar parents. Daily experience demonstrates the first position, but it takes years of observation to prove that the second is not always true. There are further, certain circumstances connected with the pursuit of the sciences of observation which tend to narrow the observer's views of the attributes of species; he begins by examining a few individuals of many extremely different kinds or species, which are to him fixed ideas, and the relationships of which he only discovers by patient investigation. He then distributes them into Genera, Orders, and Classes, the process usually being that of reducing a great number of dissimilar ideas under a few successively higher general conceptions; whilst with the history of the ideas themselves, that is, of species, he seldom concerns himself. In a study so vast as botany, it takes a long time for a naturalist to arrive at an accurate knowledge of the relations of Genera and Orders if he aim at being a good systematist, or to acquire an intimate knowledge of species if he aim at a proficiency in local Floras, and in both these pursuits the abstract consideration of the species itself is generally lost sight of: the systematist seldom returns to it, and the local botanist, who finds the minutest differences to be hereditary in a limited area, applies the argument derived from genetic resemblance to every hereditarily distinct form."

"40) It has been urged against the theory that existing species have arisen through the variation of pre-existing ones and the destruction of intermediate varieties, that it is a hasty inference from a few facts in the life of a few variable plants, and is therefore unworthy of confidence, if not of consideration; but it appears to me that the opposite theory, which demands an independent creative act for each species, is an equally hasty inference from a few negative facts in the life of certain species, of which some generations have proved invariable within our extremely limited experience. These theories must not, however, be judged of solely by the force of the very few absolute facts on which they are based; there are other considerations to be taken into account, and especially the conclusions to which they lead, and their bearing upon collateral biological phenomena, under which points of view the theory of independent creations appears to me to be greatly at a disadvantage; for according to it every fact and every phenomenon regarding the origin and continuance of species, but that of their occasional variation, and their extinction by natural causes, and regarding the *rationale* of classification, is swallowed up in the gigantic conception of a power intermittently exercised in the development, out of inorganic elements, of organisms the most bulky and complex as well as the most minute and simple; and the consanguinity of each new being to its pre-existent nearest ally, is a barren fact, of no scientific significance or further importance to the naturalist than that it enables him to classify. The realization of this conception is of course impossible; the boldest speculator cannot realize the idea of a highly organized plant or animal starting into life within an area that has been the field of his own exact observation and research; whilst the more cautious advocate hesitates about admitting the origin of the simplest organism under such circumstances, because it compels his subscribing to the doctrine of the "spontaneous generation" of living beings of every degree of complexity in structure and refinement or organization."

"On the other hand, the advocate of creation by variation may have to stretch his imagination to account for such gaps in a homogeneous system as will resolve its members into genera, classes, and orders; but in doing so he is only expanding the principle which both theorists allow to have operated in the resolution of some groups of individuals into varieties: and if, as I have endeavoured to show, all those attributes of organic life which are involved in the study of classification, representation, and distribution, and which are barren facts under the theory of special creations, may receive a rational explanation under another theory, it is to this latter that the naturalist should look for the means of penetrating the mystery which envelopes the history of species, holding himself ready to lay it down when it shall prove as useless for the further advance of science, as the long serviceable theory of special creations, founded on genetic resemblance, now appear to me to be."

"The arguments deduced from genetic resemblance being (in the present state of science), as far as I can discover, exhausted, I have felt it my duty to re-examine the phenomena of variation in reference to the origin of existing species; these phenomena I have long studied independently of this question, and when treating either of whole Floras or of species, I have made it my constant aim to demonstrate how much more important and prevalent this element of variability is than is usually admitted, as also how deep it lies beneath the foundations of all our facts and reasonings concerning classification and distribution. I have hitherto endeavoured to keep my ideas upon variation in subjection to the hypothesis of species immutable, both because a due regard to that theory checks any tendency to careless observation of minute facts, and because the opposite one is apt to lead to a precipitate conclusion that slight differences have no significance; whereas, though not of specific importance, they may be of high structural and physiological value, and hence reveal affinities that might otherwise escape us. I have already stated how greatly I am indebted to Mr. DARWIN's *rationale* of the phenomena of variation and natural selection in the production of species; and though it does not positively establish the doctrine of creation by variation, I expect that every additional fact and observation relating to species will gain great additional value from being viewed in reference to it, and that it will materially assist in developing the principles of classification and distribution" (pp. xxv–xxvi).

In the latter part of the Introductory Essay, the flora of Australia, as it was then known, is dealt with in some detail. The chief peculiarities of the flora are:

"That it contains more genera and species peculiar to its own area, and fewer plants belonging to other parts of the world, than any other country of equal extent. About two-fifths of its genera, and upwards of seven-eighths of its species are entirely confined to Australia."

"Many of the plants have a very peculiar habit or physiognomy, giving in some cases a character to the forest scenery (as *Eucalypti, Acaciae, Protea-ceae, Casuarineae, Coniferae*), or are themselves of anomalous or grotesque appearance (as *Xanthorrhoea, Kingia, Delabechea, Casuarina, Banksia, Dryandra*, etc.)."

"A great many of the species have anomalous organs, as the pitchers of *Cephalotus*, the deciduous bark and remarkable vertical leaves of the *Eucalypti*, the phyllodia of *Acacia*, the fleshy peduncle of *Exocarpus*, the inflorescence and ragged foliage of many *Proteaceae*."

"Many genera and species display singular structural peculiarities, as the calyptry of *Eucalyptus*, stigma of *Goodeniaceae*, staminal column of *Stylidium*, irritable labellum of various *Orchideae*, flowers sunk in the wood of some *Leptospermeae*, pericarp of *Casuarina*, receptacle and inner staminodia of *Eupomatia*, stomata of *Proteaceae*."

"On the other hand, if, disregarding the peculiarities of the Flora, I compare its elements with those of the Floras of similarly situated large areas of land, or with that of the whole globe, I find that there is so great an agreement between these, that it is impossible to regard Australian vegetation in any other light than as forming a peculiar, but not an aberrant or anomalous, botanical province of the existing Vegetable Kingdom. I find:—

"That the relative proportions of the great classes of Monocotyledons to Dicotyledons, of genera to orders, and of species to genera, are the same as those which prevail in other Floras of equal extent."

"That the subclasses distinguished by a greater or less complexity of the floral envelopes, or their absence, as *Thalamiflorae, Calyciflorae, Corolliflorae*, etc., are also in the same relative proportions as prevail in other Floras."

"That the proportion of Gymnospermous plants to other Dicotyledons is not increased."

"That all the Australian Natural Orders, with only two small exceptions, are also found in other countries; that most of those most widely diffused in Australia are such as are also the most widely distributed over the globe; and that Australia wants no known Order of general distribution."

"That the only two absolutely peculiar Natural Orders contain together only three genera, and very few species; they are, further, comparatively local in Australia, and are rather aberrant forms of existing natural families than well-marked isolated groups: *Brunoniaceae* being intermediate between *Goodeniaceae* and *Compositae*, and *Tremandreae* between *Polygaleae* and *Buettneriaceae*."

"That the large Natural Orders and Genera , which, though not absolutely restricted to Australia, are there very abundant in species and rare elsewhere, and for which I shall hence adopt the term Australian, stand in very close relationship to groups of plants which are widely spread over the globe (as *Epacrideae* to *Ericeae*, *Goodeniaceae* to *Campanulaceae*, *Stylideae* to *Lobeliaceae*, *Casuarinae* to *Myricae*)."

"That these Australian Orders are exceedingly unequally distributed in Australia; that there is a greater specific difference between two quarters of Australia (south-eastern and south-western) than between Australia and the rest of the globe; and that the most marked characteristics of the Flora are concentrated at that point which is geographically most remote from any other region of the globe."

"That most of those Australian Orders and genera which are found in other countries around Australia, have their maximum development in Australia at points approximating in geographical position towards those neighbouring countries. Thus the peculiarly Indian features of the Flora are most developed in north-western Australia, the Polynesian and Malayan in north-eastern, the New Zealand and South American in south-eastern, and the South African in south-western Australia."

"That of the nine largest Natural Orders, which together include a moiety of the Australian species of flowering plants, no fewer than six belong to the nine largest Natural Orders of the whole world, and five belong to the largest in India also."

"That in Australia itself, in advancing from the tropics to the coldest latitudes, or from the driest to the most humid districts, or from the interior to the seashore, or in ascending the mountains, the changes in vegetation are in every aspect analogous to what occur in other parts of the globe."

"That the relations between the epochs of the flowering and the fruiting of plants, and the seasons of the year, are the same in Australia as elsewhere, and most remarkably so, the *Orchideae* being spring flowers, the *Leguminosae* summer, the *Compositae* autumn, and the *Cryptogamia* winter."

"That the peculiarities of the Australian Flora in no way disturb the principles of natural arrangement derived from the study of the Flora of the globe apart from that of Australia; for after having attempted to consider the Australian vegetation in a classificatory point of view, shutting out of my view, as far as I could, that of other countries, I have been led to the conclusion that the authors of the Natural System—RAY, LINNAEUS, and the JUSSIEUS—might have developed the same Natural System had they worked upon Australian plants instead of upon European."

"I find further, that the classes, orders, genera, and species, may be about as well (or as ill) fixed or limited by a study of their Australian members as by those of any other country similarly circumstanced, and that there is the same vagueness as to the exact limits of natural groups, a similar inequality amongst them in numerical value and botanical characters, and an analogous difficulty in forming subclasses intermediate between classes and orders, as other Floras present. The Australian Flora, in short, neither breaks down nor improves the Natural System of plants as a whole, though it throws great light on its parts; the Australian genera fall into their places in that system well enough, though that system was developed before Australia was known botanically, and was chiefly founded upon a study of the vegetation of its antipodes."

"Thus, whether the Australian Flora is viewed under the aspect of its morphology and structure, as exhibited by its natural classification, or its numerical proportions or geographical distribution, it presents essentially the same primary features as do those of the other great continents: and it hence appears to me rash to assume that its origin belongs to another epoch of the earth's history than that of other Floras, when the proportions of its classes, etc., are identically the same with these; or that it should be attributed to a distinct creative effort, if this is manifested only in effecting morphological differences requisite to constitute species and genera in our classification, without disturbing the proportions of these; or that the local influence of the Australian climate should be essentially different from that of other countries, and yet effect no physiological change in the periods of flowering and fruiting, or produce any other functional disturbances of the vegetable organisms, or affect the agency of humidity, temperature, soil, and elevation, on plants" (pp. xxvii–xxix).

HOOKER estimates that the species of flowering plants in the Australian flora will be found eventually to be from 9000 to 10,000. The ratio of Monocotyledons to Dicotyledons is 1 : 4.6 (that for the whole earth is given as 1 : 4.9.) In the tropical Australian flora of about 2200 species the ratio is 1 : 3.5 and in the temperate flora of 5800 species it is 1 : 5.0. The ratio of Gymnosperms to Phanerogams is 1 : 184. The *Brunoniaceae* and *Tremandreae* are the only families ("natural orders") confined to Australia, but the following are almost confined: *Stackhousiae, Goodeniaceae, Stylideae, Epacrideae,* and *Casuarineae,* and the sections *Xerotideae* and *Aphyllantheae* of *Liliaceae* or *Junceae.* HOOKER'S nomenclature and concepts are here retained. Largely developed and equally or more characteristic of Australian vegetation are: *Dilleniaceae, Rutaceae, Proteaceae, Restiaceae, Thymeleae, Haemodoraceae, Buettneriaceae,* and *Droseraceae.* The nine largest families ("natural orders") are: *Leguminosae, Myrtaceae, Proteaceae, Compositae, Gramineae, Cyperaceae, Epacrideae, Goodeniaceae,* and *Orchideae.*

The number of genera of Australian flowering plants exceeds 1300, and each genus has on the average about 6 species. *Acacia* has over 200 species and every one of the following genera over 100: *Eucalyptus, Melaleuca, Leucopogon, Stylidium, Grevillea,* and *Hakea.*

The tropical Australian flora is dealt with at some length. It extends farther south on the west than on the east coast but it can be concluded in general terms that the tropical and temperate floras, both east and west, blend between 26° and 29° S. The statement is made that "the number of species in tropical Australia appears to be extremely small." It is concluded that: "The diminution of vegetable forms in advancing from temperate to tropical Australia is to a great extent due to the rarity or absence of Orders which, though more typical of hot latitudes in other parts of the globe, abound in the temperate regions only of Australia" (p. xli). The relations of the tropical flora of Australia with that of India is shown by lists of species and genera.

The extra-tropical flora of Australia is also analyzed.

"In studying the extra-tropical Flora of Australia, the first phenomenon that attracts attention is the remarkable difference between the eastern and western quarters, to which there is nothing analogous in the tropical region. What differences there are between eastern and western tropical Australia are confined to more Asiatic forms in the latter, and more Polynesian and temperate Australian ones in the former; this is analogous to that preponderance, to which I shall hereafter allude, of the South African types in south-western Australia, and of New Zealand and Antarctic ones in south-eastern; but offers nothing analogous to the fact that the species, and in a great extent the genera, of south-western Australia differ from those of south-eastern, though these species and genera belong to the same Natural Orders, and in many cases to peculiarly Australian Orders or divisions of Orders."

"I have endeavoured to estimate this difference by tabulating the genera and species of each country, and though the results must, in the present state of our knowledge, be very vague, they may serve to give an approximate idea of the amount of difference, which it is all the more important to do because I believe the phenomenon to be without a parallel in the geography of plants. These Floras I estimate as containing about:—

South-western		*The South-eastern Flora, including Tasmania.*	
Natural Orders . .	90	Natural Orders .	125
Genera	600	Genera	700
Species	3,600	Species	3,000"

"As far as I can make out, about one-fifth of the south-eastern species are found beyond that area; but only-one tenth of them are found in south-western Australia" (pp. l–li).

Lists of genera, with numbers of species, characteristic of south-eastern and south-western Australia respectively, indicate:

"*1*) How greatly larger the genera of the south-western Flora are, there being 80 genera with upwards of 10 species in its column, and only 55 in the south-eastern. *2*) That the 55 genera of the south-eastern Flora contain about 1,206 species, and the 55 highest of the south-western 1,727 species. *3*) That of these 55 south-western genera 36 do not appear at all in the south-eastern list, and 17 (marked with a *) are absolutely confined to the south-west, or almost so."

"Altogether, I find the proportion of genera to species in the south-western Flora to be 1: 6, and in the south-eastern 1: 4. This increased number of genera in south-eastern Australia over the south-western is mainly due to the presence of more Antarctic, European, New Zealand, and Polynesian genera in the south-east, to which I shall hereafter allude."

"The proportion of species belonging to peculiar or endemic genera in the south-west is about one-third of the whole, and in the south-east one-sixth."

"The proportion of species common to other countries in the south-west is about one-tenth of the Flora, and in the south-east one-sixth."

"There are about 180 genera, out of about 600, in south-western Australia that are either not found at all in south-eastern, or that are represented there by a very few species only, and these 180 genera include nearly 1,100 species."

"Of generally diffused Australian genera that are absent in the south-west, I find *Viola*, *Polygala*, *Epacris*, *Lycopus*, *Ajuga*, *Smilax*, and *Eriocaulon*; and of European genera which occur in that quarter, but which I have not seen from elsewhere in Australia, are *Echinospermum*, *Eritrichium*, *Orobanche*, *Althenia*, and *Lepturus*, several of which I suppose to be introduced, and, if so, will soon be found in other colonies."

"This curious case of great differences in the genera and species of the two quarters of a small continent, accompanied by an increased number of species in the smaller and more isolated quarter of the continent, which is, further, by far the most uniform in physical conditions, will no doubt eventually be found to offer the best means of testing whatever theory of creation and distribution may be established. In the meantime, the theories which I have sketched in the early pages of this Essay cannot, in the present state of our geological knowledge of Australia, be brought to bear fully upon it. That no Natural Order, but that many genera, and a whole Flora of species, should be created in the smaller and more isolated area of western Australia, different from what eastern Australia presents, seems at first sight favourable to the idea that these are derivative genera and species, formed during the gradual migration of certain of the Orders and Genera of the east towards the west. But on the other hand, this massing of most of the peculiar features of the Australian Flora in the west, unmixed there with Polynesian, Antarctic or New Zealand genera, is an argument for regarding western Australia as the centrum of Australian vegetation, whence a migration proceeded eastward; and the eastern genera and species must in such a case be regarded as the derivative forms. Had we any idea of the comparative geological age of species in that area, and if so, we may connect the richness in species of the western Australian Flora with its singular uniformity of character, for it is purely Australian, without admixture of any other element. As this excessive multiplication must, under the theory of creation by variation, have occupied a great length of time, it seems to be more natural to assume, on purely botanical grounds, that the western Australian Flora is the earliest, and sent colonists to the eastern quarter, where they became mixed with Indian, Polynesian, etc., colonists, than that the western Flora was peopled by one section only of the inhabitants of the eastern quarter."

"So much for the botanical aspect of the question. The geological one suggests a different explanation. That part of the Australian continent which alone is clothed with any considerable amount of vegetation, may be likened to a horse-shoe of more or less elevated land, with its convexity to the north and a vast enclosed central depressed area, that opens to the sea on the south and advances north almost to the Gulf of Carpentaria. According to Mr. JUKES's clever 'Sketch of the Physical Structure of Australia,' this central and southern area was recently an oceanic bay, and existing species of Mollusca are found on its surface for many miles along the coast, and inland from it, in an almost unchanged condition. To the east of this depressed area, the mountains are far loftier and the rocks of a much greater age than to the west of it; and were the question of the age of the Floras comprised in that of the rocks they inhabit, little doubt would be entertained that the western one was modern and derivative; but in no other part of the world are recently-formed lands tenanted exclusively by endemic plants, nor do they present assemblages of very local species; on the contrary, they are inhabited by many individuals of a few species derived from surrounding countries, of which some few are so altered as to be distinguished as varieties or even species; and we cannot therefore accept the geological evidence as good for explaining the botanical phenomena."

"There is another way of viewing the whole question, but one so purely speculative that I hesitate to put it forward. It is that the antecedents of the peculiar Australian Flora may have inhabited an area to the westward of the present Australian continent, and that the curious analogies which the latter presents with the South African Flora, and which are so much more conspicuous in the south-west quarter, may be connected with such a prior state of things" (pp. liii–lv).

The flora of Tasmania is identical in all its main features with that of Victoria and especially with that of the mountainous parts of the latter state (in HOOKER's time "colony"). Out of 1063 species, only 280 have not been found on the Australian continent.

"It will probably be conceded that Tasmania once formed a continuous southward extension of Victoria, and that as Britain was peopled with continental plants before the formation of the Channel, so Tasmania and Victoria possessed their present Flora before they were separated by Bass' Straits; but if the effects of segregation and natural selection have done

12

so little towards modifying the Floras of the opposite shores during the immense epoch that has intervened since the earliest formation of Bass' Straits, we are all the more puzzled to account for the complete change of the south-western Flora, which is isolated by no such barrier from the south-eastern."

"There are only 592 flowering plants peculiar to Tasmania and Australia, or 860 if those peculiar to Tasmania are included, so that fully one-fifth of the Flora is extra-Australian; whereas only one-sixth of the south-eastern Flora and one-tenth of the south-western are extra-Australian. Considering the before-mentioned isolation of Tasmania, this is certainly a most remarkable fact, and requires a close scrutiny."

"Turning to the genera again, I find that out of the whole (394), only 22 are absolutely peculiar to Tasmania; or adding these to the 122 which are exclusively Australian and Tasmanian, I find only 144 in all. In other words, considerably more than two-thirds of the Tasmanian genera are found in other countries besides Australia; whereas in south-western Australia much less than half the genera are extra-Australian, in south-eastern somewhat more than half, and in the whole Australian Flora, between one-half and two-thirds" (p. lxxxiv).

There is a remarkable rise in the proportion of European forms in Tasmania. The New Zealand flora is another which enters proportionally much more largely into the Tasmanian than into the Australian, nearly 200 of the genera and 170 of the species of Tasmania being common to New Zealand. There is also a larger proportion of "antarctic" plants, "nearly 100 genera and 56 species being common to this island and the groups south of New Zealand, Fuegia, the Falkland Islands, etc." (p. lxxxv).

"Another interesting subject of detail, requiring fuller materials, is the alpine Flora of Tasmania, upon which MUELLER's Victorian Alps collections have thrown so much light. I find, on a rough estimate, that there are 200 alpine and subalpine species in Tasmania (of which half are alpine); considering as such those which are most prevalent in or confined to altitudes above 3,000 feet: of these 30 are probably altered forms of lowland plants; 120 are of Australian genera (10 of them are probably varieties); about 10 are of New Zealand genera; 55 are of European genera (17 of them probably varieties); and 25 are Antarctic forms."

"This proportion of varieties amongst the alpine and subalpine plants, amounting as it does to 15 per cent., is very large; the proportion amongst the lowland plants being considerably under 10 per cent. The small proportion of varieties amongst the alpines belonging to Australian genera compared with those of European genera is also worthy of notice, as an exemplification of an observation made by Mr. DARWIN, that the species of widely distributed genera are more variable than those of local genera" (p. lxxxv).

New Zealand has representatives of many Australian genera but some of the most extensive and widely distributed are wholly absent, as are *Eucalyptus, Acacia, Stylidium, Casuarina, Callitris, Xyris, Xerotus, Thysanotus, Hibbertia, Pleurandra, Banksia, Dryandra, Grevillea,* and *Hakea.* "It is even more remarkable that most of the highly characteristic Australian Orders are wholly or nearly absent in New Zealand" (p. lxxxvii).

On the other hand, the points of resemblance between the floras of Australia and New Zealand are numerous.

"In the first place, there is no New Zealand Order absent from Australia except *Coriarieae, Brexiaceae,* and *Chloranthaceae,* which are single genera rather than Orders. Of the 282 genera of Phaenogams in New Zealand, 240 are also Australian, and 60 are almost confined to these two countries. The greatest amount of generic affinity exists in three of the largest Orders in each, *viz. Compositae, Orchideae,* and *Gramineae,* which may be considered generically identical in both. To this category of resemblances also belong the antarctic genera and representative genera, many of which are also found in America, and which will be hereafter considered. Of these 240 genera, by far the larger proportion are confined to eastern Australia, not one being exclusively western Australian."

"Descending to species, I find that 216, or one-fourth of the New Zealand Phaenogams, are natives of Australia, and of these 115 are confined to these two countries. Of the remaining 101, 77 are common to America, 75 to India, and 52 to Europe. The comparatively small number of these that are common to India, and greater number common to America, is a remarkable fact, considering the relative position of these countries; and the large number of European genera is no less so."

"Another interesting anomaly is, that of the 115 species peculiar to Australia and New Zealand, only 26 belong to genera peculiar to those countries and only 6 to the long list of Australian genera that contain upwards of 20 species each. Again, upwards of 20 of these 115 are scarce and chiefly alpine plants in both countries, occupying comparatively very small areas; whereas of the 101 that are found in other lands besides Australia and New Zealand, only 5 or 6 are alpine, and most of these are antarctic also."

"Thus, under whatever aspect I regard the Flora of Australia and New Zealand, I find all attemps to theorize on the possible causes of their community of feature frustrated by anomalies in distribution such as I believe no two other similarly situated countries in the globe present. Everywhere else I recognize a parallelism or harmony in the common features of contiguous Floras, which conveys the impression of their generic affinity at least being effected by migration from centres of dispersion in one of them, or in some adjacent country. In this case it is widely different. Regarding the question from the Australian point of view, it is impossible in the present state of science to reconcile the fact of *Acacia, Eucalyptus, Casuarina, Callitris,* etc., being absent in New Zealand, with any theory of transoceanic migration that may be adopted to explain the presence of other Australian plants in New Zealand; and it is very difficult to conceive of a time or of conditions that could explain these anomalies, except by going back to epochs when the prevalent botanical as well as geographical features of each were widely different from what they are now. On the other hand, if I regard the question from the New Zealand point of view, I find such broad features of resemblance, and so many connecting links that afford irresistible evidence of a close botanical connection, that I cannot abandon the conviction that these great differences will present the least difficulties to whatever theory may explain the whole case" (pp. lxxxviii–lxxxix).

In treating of "antarctic" plants in the Australian vegetation, Hooker makes clear his reasons for using the term "antarctic".

"From the geographical position of Australia, no less than from the altitude of its southern mountains, it is well placed for the maintenance of those types of vegetation which I have denominated Antarctic. These, it must be remembered, are not so called because they really inhabit the country of that name beyond the Polar circle, but because in a botanical point of view, no less than in position relative to the south temperate Flora, they represent the Arctic Flora. They might indeed almost be called alpine plants, for many which are found at the level of the sea in the so-called Antarctic islands, also ascend the mountains of more genial latitudes. An alpine vegetation, however, in the tropics especially, is supposed to commence only where the forest is replaced by low brushwood; whereas, owing to the uniformity and humidity of the high southern latitudes, an arboreous vegetation there encroaches upon the limits of perpetual ice. In the longtitude of Cape Horn, on the mountains of Fuegia, of the Middle Island of New Zealand, and of Australia, the belt of country occupied by low and chiefly herbaceous plants, that intervenes between the arboreous vegetation and the extinction of phaenogamic life, is a very narrow one indeed compared with what analogous regions the Alps, Andes, Himalaya, or Arctic latitudes present." (p. lxxxix).

A list of "antarctic" plants is given.

"The most curious point in this list is the number of European species it contains, amounting to seventeen, of which most are British; there are besides two other species which inhabit the north temperate zone of the New World, *Triglochin triandrum* and *Crantzia lineata*; *Apium australe* is in some of its states with difficulty distinguished from *A. graveolens*."

"The genera that are most characteristic of the Antarctic regions amongst them are,—*Colobanthus, Acaena, Donatia, Nertera, Forstera, Leptinella, Ourisia, Drapetes, Fagus, Oreobolus,* and *Carpha*. Only one (*Lomatia*) can be said to betray any generic affinity between the peculiar Flora of Australia and the Antarctic regions; though *Forstera*, as belonging to *Stylidieae*, may be classed with Australian representatives" (p. xci).

With the South African flora, Australia has family but very few generic affinities.

"The most conspicuous characters that extratropical South Africa presents in common with Australia, are the abundance of species of the following Orders, many of which being shrubby, give in certain districts of each country a character to the landscape."

Proteaceae	*Polygaleae*	*Rutaceae*
Compositae	*Restiaceae*	*Thymeleae*
Irideae	*Epacrideae, Ericeae*	*Santalaceae*
Haemodoraceae	*Decandrous Papilionaceae* and	*Anthospermous Rubiaceae.*"
Buettneriaceae	tribes *Podalyrieae* and *Loteae*	

"All these Orders are far more abundantly represented in Australia (especially south-western) and South Africa than in any other part of the world, added to which by far the greater number of the known genera and species of *Proteaceae* and *Restiaceae* are confined to these two countries. Other marks of affinity are the *Cycadeae*, the genus *Encephalartos* (to which MUELLER reduces *Macrozamia*) being common to both; *Cyphiaceae* (according to BROWN a suborder of *Goodeniaceae*) are almost confined to South Africa. Numerous terrestrial *Orchideae*, *Droseraceae*, *Zygophylleae*, *Liliaceae*, *Smilaceae*, and *Capparideae*; the genera *Pelargonium* and *Mesembryanthemum*, besides *Metrosideros*, *Acaena*, *Tetragonia*, *Weinmannia*, *Sarcostemma*, *Sebaea*, *Callitris*, *Anguillaria*, *Restio, Carpha, Uncinia*, and *Ehrharta*. The rarity in both of *Aroideae, Laurineae*, and all *Rubiaceae* except the *Anthospermeae*, is also worthy of notice. With regard to the Natural Orders enumerated above, their genera are almost unexceptionally different in the two countries. I find that of 1,000 South African genera of flowering plants, only about 280 are Australian; of these about 160 are also common to Europe, and 130 to India, leaving *Callitris, Encephalartos, Restio, Hypolaena*, and *Anguillaria*, confined to South Africa and Australia, and 10 more common to these countries, together with New Zealand and extratropical America."

"On the other hand, South Africa contains upwards of 220 European genera, of which 80 are not Australian, and of these upwards of 60 are north temperate forms. We have hence the very curious fact that in point of numbers Australia represents generically the European Flora better than South Africa does; but that the South African Flora contains a larger proportion of very northern European genera (not species) than Australia does. This no doubt because many of the so-called European genera of Australia are more properly Asiatic, and spread thence in both directions, towards Europe and towards Australia."

"Before dismissing this subject, it is as well to glance at the differences between these Floras, which may shortly be summed up. South Africa abounds in *Campanulaceae*, which are very rare in Australia, where the very closely allied Orders *Stylidieae* and *Goodeniaceae* abound. The true *Ericeae*, which swarm in certain districts of South Africa, are all but wholly absent in Australia, being represented there by their suborder *Epacrideae*. Succu-

lents are, comparatively, extremely rare in Australia, which almost wholly
wants those conspicuous features of South African vegetation the *Crassulaceae*, *Ficoideae*, fleshy *Asclepiadeae*, *Liliaceae* (Aloes), and *Euphorbieae*" (pp.
xcii–xciii).

Lists of genera and species common to Australia and Europe include 227 genera and 148 species.

"The last observation I shall make with reference to this subject is, that
the existing European Flora does not contain one Australian representative,
nor betray the remotest direct botanical affinity with the Australian. I have
elsewhere indicated (p. xxi) that there is evidence of what are now Australian
plants having once inhabited Europe. In north-eastern Asia there are however a few Australian forms, of which the *Haloragis*, *Stylidium*, and *Baeckia*
of China, the *Microtis* of Bonin, *Stackhousia* and *Thysanotus*? of Philippine
Islands, *Thelymitra* of Java, and *Proteaceae* of Japan are examples. Connecting
these again is the singular assemblage of Australian forms on the lofty mountain Kini Balou in Borneo, and which consists of species of *Drimys*, *Leptospermum*, *Leucopogon*, *Coprosma*, *Didiscus*, *Drapetes*, *Euphrasia*, *Phyllocladus*,
Dacrydium, and an Irideous and Restiaceous plant, both apparently allied to
Australian genera" (p. c).

The discussion on the fossil flora of Australia is based on an account by J. B. JUKES. From this HOOKER concludes "that the extinct
Flora of Australia was not entirely different from that now existing",
that Australia was dry land during the "Oolitic and Cretaceous"
periods, and was largely "submerged during the Tertiary epoch, when
it presented the appearance of two long islands, or chains of islands,
one, the larger, representing the elevated land of eastern Australia
and Tasmania, the other that of south-eastern Australia, together with
subsidiary groups in the western and northern parts of the continent."
HOOKER presents some arguments in favour of the antiquity of the
Australian flora as compared with the European and then discusses
the origin, in Australia, of European, Antarctic, and Asiatic types. In
view of remarks made by him in other phytogeographical works, two
extracts must be given.

"When I take a comprehensive view of the vegetation of the Old World, I
am struck with the appearance it presents of there being a continuous current of vegetation (if I may so fancifully express myself) from Scandinavia
to Tasmania; along, in short, the whole extent of that arc of the terrestrial
sphere which presents the greatest continuity of land. In the first place,
Scandinavian genera, and even species, reappear from Lapland and Iceland
to the tops of the Tasmanian alps, in rapidly diminishing numbers it is true,
but in vigorous development throughout. They abound on the Alps and
Pyrenees, pass on to the Caucasus and Himalaya, thence they extend along

the Khasia mountains, and those of the peninsulas of India to those of Ceylon and the Malayan Archipelago (Java and Borneo), and after a hiatus of 30°, they appear on the alps of New South Wales, Victoria, and Tasmania, and beyond these again on those of New Zealand and the Antarctic Islands, many of the species remaining unchanged throughout! It matters not what the vegetation of the bases and flanks of these mountains may be: the northern species may be associated with alpine forms of Germanic, Siberian, Oriental, Chinese, American, Malayan, and finally Australian and Antarctic types; but whereas these are all, more or less, local assemblages, the Scandinavian asserts his prerogative of ubiquity from Britain to beyond its antipodes" (p. ciii).

"These considerations quite preclude my entertaining the idea that the Southern and Northern Floras have had common origin within comparatively modern geological epochs; on the contrary, the European and Australian Floras seem to me to be essentially distinct, and not united by those of intervening countries, though fragments of the former are associated with the latter in the southern hemisphere. For instance, I regard the Indian plants in Australia to be as foreign to it, botanically, as the Scandinavian, and more so than the Antarctic; and that to whatever lengths the theory of variation may be carried, we cannot by it speculate on the Southern Flora being directly a derivative one from the existing Northern. On the contrary, the many bonds of affinity between the three southern Floras, the Antarctic, Australian, and South African, indicate that these may all have been members of one great vegetation, which may once have covered as large a southern area as the European now does a Northern. It is true that at some anterior time these two Floras may have had a common origin, but the period of their divergence antedates the creation of the principal existing generic forms of each. To what portion of the globe the maximum development of this Southern Flora is to be assigned, it is vain at present to speculate; but the geographical changes that have resulted in its dismemberment into isolated groups scattered over the Southern Ocean, must have been great indeed. Circumscribed as these Floras are, and encroached upon everywhere by northern forms, their ultimate destiny must depend on that power of appropriation in the strife for place which we see in the force with which an intrusive foreign weed establishes itself in our already fully peopled fields and meadows, and of the real nature of which power no conception has been formed by naturalists, and which has not even a name in the language of biology. Everywhere, however, we see the more widely distributed, and therefore least peculiar forms of plants, spreading, and the most peculiar dying out in small areas, and the progress of civilization has introduced in man a new enemy to the scarce old forms, and a strong ally of those already common: nor can it be doubted but that many of the small local genera of Australia, New Zealand, and South Africa, will ultimately disappear, owing to the usurping tendencies of the emigrant plants of the northern hemisphere energetically supported as they are by the artificial aids that the northern races of man afford them" (pp. civ–cv).

* * *

HOOKER raised very clearly the problems of the botanical relationships between the southern parts of the southern continents and associated or intermediate islands. These problems still intrigue plant geographers. The Antarctic continent itself has probably no, or at most very few, species of seedbearing plants at the present time. South America, South Africa, Australia, and New Zealand are now separated one from another by wide stretches of ocean. Their floras are on the whole very different one from another but there are a number of very marked and peculiarly selected linkages particularly in common genera. To explain the ranges of these exercised the mind of HOOKER and those after him who have investigated the wider problems of distributions in the Southern Hemisphere. As usual, there are two possibilities: transport of disseminules over wide stretches of ocean or dispersal over former land connections. The majority of more recent writers, like HOOKER, accept the latter, but still acknowledge difficulties and uncertainties and differ much in the degree of detail they give to their hypotheses and in the details themselves. That what is now the land of the Antarctic Continent had a milder climate suited to the growth of vascular plants in the Mesozoic and (at least part of) the Cainozoic eras appears to be established by the fossil record. Great climatic changes have to be accepted.

A brief consideration must now be given to the more or less divergent views of recent authors. We may start with R.N. RUDMOSE BROWN who has dealt with Antarctic plant life in a series of papers (BROWN, 1906, 1912, 1923, 1928). He is concerned mainly with the Antarctic proper and, therefore, has a more limited field than HOOKER and has to consider mainly *Algae*, *Lichenes*, and *Bryophyta*. He (1912, pp. 14–5) makes the following statements: "While our knowledge of Antarctic flora is certainly incomplete, all the known facts point to a Fuegian origin. Not only does an analysis of the distribution of the constituent elements indicate this, but the relative greater abundance of species in Graham Land and vicinity than in Victoria Land, as well as the absence of New Zealand forms, shows that the flora of the Antarctic is due to an emigration of species from Fuegian lands. I have discussed above (pp. 6 and 7) the ways in which seeds might cross Drake Strait. Winds and birds must have done the work of giving Antarctica its present flora, *via* Graham Land from Fuegia, and thence it must have spread westward *via* the coast to Victoria Land, but naturally only a small proportion of the species were carried so far. However, it is quite possible that by the same agencies a certain number of mosses and lichens may have reached Wilkes Land and Wilhelm Land from Kerguelen and Heard Island, while South Georgia and the South Sandwich group may have contributed to Coats Land and the coast eastward towards Enderby Land. The floras of all these subantarctic

islands from the Falklands eastward to Kerguelen have been shown
to be related to one another, and to have strong Fuegian affinities;
and Dr. COCKAYNE has pointed out the relationship between the flora
of Kerguelen and that of Macquarie Island."

"In a later part of this paper (pp. 17–20) is a fuller discussion of these
islands and their floras; but this close relationship with Fuegia that
they all exhibit, means that emigration of a species from any of these
islands to Antarctica amounts to emigration from Fuegia by a somewhat circuitous route. No other lands are near enough to Antarctica
to have affected its flora."

THISTLETON-DYER (1912) favoured a northern origin and southern
spread of plant life. He points out that the distribution of existing
taxonomic units cannot be understood correctly without reference
to fossil records and that the more these are investigated the more
do they point to a northern origin of the groups concerned.

HEDLEY (1912) took many of his examples from the animal kingdom. His position is approximately the opposite of that of THISTLETON-
DYER: "that the community of austral life is explicable only by former
radiation along land-routes from the south polar regions." In the
Tertiary, Antarctica had a warm climate. On the high plateau the ancestors of plants now found in Kosciusko and Kerguelen took refuge
while tropical migrants found a congenial climate on the coast. The
land link was maintained during the period of refrigeration, and from
the Antarctic focus first the subtropical, then the temperate, lastly the
"alpine" forms were expelled, each to gain a fresh footing in lower
latitudes.

The numerous works of SKOTTSBERG demand attention. TURRILL
(1919) summarized the botanical results of the Swedish South American and Antarctic expeditions. It is, therefore, not necessary to refer
in detail to the publications, mostly by SKOTTSBERG, considered in that
summary. Additional papers are SKOTTSBERG 1915, 1931, 1936, and
1940. One quotation from TURRILL (1919, p. 279) must suffice to sum
up SKOTTSBERG's earlier and important contributions to the subjects
now under discussion.

"SKOTTSBERG has undoubtedly added valuable confirmation to the
position taken up by HOOKER in the introductory essay to the "Flora
Antarctica." The southern part of the American continent is the head-
quarters of a botanical region whose flora is still found in isolated
groups of islands extending 5000 miles to the east, some being nearer
to the African and Australian continents than to the American. The
existence must be accepted of an old Antarctic or Subantarctic Flora,
at one time more continuous, or at least with its now separate parts
having had a common origin. In reading SKOTTSBERG's works one is

impressed by the effect which the last Ice Age or series of Ice Ages had on the vegetation. The north and south direction of the Andes allowed the southern flora to retreat northwards in South America, and to return again when better conditions prevailed. A comparison between the flora of South Georgia which was, at least once, completely glaciated and that of Kerguelen, which was never completely glaciated, is most instructive in this connection."

In later papers, SKOTTSBERG has maintained the theory of an Antarctic continent whence plants emigrated northwards. In his 1940 paper he said that an earlier calculation (now in part out-of-date) of 80 genera, classified into about 50 different families, indicated the characteristic more or less circumpolar subantarctic flora which, if it did not originate in Antarctica certainly had a secondary centre there. Examples are *Araucaria, Drimys, Laurelia, Nothofagus, Eucryphia, Oreobolus*, and *Colobanthus*. Though Antarctica is now mainly ice-covered, the lichen and moss flora is richer than was formerly supposed and there is definite evidence of former vegetation, including woody plants, in various fossil remains. Moreover, the isolation of Antarctica is not so pronounced as has been imagined. The "Scotia-bow" links South America and Western Antarctica and other past connections may have been between Victoria Land by way of the Macquarie, Auckland, and and Campbell Islands to New Zealand and from some area farther westwards to Tasmania and New Zealand. SEWARD had already suggested a land connection between a greater Kerguelen Land and Antarctica. Another connection suggested is with South Africa.

SKOTTSBERG discussed WEGENER's theory with regard to Antarctica. He noted that the former land connections which (many) biologists regard as necessary are easily, perhaps too easily, brought about or made in theory by accepting one former continental land mass. The question is extremely complicated but the present day composition and distribution of the austral and subantarctic flora may be explained without the violent changes which the WEGENER hypotheses import. The vertical movements which are required for a reconstruction of the Scotia-bow or curve and the other land connections, are not of such an order of size that they would modify the theory of the permanence of the Pacific Ocean.

Accepting that the subantarctic continent was formerly connected northwards its flora wandered far afield and SKOTTSBERG considered it is to be met with in the high mountains of Java, Borneo, and the Philippines and can be traced by way of Fiji to Samoa, Tahiti, the Marquesas Islands and Hawaii. Along the Andes it reached the equator, and its extreme outpost is situated in California. In Africa it found its way to the highlands of Abyssinia, Madagascar and the Mascarene Islands.

But it is very weak in these sectors and has a less pronounced temperate character. There is every reason to assume that Africa was separated from its Antarctic connections earlier than were America or
Australia-New Zealand.

HILL (1929) dealt with Antarctica and problems in geographical
distribution and reached the following conclusion: "That there was
a large and extensive land mass, which we may call Antarctica seems
certain, from the evidence afforded by a study of the living and fossil
floras, and it seems equally certain that from this lost continent, which
enjoyed from time to time a temperate or even a subtropical climate,
plants were distributed to South America on the one hand and to
Australia and New Zealand on the other, either by means of direct
land connections or by means of a chain of Islands." (p. 1486).

Du RIETZ (1940) published a long paper with the title "Problems of
Bipolar Plant Distribution." Questions of wider import than those we
are now considering are reviewed by Du RIETZ. This author rejects
WEGENER's theory of continental drift and replaces it by a hypothesis
of ENQUIST's. This is an isostatic hypothesis that "presupposes mobil-
ity and unstable equilibrium between ocean blocks and land blocks.
Consequent on this are variations in the sea-level: transgressions and
retrogression." Land areas may turn into deep sea bottom—the opposite cannot occur. "The originally continental part of the earth's
surface was considerably greater than at present. The areas of the
oceans were considerably restricted; it is possible that they amounted
only to half their present size. Their average depth, on the other hand,
was consequently much greater than to-day. Owing to successive eruptions of magma, and owing to the displacements of these eruptives
along the earth's surface that were rendered possible by the original
contrasts in the cross-section, the conditions were changed during the
course of geologic time. The present state is a stage in a development,
which will proceed along the same lines; less and less contrast between
the mean level of continents and sea-floors, constant shrinking of the
continents, while the abnormal differences in density between crust
and substratum will gradually be abolished. In the end a flat, uniform
earth crust built up of the lightest material of the globe will rest upon
a magma whose density increases towards the centre. This earth crust
will then be uniformly covered by the waters of the oceans."

Whatever be the verdict of geophysicists and geologists it would
appear that ENQUIST's hypothesis may be elastic enough to cover the
legitimate demands of phytogeographers for now absent land connections while avoiding the difficulties of WEGENER's views, especially
those following acceptance of wandering poles. At any rate, Du
RIETZ sums up as follows: "To explain the facts of bipolar plant dis-

tribution it seems necessary to look for epeirogenetic [epeirogeny, or continent-making, is, according to GEIKIE, Text Book of Geology, ed. 4 (1903) p. 392, footnote, "rather displayed in slow secular deformation of the crust"] transtropical highland bridges older than the mountain chains of the Alpine Orogen. Such highland bridges may have existed not only in Africa, but also bordering the transtropical Alpine geosynclines (*i.e.* the Andean and the Malaysian geosynclines), partly passing over present deep sea bottom" (p. 272).

The flora of the Falkland Islands has been excellently illustrated by VALLENTIN (1921) and both the flora and vegetation carefully investigated by SKOTTSBERG (*see* summary and references *in* TURRILL, 1919). Other researches are those of BIRGER (1906-7), BOYSON (1924), DAVIES (1939), and DUSÉN (1898). A valuable concise recent account is that by SKOTTSBERG (1942). The most striking feature of the scenery is the absence of trees due to exposure to almost daily winds. The climate is ideal for peat formation, to which all the main plant communities give rise. Fossil remains of at least two coniferous trees have been discovered. The flora is poor: about 145 phanerogams and 15 pteridophytes. The affinity is with the Magellanian flora of which, indeed, it is a part. "Two sources of origin can be distinguished: the old Antarctic flora, now scattered over the circumpolar Subantarctic zone, and the Andean. More than 40 Subantarctic species are found in Falkland" (p. 23).

HOOKER's personal experiences in temperate South America, apart from the Falkland Islands, were limited to the south-western part, in the Magellan Straits area and the Chonos Archipelago. A summary of botanical exploration in Chile and Argentina was published by TURRILL (1920). The southern political boundaries, between Chile and Argentina, are not botanical boundaries and the following will be found to be useful references to Fuegia and Patagonia: BEETLE (1943), BROCKMANN-JEROSCH (1928), DAVIES (1940), DONAT (1931, 1932, 1934, 1935), DUSÉN (1903, 1905), DUSÉN and NEGER (1908), GOOD (1933), GOODSPEED (1945), HAUMAN (1928), HAUMAN, BURKART, PARODI, and CABRERA (1947, and bibliography here), MACLOSKIE (1904–6, 1906), MACLOSKIE and DUSÉN (1914), NEGER (1901), REICHE (1907), ROTHKUGEL (1916), SKOTTSBERG (1931), and TURRILL (1919).

The most complete work on the plant life of Chile is that of REICHE (1907) and though this is now more than forty years old some references must be made to the accounts given there of the areas visited by HOOKER and of conclusions reached on problems discussed by HOOKER. The vegetation of Magellan districts, including the Chonos Islands, is dealt with in pages 230–267. The connections between the floras of Chile and New Zealand are considered in pages 299–302, with lists of families, genera, and species. REICHE (pp. 310–1) mentions as alternative hypotheses the former existence of an antarctic continent radiating migration routes northwards, to Chile and New Zealand especially, and the dispersal of fruits and seeds over wide areas of ocean, without reaching any very definite conclusions.

DIELS (1906), in considering very fully the plant life of Western Australia, refers to the connections between Australia and other parts of the world and comments upon HOOKER's views on the relationships of southern floras. DIELS recognizes three floristic elements in the Australian flora: the Antarctic, Malayan, and Australian. The first of these is that which mainly concerns us here. It is limited to the south east corner of the country and is only rich in kinds in the mountains, but in the "alpine region" its role is significant. The Antarctic element scarcely extends beyond the south east of Australia (p. 32). HOOKER probably over emphasized the floristic community of Australia and S. Africa. There are very deep-seated differences between the floras of the two areas. For the most part the resemblances are due either to convergences or to a common source in an old flora of the Southern Hemisphere (p. 371). The main mass of the Australian flora has been regarded as forming an old Panaustralian element but that aggregated in Western Australia has also many types derived by immigration (p. 384).

A good deal of modern work on the phytogeography of Australia is summed up by CROCKER and WOOD (1947) and their list of references should be particularly noted. Of the elements constituting the flora of Australia, the Australian is the most prominent in the Southern regions but it probably developed when Australia as a whole had a temperate and more uniform climate. The evidence is strongly in favour of a pan-Australian flora in the early Tertiary when the continent was reduced to an almost perfect peneplain. The uniformity was broken by marine transgressions, volcanic activity, and earth movements which reached their maximum in the late Tertiary. Habitat diversity was increased in the Pleistocene by the generally high rainfall causing erosion. A post-Pleistocene period of aridity followed and the severe desiccation destroyed a considerable portion of the pre-arid flora. Remnants survived in numerous refuges and later re-colonized vast areas when climatic conditions slightly improved. "In a broad way the flora of the arid regions is a blending of the Indo-Melanesian element from the north and north-east with the Australian element from the south-west, south and south-east, which has occurred subsequent to the Great Australian Arid Period". The associations are very young and their distribution within a climatic zone has been determined chiefly by edaphic conditions. The richness of western Australia in endemics is explained by its isolation from the south-east in Miocene times. Edaphic barriers are particularly emphasized in this isolation.

Kerguelen:— In the Flora Antarctica, HOOKER recorded what was known to that date of the flora of Kerguelen, and there described most of the peculiar flowering plants and some of the cryptogams of which it is constituted. In a later paper he gives a more complete account, under the general title "An account of the botanical collections made in Kerguelen's Land during the Transit of Venus Expedition in the years 1874–5" (Phil. Trans. Roy. Soc. 168: 1–15, 1879), based especially on MOSELEY's collection.

"The three small archipelagos of Kerguelen Island (including the Heard Islands), Marion and Prince Edward's Islands, and the Crozets, are individually and collectively the most barren tracts on the Globe, whether in their own latitude or in any higher one, except such as lie within the Antarctic Circle itself; for no land even within the N. Polar area presents so impoverished a vegetation."

"The chief interest attached to the flora of these archipelagos lies in the indication it affords of their being, in all probability, the remains of a much larger land area, which, though peopled with plants mainly from the southern extreme of S. America, 4,000 miles to the westward, possessed an endemic flora of its own, which included forest trees of considerable dimensions" (pp. 2–3).

HOOKER shows that the affinities of the flora are Fuegian and he classifies the Phanerogams of Kerguelen as follows:

"1 Endemic genus, which has no near ally — *Pringlea antiscorbutica*.

1 Endemic genus allied to an Andean one — *Lyallia kerguelensis*.

6 Endemic species allied to American congeners — *Ranunculus crassipes* and *Moseleyi, Colobanthus kerguelensis, Acaena affinis, Poa Cookii, Festuca kerguelensis*.

5 species common to Fuegia but not found elsewhere: *Ranunculus trullifolius, Azorella Selago, Galium antarcticum, Festuca erecta, Deschampsia antarctica*.

6 species common to America, and also to New Zealand and the islands south of it. *Tillaea moschata, Montia fontana,* Callitriche obtusangula,* Limosella aquatica,* Juncus scheuzerioides, Agrostis Magellanica*. (Most of these are aquatic or marsh plants, and those marked with an asterisk are also European and very widely dispersed).

2 species found elsewhere but not in Fuegia, *Cotula plumosa*, common to Lord Auckland's group and Campbell's Island south of New Zealand, and *Uncinia compacta*, a native of the mountains of Tasmania and New Zealand" (pp. 3–4).

"This American affinity of the Kerguelen Island flora thus clearly established by its flowering plants is very strongly manifested by its Cryptogams, amongst which, however, the only evidence of migration from South Africa occurs. This is the case of *Polypodium vulgare*, a widely distributed fern in the north temperate zone, but known in the southern only from the Cape Colony, Marion, and Kerguelen Islands; what is further curious respecting it is, that the Kerguelen Island individuals are referable to a variety with pellucid veins, hitherto known only from the Sandwich Islands" (p. 4).

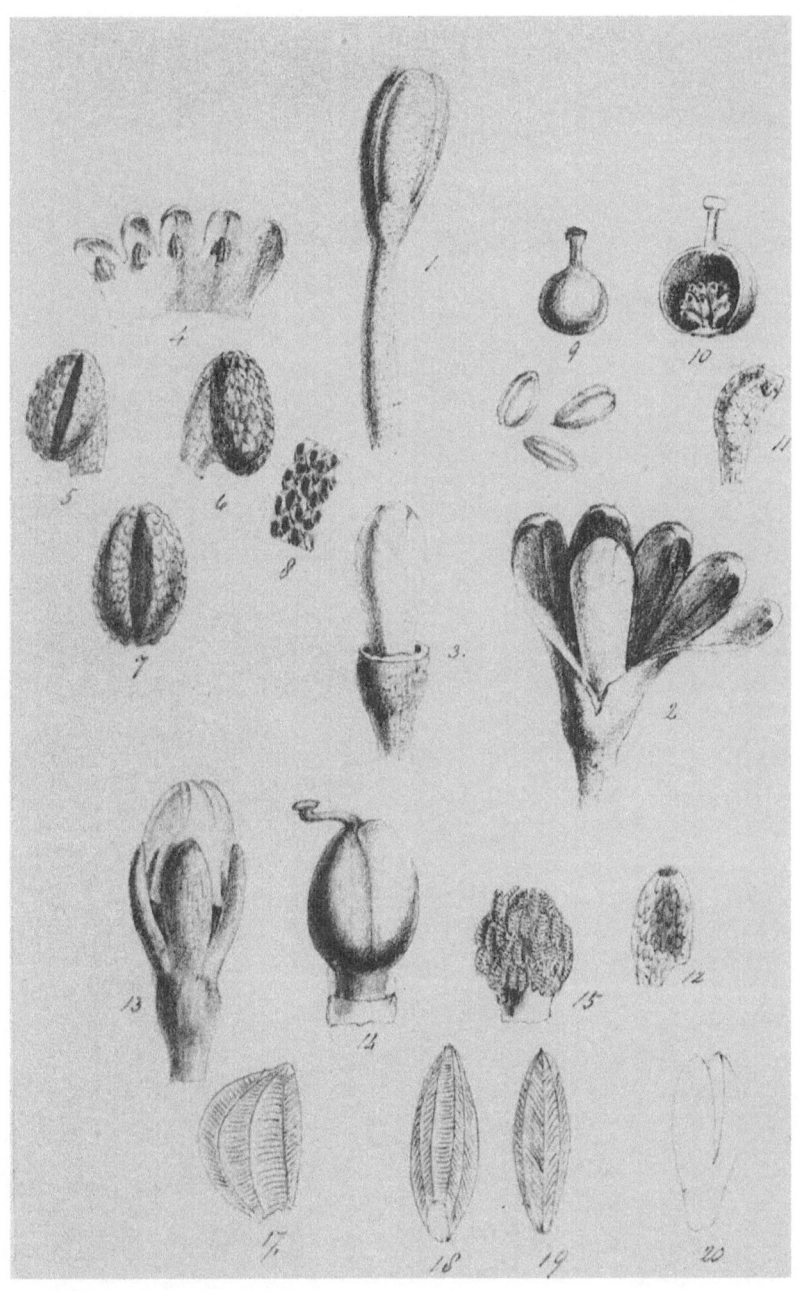

Plate 18. — FROM J. D. HOOKER'S MANUSCRIPTS AT KEW. Photograph of a page of
drawings of a plant (later determined as *Limosella aquatica* L.)
from Kerguelen Island.

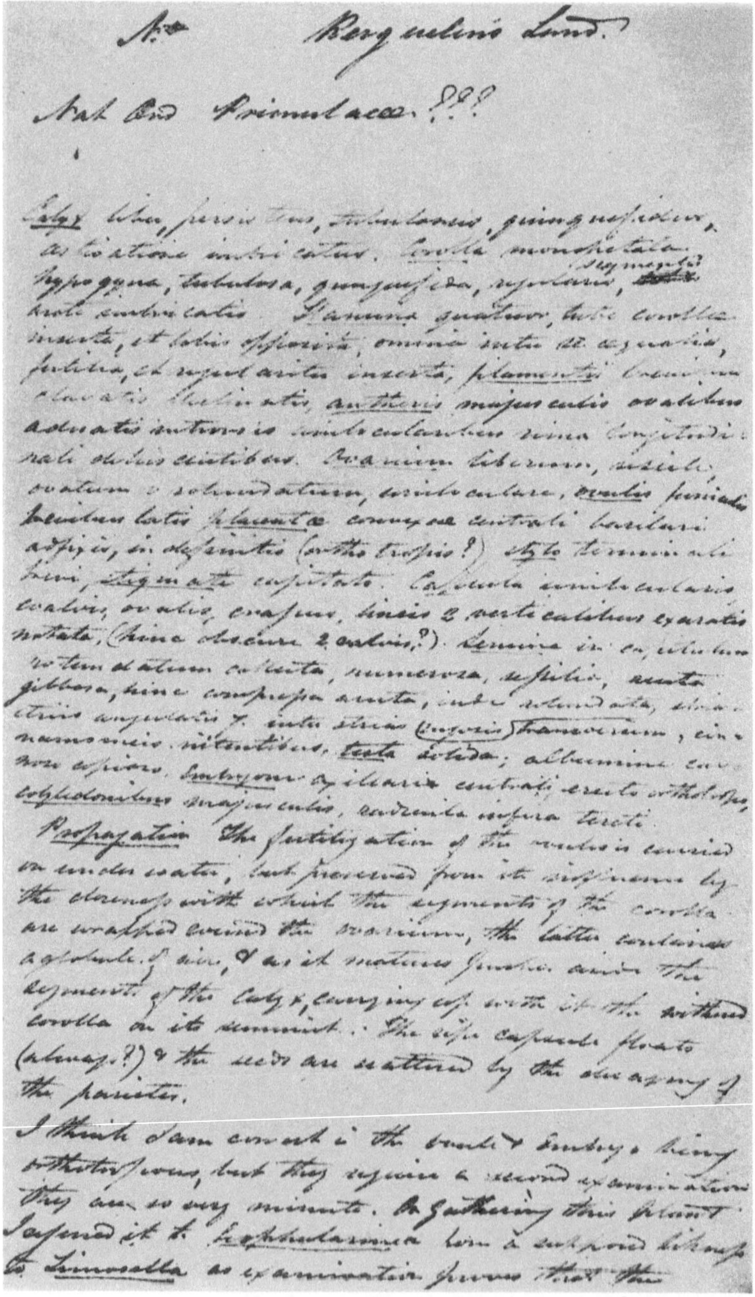

HOOKER discusses the possibilities of the plants of Kerguelen, or their ancestors, being transported from Fuegia as spores, seeds, or fruits by birds, winds, or ocean currents. He concludes that while winds could account for the transport of spores of cryptogams and ocean currents could have brought the marine algae, the transport of fruits and seeds over upwards of 4,000 miles of ocean is difficult to imagine.

"The supposition that more land formerly existed along the parallels between Fuegia and Kerguelen Island, possibly in the form of islands, remains as the forlorn hope of the botanical geographer. By such stepping stones the land birds, so numerous in the Falklands Islands (which lie in the direction of such hypothetical islands), and of which the vegetation is identical with that of colder South America, might, favoured by the prevalent westerly gales, have passed from thence to Kerguelen Island, having adhering to them fruits and seeds. The absence of such birds from the present Avi-fauna of Kerguelen Island offers no obstacle to such a speculation, as such immigrants would on arrival speedily be destroyed by the predatory gull and petrels of the island."

"Various phenomena, of very different relative value and nature, but common to the three archipelagos, Kerguelen, the Crozets, and Marion, favour the supposition of these all having been peopled with land plants from South America by means of intermediate tracts of land that have now disappeared; in other words, that these islands constitute the wrecks of either an ancient continent or an archipelago which formely extended further westwards, and that their present vegetation consists of the waifs and strays of a mainly Fuegian flora, together with a few survivals of an endemic one" (pp. 5–6).

In a short paper (Journ. Linn. Soc. 14: 474–82, 1875) HOOKER brought together what was known of the flora of Amsterdam and St. Paul's Islands, about 800 miles N.E. of Kerguelen. In his 1879 paper he summarizes for these islands and for Tristan d'Acunha as follows:

"Their scanty vegetation is on the whole more temperate than antarctic, and approximates to that of S. Africa in containing such genera as *Phylica*, *Spartina*, and *Danthonia*. Their fern flora is very interesting; one fern only is common to Kerguelen (*Lomaria alpina*), one (*Nephrodium antarcticum*) is peculiar, though allied to a Mauritian species, and two others (*Blechnum australe* and *Asplenium furcatum*) are natives of the Cape and other countries; but what is most singular is, that neither the *Polypodium vulgare* nor *Aspidium mohrioides* have been found in either island, though the former is common to the Cape, Marion Island, and Kerguelen's Land, and the latter to the two first of these localities."

"Tristan d'Acunha, in 12° W. long, and 37° S. lat., and the adjacent islets called Nightingale and Inaccessible, all nearly in the latitude of Amsterdam Island and the Cape of Good Hope, are the only other islands whose vegetation demands a passing notice here. (For the latest account of this group *see* MOSELEY in Journ. Linn. Soc. 14 : 377). Their flora is essentially Fuegian, with an admixture of Cape genera, but with none of those characteristics of Kerguelen Island. Of Cape types, it contains a *Pelargonium* and an abundance of both the *Phylica* and *Spartina* of Amsterdam Island, together with species of *Oxalis* and *Hydrocotyle*. The Fuegian and Falkland Island plants of Tristan d'Acunha and its islets, which have not hitherto been found in the islands south and east of them, are however more numerous than are the Cape genera even, and include *Cardamine hirsuta, Nertera depressa, Empetrum nigrum*, var. *rubrum, Lagenophora Commersoniana*, and *Apium australe*, and it contains besides the strictly American genus *Chevreulia*. Two land birds, both peculiar, are common in the Tristan group, and they possess a water hen, which has a representative in Africa and S. America. I am not aware whether land birds are found in Amsterdam Island; if so they may help to account for the wonderful fact of the Tristan d'Acunha *Phylica* and *Spartina* being found also in it, though separated by 3,000 miles of ocean."

"In conclusion, I have to state that no trace of the mountain flora of S. Africa has been found in any of the southern groups of islands" (p. 8).

* * *

HOOKER included in his later account of the flora of Kerguelen the collections made during the Challenger Expedition. HEMSLEY (1885) published valuable field notes written by MOSELY but otherwise added little to HOOKER's account.

Later expeditions to Kerguelen than those whose collections were dealt with in HOOKER's papers have added very few vascular plants to the known native flora (*Hymenophyllum peltatum* was found by NAUMAN) but many non-vascular cryptogams. The results published since 1879 have, however, given lengthy descriptions of the ecology—both autecology and synecology—, vegetative anatomy, and floral structure, and important information on fossil plant remains. The establishment of certain introduced species has also been recorded.

The two chief botanical explorations to which reference must be made were those of the "Valdivia" (1898–9) and of the German South Polar Expedition (1901–3). In the former A. F. W. SCHIMPER was the botanist whose collections and notes were used in preparing the published accounts. MARDNER (1902) investigated the anatomical structure of a dozen Kerguelen plants and concluded that the characters he found were partly hygrophilous, due no doubt, he suggested, to the air moisture being in no way reduced by the strong winds. SCHENK (1905) "wrote up" SCHIMPER's field work and also (1906) prepared the taxonomic account of the German South Polar Expedition. The ecological and phytogeographical accounts of the latter were, however, prepared by WERTH (1906–11) who was principal botanist on the expedition.

The vascular flora of the Kerguelen group of islands is small— 9 vascular cryptogams and 21 phanerogams are accepted as native. The following are the endemics: *Poa cookii* Hk.f., *Festuca kerguelensis* Hk.f., *Colobanthus kerguelensis* Hk.f., *Lyallia kerguelensis* Hk.f., *Ranunculus moseleyi* Hk.f., and *Pringlea antiscorbutica* R. Br. *Lyallia* and *Pringlea* are endemic genera. HOOKER's statement that the main taxonomic affinity of the flora is with South America has been strengthened by later investigations. SCHENK and SCHIMPER believed that the floristic connections must be attributed to dissemination by winds and birds, the latter directed by the former. WERTH, on the other hand, will have none of this. He concluded that the flora of the Kerguelen group of islands, was almost completely destroyed during the Ice Age and that the present plants are not relatively young immigrants but relicts of a pre-glacial epoch in which a rich and varied vegetation covered not only Subantarctica but also Antarctica proper. On this view, only a few species survived the glaciation and persisted on steep ice-free rock-walls. These few species now form the main element of the Kerguelen flora, besides which the limited number that have reached Kerguelen since by wind, birds, or icebergs are not of great importance. E. A. DE LA RÜE (1932) has shown that Kerguelen is largely built up of basaltic lava with subordinate trachytes and rhyolites and is situated on a submarine ridge connecting it with a continent. He assigns the earliest lavas to the Mesozoic or to an even earlier epoch. There was prolonged weathering during the Tertiary when a very different vegetation existed than that now present (*see below*). Volcanic activity was renewed in the Pliocene period and was followed by the Pleistocene glaciation when the ice covered practically the whole island from the centres of the main mountain massifs where large ice-sheets still exist. Kerguelen's old connection with a continental mass, probably Antarctica, was probably by way of Heard Island, on the same submarine ridge which extends southwards to Queen Mary Land. WEGENER's hypothesis of continental drift is accepted by DE LA RÜE to explain the separation of Kerguelen from Antarctica.

SEWARD (1933, 1934) has described in detail plant fossils collected by DE LA RÜE believed to be of Tertiary (probably Oligocene) age. These include diatoms, mosses, ferns, some dicotyledonous leaves, and *Araucarites ruei*. The latter is compared most closely with *Araucaria excelsa* and the remains consist of foliage shoots, leaves, cone-scales, and wood. SEWARD points out that *Araucaria* is not well constructed for wide dispersal over oceans and sums up in favour of the Kerguelen Archipelago having been part of a continent and though leaving the question open has an obvious bias in favour of some kind of continental drift.

COOKSON (1946, 1947) examined a sample of lignite from Kerguelen and found abundant spores and pollen grains. "Among these, several varieties of winged pollens of podocarpaceous affinities appear to predominate; pteridophyta spores are rather numerous but angiospermous pollens seem to be infrequent. Some large wingless grains that suggest comparison with araucarian pollens have also been observed."

Ecologically, wind is the main determiner in the environment of Kerguelen plants and is associated with low summer temperatures and with a mean difference of only 6.5° C. between the warmest and coldest months. As SCHIMPER said the Kerguelen desert is a wind-desert. WERTH divides the climatic plant formations into three facies or belts: (1) The desert, with only very scattered plants; (2) the tundra, with uniform but discontinuous plant cover; and (3) the heath, with a complete or almost complete plant cover. Locally peculiar communities are those of rocks, marshes, and sea-strand. Except for the last two, it is the proportion and behaviour of constituent species rather than the presence or absence of this or that species that are diagnostic. The predominance of autogamy and absence of obligate entomophily are correlated with existing climatic conditions.

Raoul or Sunday Island:— Raoul or Sunday Island (lat. 29° 16′ S., long, 177° 53′ W., Naval Intelligence Division, B.R. 519B, 1944) is the largest (29.25 sq. km.) and most northern island of the Kermadec group. HOOKER, *in* Journ. Linn. Soc. Bot. 1: 125–9 (1857), dealt with its flora, as sampled in a collection made by J. M'GILLIVRAY (also spelt McGILLIVRAY and MACGILLIVRAY), in a paper with the title "On the botany of Raoul Island, one of the Kermadec group in the South Pacific Ocean." The following extract gives the gist of his phytogeographical conclusions.

"The most interesting circumstance connected with the vegetation of Raoul Island is the identity of most of the flowering plants, and all but one of the ferns, that have been collected upon it, with those of New Zealand. The great extent of intervening ocean (450 miles), and the small size of the islands, would appear to render it extremely difficult to account for this similarity of vegetation by transport; added to which, the prevailing winds blow from the north-west, and the oceanic currents set in the same direction."

"It is also worthy of remark, that of the nine species that are not natives of New Zealand, four are new, and three are nearly allied to New Zealand plants; whilst of those five that are not new, three are widely diffused throughout the tropical and subtropical Pacific islands, and would appear not to be capable of enduring the cold of New Zealand; these are the *Metrosideros polymorpha*, *Piper latifolium*, and *Omalanthus nutans*."

"The absence of any Ferns (with a single exception) but such as are natives of New Zealand, is, however, a far more striking fact, both because the list is a large one for so small an island (twenty-two species), and because, if their presence is to be accounted for wholly by trans-oceanic transport of these species, the question at once occurs, why has there been no addition of some of the many Fiji or New Caledonian Island ferns, that are common tropical Pacific species, the Fiji Islands being only 700 miles north of the Kermadecs, and New Caledonia 750. The only fern which is not a native of New Zealand, is the Norfolk Island *Asplenium difforme*."

"Still more remarkable is the total absence in the collection of any of the plants peculiar to Norfolk Island, for Raoul Island is in the same latitude as Norfolk Island, is exactly the same distance from New Zealand, and the winds and currents set from Norfolk to Raoul Island: in short, though the northern extreme of New Zealand, Norfolk Island and Raoul Island form an equilateral triangle, with the exception of *Asplenium difforme*, there is not a single fern of Norfolk Island found in Raoul Island that is not also found in New Zealand; whilst of the twenty flowering plants of Raoul Island, no less than six are absolutely peculiar to New Zealand and Raoul Island, and with the exception of the tropical, widely diffused Pacific species, there are no phaenogamic plants or ferns confined to Norfolk Island and Raoul Island. It is further remarkable that of the Raoul Island ferns, *Cyathea medullaris* and *Pteris falcata* have not been found in Norfolk Island."

"There is no doubt that a complete flora of Raoul Island would modify these results; but there can also be no doubt that it would confirm these indications of its affinities being most strong with that of New Zealand, and feeble to a very unaccountable degree with the floras of those other groups with which it might be expected to possess a very strong relationship."

"Of the twenty flowering plants, three are noticed by the collector as being possibly introduced by man, *viz. Sicyos angulatus*, *Gnaphalium luteoalbum*, and *Oplismenus aemulus*, all of which were found to affect cultivated ground. These are, however, so widely distributed in the South Pacific Islands, New Zealand, and Australia, that it is quite as probable as not that they are truly wild in the Kermadec group, and only grow in more abundance upon prepared soil. All have, however, appendages that would favour their transport, as the glochidiate setae of the fruit of the *Sicyos*, the awn of the glume of *Oplismenus*, and the pappus of *Gnaphalium*."

"With regard to the remaining seventeen flowering plants, I recognize special adaptations for transport in the following two only:— *Bidens leucantha*, in the barbed setae of the fruit, and *Lagenophora petiolata*, in the viscid fruit. Of the rest none seem in any way adapted for transport, unless the minute and numerous seeds of the *Lobelia*, *Acianthus*, and *Metrosideros* be so regarded" (pp. 126–7).

In HOOKER's Kew Journ. Bot. 7: 151–2 (1855) there is a short reference to Sunday Island by W. G. MILNE. Later accounts of the flora of the Kermadec Islands are those of CHEESEMAN (1888) and OLIVER (1910, 1911). A summary of their results is given by COCKAYNE (1928, pp. 324–8). The more modern investigations have greatly increased the known flora which was recorded (in 1928) as consisting of

117 vascular plants (pteridophytes 38, monocotyledons 25, dicotyledons 54). Of seed-bearing plants the largest families are: *Gramineae* 13, *Compositae* 9, and *Cyperaceae* 6. Only six genera contain three species or more, the largest being *Asplenium* with five species. The phytogeographical elements of the flora are: endemic 15 species (9 closely related to New Zealand, 3 to Polynesian, 2 to Norfolk Island, and 1 to New Hebridean species); New Zealand 89 (of which 65 are also Australian and 45 occur in Norfolk Island or Lord Howe Island or both); Australian (excluding those also in New Zealand) 7. These figures support Hooker's view that the flora has very strong affinities with that of New Zealand, but the Polynesian element is stronger than was allowed for by HOOKER, and OLIVER also notes that it is evident that a strong ocean current flows from New Zealand in a north-easterly direction.

The plant communities of Raoul Island have been greatly modified by the action of introduced goats. The climate is of a humid subtropical type and the volcanic tuff soil extremely porous. Relatively small areas are occupied by coastal communities (rock vegetation, "*Mariscus* slopes," coastal scrub, sand dunes, and gravel-flat vegetation), inland rock communities, and swamps. Forest occupies by far the greater part of the island and may be "wet" or "dry". COCKAYNE (1928) summarizes the main features as follows:

"*Wet forest* is composed of a mixture of trees none of which is dominant. The principal trees are:— *Ascarina lanceolata, Melicytus ramiflorus, Nothopanax kermadecensis, Suttonia kermadecensis* and *Coprosma acutifolia. Metrosideros villosa*, of huge dimensions, occurs in places. Frequently one species or another makes a pure stand. *Rhopalostylis Cheesemanii* and *Cyathea kermadecensis* are generally extremely plentiful. The abundance of epiphytic ferns is a striking feature."

"*Dry forest* is characterised by the dominance of *Metrosideros villosa*. The average height is some 20 m. There are 3 tiers of vegetation. Different combinations form the lowermost tier in different localities, *e.g. Polystichum aristatum* 1 to 2 m. high may make an impenetrable mass, or *Nephrolepis cordifolia* occupy wide areas, its matted roots spreading over the ground or fallen logs or climbing the tree-fern trunks, or again the undergrowth be merely a dense mass of the stems of *Macropiper*. The second tier may consist of small trees, the palm and tree-ferns, the first-named being especially *Melicope ternata, Boehmeria dealbata, Coriaria arborea, Corynocarpus laevigata, Melicytus ramiflorus, Suttonia kermadecensis, Myoporum laetum* and *Coprosma acutifolia*. In places the palm forms colonies (Fig. 94). The third tier consists of the *Metrosideros* together with *Corynocarpus* and *Myoporum* which almost equal it in stature. The forest-roof is fairly dense" (p. 328).

Chapter IX

MISCELLANEOUS AND GENERAL

Mention must be made of the very full essay-review, by J. D. Hoo-KER, of A. DE CANDOLLE's Géographie Botanique Raisonnée, published in HOOKER's Journ. of Bot. (Kew Journ. Bot.) 8: 54–64, 82–8, 112–21, 151–7, 181–91, 214–9, 248–56 (1856). The importance of this review is twofold. First, it shows how very fully HOOKER studied DE CAN-DOLLE's methods and theories, and, though he sometimes disagrees with many of his conclusions, they undoubtedly influenced some of his later work. Secondly, in the final instalment, under the heading *"Concluding remarks,"* HOOKER discusses at some length his view on the nature of species. DE CANDOLLE believed "that the majority of species were created such as they now exist," but HOOKER adds "there is not a shadow of a proof of this." The arguments for trans-mutation of species are put forward and include: "the fact that species are variable", that "this theory of transmutation accounts better for the aggregation of Species, Genera, and Natural Orders in geographi-cal areas, and for their limitation", and that transmutation involves "less of the marvellous at first sight" than does special creation. On the other hand, HOOKER says, "unfortunately transmutation brings us no nearer the origin of species, except the doctrine of progressive de-velopment be also allowed, and, as we can show, the study of plants affords much positive evidence against progressive development, and none in favour of it" (p. 252).

Two sets of what in 1856 he regarded as botanical facts were against HOOKER accepting "progressive development." These were the evidence of the fossil record and the high place (in structure and therefore in systematic arrangement) he assigned to the *Coniferae*.

"That the various Dicotyledonous Natural Orders of the Chalk afford no proof of being higher nor lower in development than those of the present day, but much of their being equal in rank; that the *Cycadeae* of the Lias and Oolite are certainly as highly organized as their existing allies; that the *Coniferae* are too imperfect to afford the smallest evidence of their relative development; that the Ferns of the Oolite and Coal are as highly organized as those of the present day; and that the *Lycopodiaceae* of the Carboniferous epoch are, in general structure, the same with those now existing, but were very much more highly developed in stature and organization."

"It is further to be remarked that the above Natural Orders embrace some of the most highly organized in the vegetable kingdom; though with regard to that which we consider as amongst the very highest, namely the *Coniferae*, the evidence is the most incomplete as to the perfection of its members, as compared with those now existing" (p. 253).

One has to acknowledge that the origin and early development of the angiosperms and of angiospermous families is still largely a matter of faith, in that relevant palaeobotanical evidence is yet to be obtained. Regarding the *Coniferae* (and other gymnosperms) we now know that HOOKER was wrong in the "high" position he assigned to them, though there may still be long philosophical debates as to just what is meant by "higher or lower".

A final quotation shows clearly the half-way, compromising, or "on-the-fence" position HOOKER took up in 1856.

"The only other theoretical point to which we shall allude, is the appearance of species (whether as creations or transmutations) in one spot or in many spots. Here again we have no evidence to guide us, and can only assume a position, as in the former cases, upon the broadest view of the facts of distribution; now it is undisputed that the most prominent feature in distribution is that, as a general rule, species are grouped in more or less restricted areas, after a manner that is quite consistent with the hypothesis of their having spread from one spot. The exceptions may be very numerous, and the question remains, how may those cases be most easily accounted for which cannot be explained by migration; if by a double creation of the same species, we wander further into the regions of pure hypothesis; but if by transmutation, we may assume that the power that species have of forming races, has been developed at two or more spots instead of one. This demands less of the marvellous than the hypothesis of a double creation, and allows more latitude for variation: for whereas it is adding miracle to miracle to assume the same species to be created not only at two or more spots, but at two or more times, and under two or more forms; it is but extending one law now in operation to suppose that this would happen if transmutation thus gave origin to races and species; for the conditions that induce the change, and hence the race, need not have occurred at the same time at two or more spots, nor when they did occur would they act with equal power or upon exactly similar individuals, whence the individual races would not be altogether similar."

"We have thus endeavoured to put the argument in favour of transmutation in as strong a light as we believe it to be capable of bearing in the existing state of our knowledge. For our own part we confess that we see no more means of forming an opinion on the subject of the origin of species, than we do of the origin of time; whether they are all suffering transmutation or not, appears to be immaterial as regards the progress of botanical science; on the one hand we cannot treat practically of the species of plants, either systematically or physiologically, save under the assumption that most are

MISCELLANEOUS AND GENERAL

hereditarily permanently distinct; and on the other, we cannot study any species or organ physiologically or morphologically without being strongly impressed with the fact that variability is an ever-operating law."

"Species of plants are so far constant as to admit of their being treated upon the whole as if they were permanent creations; and though so plastic under altered conditions, they are capable of better and more natural systematic arrangement and circumscription by characters than minerals, climates or diseases. The difference between the views of those who advocate the theory of the creation of species by transmutation, and those who believe in a special creation, is very wide perhaps, but not so wide as to allow of their employing different methods towards the advancement of Botany in any one of its departments. For ourselves, we believe that fully one-half of the registered species of plants are reducible to races or varieties; with regard to the other half, whatever their origin may be, they are, in comparison, permanently distinct as species. That these species do run into varieties; that two or more of them may have originated in an altered state of some pre-existing form, or may in the course of ages assume still other forms, is perfectly intelligible; but for any such species so to change as to assume all the characters of another within the limits of our experience, is for Nature to break one of her best-established laws: *Natura nihil facit per saltus*" (pp. 254–6).

Geographical Distribution:— At York in 1881, HOOKER was president of the Geographical Section of the British Association for the Advancement of Science and took for his presidential address the subject of geographical distribution. Much of the address is concerned with the history of the subject and with the bearing of topography, geology, and meteorology in explaining the causes of the known distribution of plants. The phytogeographical work of LINNAEUS, HUMBOLDT, FORBES, and others is reviewed and the importance of Darwinian theories is fully acknowledged.

"Before the publication of the doctrine of the origin of species by variation and natural selection, all reasoning on their distribution was in subordination to the idea that these were permanent and special creations; just as, before it was shown that species were often older than the islands and mountains they inhabited naturalists had to make their theories accord with the idea that all migration took place under existing conditions of land and sea. Hitherto the modes of dispersion of species, genera and families, had been traced; but the origin of representative species, genera and families, remained an enigma; these could be explained only by the supposition that the localities where they occurred presented conditions so similar that they favoured the creation of similar organisms, which failed to account for representation occurring in the far more numerous cases where there is no discoverable similarity of physical conditions, and of their not occurring in places where the conditions are similar. Now under the theory of modification of species after migration and isolation, their representation in distant localities is only a question of time and changed physical conditions. In fact, as DARWIN well sums up, all the leading facts of distribution are clearly explicable under this theory; such as the multiplication of new forms; the importance of barriers in forming and separating zoological and botanical provinces; the concentration of related species in the same area; the linking together under different latitudes of the inhabitants of the plains and mountains, of the forests, marshes, and deserts, and the linking of these with the extinct beings which formerly inhabited the same areas; and the fact of different forms of life occurring in areas having nearly the same physical conditions."

"With the establishment of the doctrine of the orderly evolution of species under known laws, I close this list of those recognised principles of the science of geographical distribution, which must guide all who enter upon its pursuit. As HUMBOLDT was its founder, and FORBES its reformer, so we must regard DARWIN as its latest and greatest lawgiver. With their example, and their conclusions to guide, advance becomes possible whenever discovery opens new paths, or study and reflection retraverse the old ones" (p. 7 of separate, p. 733 of Report).

The existence of fossil floras in the Arctic, the conclusions of ASA GRAY on the relationships of the flora of North America, and the researches of BLYTT on the Norwegian flora are considered. The section dealing with the floras of the Southern Hemisphere is so largely based on HOOKER's own investigations that it may be accepted as a summary of first hand experiences and the following quotation is, therefore, well worth giving:

"Turning now to the southern hemisphere, the phenomena of distribution are much more difficult of explanation. Geographically speaking there is no Antarctic flora except a few lichens and seaweeds. The plants called Antarctic, from their analogy with the Arctic, are very few in number, and nowhere cross the 62° of south latitude. They are, in so far as they are endemic, confined to the southern islands of the great southern ocean, and the mountains of South Chili, Australia, Tasmania, and New Zealand; whilst the few non-endemic are species of the nearest continents, or are identical with temperate northern or with sub-arctic or even Arctic species. Like the Arctic flora, the Antarctic is a very uniform one round the globe, the same species, in many cases, especially the non-endemic, occurring on every island, though there are sometimes thousands of miles of ocean between the nearest of these. And, as many of the island plants reappear on the mountains above mentioned, far to the north of their island homes, it is inferred on these grounds, as well as on astronomical and geological, that there was a glacial period in the southern temperate zone as well as in the northern."

"The south temperate flora is a fourfold one. South America, South Africa, Australia, and New Zealand contain each an assemblage of plants differing more by far amongst themselves than do the floras of Europe, North Asia, and North America; they contain, in fact, few species in common, except the Antarctic ones that inhabit their mountains. These south temperate plants have their representative species and genera on the mountains of the tropics, each in their own meridian only, and there they meet immigrants from all latitudes of the northern hemisphere. Thus the plants of Fuegia extend northward along the Andes ascending as they advance. Australian genera reappear on the lofty mountain of Kini-balu in Borneo; New Zealand ones on the mountains of New Caledonia; and the most interesting herbarium ever brought from Central Africa, that of Mr. JOSEPH THOMSON, from the highlands of the lake districts, contains many of the endemic genera, and even species of the Cape of Good Hope. Nor does the northern representation of the south temperate flora cease within the tropics; it extends to the middle north temperate zone; Chilian genera reappearing in Mexico and California; South African in North Africa, in the Canary Islands, and even in Asia Minor; and Australian in the Khasia Mountains of East Bengal, in East China and Japan."

"So too there is a representation of genera in the southern temperate continents, feeble numerically compared to what the north presents, but strong in other respects. This is shown by the families of *Proteaceae*, *Cycadeae* and *Restiaceae*, abounding in South Africa and Australia alone, though not a single species or even genus of these families is common to the two countries; by New Zealand, with a flora differing in almost every element from the Chilian, yet having a few species of both calceolaria and fuchsia, genera otherwise purely American; whilst as regards Australia and New Zealand, it is difficult to say which are the most puzzling, the contrasts, or the similarities which their animal and vegetable productions present."

"These features of the vegetation of the south temperate and Antarctic regions, though they simulate those of the north temperate and Arctic, may not originate from precisely similar causes. In the absence of such evidence as the fossil animals and plants of the North affords, there is no proof that the Antarctic plants found on the south temperate alps, or the south temperate plants found in the mountains of the tropics, originated in the south; though this appears probable from the absence in the south of so many of the leading families of plants and animals of the north, no less than from the number of endemic forms the south contains. These considerations have favoured the speculation of the former existence, during a warmer period than the present, of a centre of creation in the Southern Ocean, in the form of either a continent or of an archipelago, from which both the Antarctic and Southern endemic forms radiated. I have myself suggested continental or insular extension as a means of aiding that wide dispersion of species over the Southern Ocean, which it is difficult to explain without such intervention; and the discovery of beds of fossil trunks of trees in Kerguelen's Island, testifies to that place having enjoyed a warmer climate than its present one."

"The rarity in the existing Archipelago (Kerguelen's Island, the Crozets, and Prince Edward's Island) of any of the endemic genera of the south temperate flora, or of representatives of them, is, however, an argument against such land, if it ever existed, having been the birthplace of that flora;

and there are two reasons for adopting the opposite theory, that the southern flora came from the north temperate zone. Of these, one is the number of northern genera and species (which, from their all inhabiting north-east Europe, I have denominated Scandinavian), that are found in all Antarctic and south temperate regions, the majority of them in Fuegia, the flora of which country is, by means of the Andes, in the most direct communication with the northern one. The other is the fact I have stated above that the several south temperate floras are more intimately related to those of the countries north of them than they are to one another" (pp. 9–11 of separate pp. 735–7 of Report).

HOOKER comments upon THISELTON DYER's proposition "that the floras of all the countries of the globe may be traced back at some time of their history to the northern hemisphere, and that they may be regarded in point of affinity and specialisation as the natural results of the conditions to which they have been subjected during recent geological times, on continents and islands with the configuration of those of our globe." SAPORTA held a similar view: "that the Polar area was the centre of origination of all the successive phases of vegetation that have appeared on the globe, all being developed in the north; and that the development of flowering plants was enormously augmented by the introduction during the latter part of the secondary period of flower-feeding insects, which brought about cross-fertilisation" (p. 11 of separate, p. 737 of Report).

The address concludes with allusions to DE CANDOLLE's 'Géographie Botanique' and WALLACE's 'Island Life.'

The address was published in full in the "Report of the fifty-first meeting of the British Association for the Advancement of Science held at York in August and September 1881", London, 1882.

HOOKER outlined in his presidental address to the Royal Society in 1878 (Proc. Roy. Soc. 28: 51–5, 1879), rather more fully, SAPORTA's theories on "L'Ancienne Végétation Polaire" (The Kew copy of this paper records "Extrait des Comptes-Rendus du Congrès International des Sciences Géographiques," Paris, 1877.). The following extract is from HOOKER's address:

"That the polar area was the centre of origination for the successive phases of vegetation that have appeared in the globe is evidenced, under Count SAPORTA's view, by the fact that all formations, Carboniferous, Jurassic, Cretaceous, and Tertiary, are alike abundantly represented in the rocks of that area, and that, in each case, their constituents closely resemble that of much lower latitudes. The first indications of the climate cooling in these regions is afforded by *Coniferae*, which appear in the polar lower Cretaceous formations. These are followed by the first appearance of Dicotyledons with deciduous leaves, which again marks the period when the summer and winter season first became strongly contrasted. The introduction of these (deciduous-leaved trees) he regards as the greatest revolution in vegetation that the world has seen; and he conceives that once evolved they increased, both in multiplicity and in diversity of form, with great rapidity, and not in one spot only, and continued to do so down to the present time."

"Lastly, the advent of the Miocene period, in the polar area, was accompanied with the production of a profusion of genera, the majority of which have existing representatives which must now be sought in a latitude 40° farther south, and to which they were driven by the advent and advance of the glacial cold; and here Count SAPORTA's conclusions accord with those of Professor A. GRAY, who first showed, now twenty years ago, that the representatives of the elements of the United States Flora previously inhabited high northern latitudes, from which they were driven south during the Glacial period."

"Perhaps the most novel idea in Count SAPORTA's Essay is that of the diffused sunlight which (with a densely clouded atmosphere), the author assumes to have been operative in reducing the contrast between the polar summers and winters. If it be accepted it at once disposes of the difficulty of admitting that evergreen trees survived a long polar winter of total darkness, and a summer of constant stimulation by bright sunlight; and if, further, it is admitted that it is to internal heat we may ascribe the tropical aspect of the former vegetation of the polar region, then there is no necessity for assuming that the solar system at those periods was in a warmer area of stellar space, or that the position of the poles was altered, to account for the high temperature of Pre-Glacial times in high northern latitudes; or, lastly, that the main features of the great continents and oceans were very different in early geological times from what they now are" (pp. 53–4).

Island Flora:— HOOKER's famous "Lecture on Insular Floras" was delivered before the British Association for the Advancement of Science at Nottingham, 27 August 1866. It was published in the Gardener's Chronicle Jan. 1867 (a reprint exists at Kew) and as a separate pamphlet in 1896 to which the page figures refer. He uses the term "Insular" Floras in a restricted manner for "the Floras of those islets that rise as mere points of land from out the broad breasts of the great oceans. With few exceptions all are volcanic, all mountainous, and so small that no man has realised their smallness who has not sailed in search of them." Our author points out that:

"The relationships between these oceanic island Floras are of two kinds, that must not be confounded: one a relationship of analogy between themselves, due to physical conditions common to them all—to their climate, exposure, limited area, distance from continents, etc. Thus they are rich in Ferns, Mosses, and other Flowerless plants; and they possess many evergreen, but comparatively few herbaceous plants, and fewer or no indigenous annuals. Plants which are herbs on continents, often either themselves become shrubby in islets, or are represented by allied species that are shrubby or arboreous. Species are few in proportion to genera, and genera in proportion to orders. The mountains, however lofty, present few alpine or sub-alpine species; and the total number of species is usually small compared with what continental areas of equal size and similar conditions contain. The other is a relationship of affinity, a *bona fide* kinship, which the Floras of islands display in common with one another or with certain continents: as is shown by Madeira, the Azores, and Canaries containing many plants in common that are not found on any continent; by the Canarian Flora being in the main a Mediterranean one; the St. Helenan being an African, and so forth" (pp. 5–6).

In the first, essentially descriptive, part of his lecture HOOKER deals with the floras of seven oceanic islands or island groups: Madeira, Canary Islands, Azores, Cape Verde Islands, St. Helena, Ascension, and Kerguelen.

In the Madeiran Group (Madeira, Porto Santo, and the Dezertas) the first apparent character of the vegetation is European.

"But though the vegetation is European in the main, it is not so altogether, and even its European features soon arrange themselves in a botanist's eye under different categories, very much as follows: The majority, including almost if not all the annuals, we find to be identical with European plants, and undistinguishable from them; others differ from European plants by slight but certain characters, as varieties we say; a third class are specifically different from European, and yet seem to hold a place corresponding to what their nearest allies occupy in Europe—these are representative species; and a fourth class comprises plants that are evidently allied to European, but belong to different genera—these constitute representative genera."

"Now, it is a curious fact, that when we tabulate these classes, we find that to a great extent they form a graduated series, not only in systematic order and structure, but in point of numbers; in other words, the plants identical with those of Europe are both the most numerous in species, and the species are most numerous in individuals: then come the varieties— some are scarcely perceptibly different from European plants, others constantly, and these are less numerous and less common. Then come the distinct species: of these some would be called varieties by many botanists, and others good species by all; these are still less common. Lastly, of the different genera, some constitute what all botanists call a good genus, others would, with some botanists, take rank as slight modifications of European genera; these are both the fewest in number and most local in distribution, many indeed being confined to single spots, or even represented by single plants" (pp. 7–8).

"On penetrating to the rocky and precipitous interior, whether of the main island or the smaller, we find many indigenous trees and shrubs that are not only foreign to Europe, but are allied to American, to African, and to Asiatic plants: thus we have trees of *Clethra* and *Persea*, genera found in no continent but America; of *Apollonias* and others found elsewhere on a continent only in the East Indies; and of *Dracaena* and *Myrsine*, that betray an African affinity. As these non-European plants inhabit the Canaries and Azores also, they have been called *Atlantic types*, under which name I shall speak of them."

"Lastly, when we ascend the mountains of Madeira above 4000 feet and up to their summits (6000), we find little or none of that replacement of the species of a lower level by those of a higher northern latitude, with which we are so familiar in ascending any continental mountains of equal or less height. Plants become fewer and fewer as we ascend, and their places are not taken by boreal ones, or by but very few."

"Here then are various botanical features in respect of which Madeira and its satellite islands, Porto Santo and the Dezertas, differ very much from continental areas of equal extent and elevation, or from islands lying near the coast of a continent, the Floras of which are therefore continental" (pp. 8–9).

HOOKER then points out that Great Britain contains twice as many flowering plants as the Madeiran group, but these are almost entirely identical with plants of continental Europe.

"And so it is with any other corresponding area in Europe: none present a similar assemblage of Asiatic and American plants, nor an equal number of peculiar varieties, species, and genera, as the Madeiran group does, nor so many peculiar plants represented by so very few specimens; and nowhere do we find the rocky islets on the coast of a continent to be tenanted by numerous singular genera, species, and varieties, which are to be found nowhere else on the surface of the globe. What should we say, for instance, if a plant so totally unlike anything British as the *Monizia edulis* (an Umbelliferous plant, with a stem like an inverted elephant's trunk, crowned with a huge tuft of Parsley-like foliage), were found on one rocky islet of the Scillys, or another Umbelliferous plant (*Melanoselinum*) with a slender trunk like a Palm, on one mountain of Wales; or if the Isle of Wight and Scilly Islands had varieties, species, and genera too, differing from anything in Britain, and found nowhere else in the world!"

"Of all the above peculiarities, it is those very rare and local plants that are isolated as genera, and in geographic distribution, that arrest the inquirer's attention and force him to speculate. We *must* ask ourselves, were these almost unique isolated individuals created as complete highly specialised organisms, or are they modifications of allied plants, owing their strange forms and special attributes to centrifugal variation operating through countless ages? and however they have originated, are we to regard these solitary representatives of such strange forms of vegetation as the first of their several races, destined, mayhap, to increase, and become in future as common as they are now rare; or the last of their races, which, but for the rapid advance of modern science, would have passed away, along with those countless forms of animal and vegetable life that once peopled the globe, but whose forms and structures will never be revealed to us?"

"Considerations, which I cannot here enter into, warrant our belief that such plants on oceanic islands are, like the savages which in some islands have been so long the sole witnesses of their existence, the last representatives of their several races; and the question involuntarily arises—How did this come about?"

"Excluding the direct agency of man, and of animals introduced by man, I believe that a principal cause of the rarity or extinction of old species on oceanic islands is the subsidences they have all experienced. This sinking of the island operates in various ways. (*1*) It reduces the number of spots suitable to the habits of the plant. (*2*) It accelerates that struggle for existence which must terminate in the more hardy or more prolific displacing the less hardy or less prolific. (*3*) It reduces both the numbers and kinds of insects to whose activity the fertilising process in plants, and hence their propagation, is so largely due; and not only does it reduce the number and kinds of insects, but the destruction falls heaviest on the winged kinds, which, as has lately been shown, are almost exclusively the agents in this process; for these, as the area becomes contracted, are blown out to sea and lost in greater proportion than the wingless. Nor is this mere conjecture. Mr. WOLLASTON's careful entomological researches in Madeira and the Canaries prove that winged insects exist in wonderfully smaller proportions to wingless, in these islands, than on the continents, and I can extend this observation to other oceanic islands that I have visited" (pp. 9–11).

Man has much modified the vegetation of Madeira by the destruction of the forests, cultivation, and the introduction of exotics. With regard to Porto Santo:

"In about the year 1418 a mother rabbit and her brood were landed, and increased so rapidly, that they not only consumed the native vegetation, but the cultivated, and actually drove the settlers from the island" (p. 11).

"The Canary Islands contain upwards of 1000 native species, of which fully one-third are absolutely peculiar to the group, and these admit of almost precisely the same classification as the Madeiran plants. Thus, the mass of the plants are identical with Mediterranean species; then follow, in numerical importance, those that are representative, as slight or well marked varieties, or congeners or co-ordinates of the Mediterranean genera and species. After these come, and in great force, the Atlantic plants, including no less than forty of the Madeiran shrubs and trees that are not found in Europe or Africa, and as many representatives of Madeiran genera, species, and varieties, together with a number of allied ones more nearly related to African, Indian, and American plants than to European. Lastly we have, as was to be expected, a sprinkling of African plants, belonging to that division of the African Flora, which, being different from the Mediterranean on the one hand and from the Equatorial African on the other, extends from Western Asia through Arabia and across the Sahara to Cape Blanco—a Flora conterminous in longitude with the distribution of the domesticated camel, which is used as a beast of burthen even in the Canary Islands. This Flora I would call the Arabo-Saharan."

"The lofty mountains of the Canaries, though upwards of 11,000 feet high, contain no alpine plants, and as in the case of the Madeiran group, many of the most peculiar forms are extremely rare and local. Lastly, the Floras of the several islets of the group differ much from one another. The two easternmost, Lancerote and Forteventura, especially, thus standing in the same relation to the others, that Porto Santo and the Dezertas do to the main island of Madeira" (pp. 12–3).

A brief reference is made to the Salvages, almost midway between Madeira and the Canaries. They contain an Atlantic flora, intermediate between that of Madeira and the Canaries, but most closely related to the latter. "The Salvages hence appear to be the peaks of a submerged island that once occupied an important botanical as well as geographical position in the Atlantic Ocean, more or less closely linking the Canaries with Madeira" (p. 13).

HOOKER next turns to the Azores and notes that they demand special notice because of their remoteness from any continent.

"Of flowering plants 350 species have been collected from the principal islands, a very small number considering their extent, but enough to give us a clear insight into the nature of the Azorean Flora. Of these some thirty are peculiar species of well-marked varieties, representatives for the most part of Madeiran or European plants. About thirty are Atlantic types, common to the Azores and Madeira, or to the Azores and the Canaries, or to all; the rest are Portuguese and Spanish plants. Thus, though the absolute number of plants foreign to Europe is even less than in the Canaries and Madeira, these hold a far more important position in the whole Flora, from including so many of those peculiar Atlantic trees and shrubs that link all these three groups into one well-marked though fragmentary Flora."

"Though so much further north than Madeira, the Azores contain scarcely any more boreal plants than Madeira, or even than the Canaries; and such as it does possess are likewise found in the mountains of the Spanish Peninsula. The most notable are the common Ling or Heather (*Calluna vulgaris*), and the beautiful St. Dabeoc's Heath, which is elsewhere found only in the extreme west of Ireland, and in the Pyrenean region. A third is *Littorella lacustris*, a little water plant that inhabits a mountain lake, probably the crater of an extinct volcano, much frequented by migratory water-fowl."

"As in the other groups, there is here a considerable difference between the Floras of the separate islets; and one of the most conspicuous and beautiful plants in the Azores, the *Campanula Vidalii*, is (so far as is known) absolutely confined to a single sea-girt rock off the east coast of Flores" (p. 14).

"Considering how far removed the Azores are from Europe, and how much nearer they are to America than Madeira and the Canaries are to that continent, it might appear strange that the group contains scarcely any American plants not found in the other groups. But such is the case, and more than this; for even the *Clethra* of the Canaries and Madeira, a genus found nowhere else out of America, does not inhabit the Azores."

"The only trace of American influence on the Azorean Flora that I can substantiate, is in a species of the Umbelliferous genus *Sanicula*. Of this genus a common European species is spread almost all over the globe, inclusive of Madeira and the Canaries, but exclusive of the Azores; whereas another species of this same genus takes its place in the Azores, and this species is most closely allied to an American one."

"It is a significant fact, that the minute seed-vessels of *Sanicula* are provided with hooked bristles, suggesting the probability that these were originally transported by birds across the Atlantic" (p. 15).

"The Cape de Verde, situated far within the tropics, 800 miles south of the Canaries, and 300 distant from the African coast are dealt with rather briefly."

"I visited this group in 1839, and found the Flora of the lowlands to be purely African and Arabo-Saharan in character, but on ascending the mountains, I met with a few plants very characteristic of the Canaries and Madeira. The Rev. Mr. LOWE has during the last two winters diligently botanised this group, with most interesting results. He finds, as I did, that the mass of the Flora is African, and that the mountains contain many Canarian types; but that all these are the types that have representatives in the Mediterranean region, whilst of those peculiar Canarian, Madeiran, and Azorean plants that have no near allies or representatives in Europe, not one is found in the Cape de Verde, with the single exception of the Dragon's-blood tree."

"Also, ascending above the tropical zone to 5000 feet and upwards, many of the same middle-European plants are found, that appear at correspondingly lower elevations in Madeira, the Canaries, and Azores, and I may add that these are also found on the lofty mountains of Equatorial Africa and Abyssinia."

"We have thus in the Cape de Verde Islands a certain relationship with the Canaries and Madeira almost to the exclusion of the Azores; but it is a feeble one, and so blended with that of the African continent, and especially of the Mediterranean region, as to suggest other considerations than what concern us here" (p. 16).

HOOKER also visited St. Helena. He gives details of the destruction of the natural vegetation.

"When discovered, about 360 years ago, it was entirely covered with forests, the trees drooping over the tremendous precipices that overhang the sea. Now all is changed, fully five-sixths of the island are utterly barren, and by far the greater part of the vegetation that exists, whether herbs, shrubs, or trees, consists of introduced European, American, African, and Australian plants. The indigenous Flora is almost confined to a few patches towards the summit of Diana's Peak, the central ridge, 2700 feet above the sea" (p. 17).

The destruction was due mainly to goats introduced in 1513 and after these were all killed, about 1810, introduced plants, according to

HOOKER, won in the struggle for existence against the indigenous flora "and wherever established, they have actually extinguished the indigenous Flora."

On the basis of collections made by BURCHELL and ROXBURGH, HOOKER estimates that about 45 indigenous species inhabited the island before the goats were destroyed and alien plants introduced by BEATSON. "All are shrubs, trees, or perennial plants; not one is an annual (though introduced annual plants abound, both tropical and temperate). Forty of them are absolutely confined to the island, and five are tropical weeds or seaside plants of very wide distribution.

"These forty are absolutely peculiar to St. Helena and, with scarcely an exception, cannot be regarded as very close specific allies of any other plants at all. No less than seventeen of them have been referred to peculiar genera, and of the others, all differ so markedly as species from their congeners, that not one comes under the category of being an insular form of a continental species. Many of them are excessively scarce, being now found in very small numbers, and on single rocks; not a few have never been gathered since Dr. BURCHELL's visit, some are certainly now extinct, as the beautiful Ebony tree, and probably nearly one-fifth have totally disappeared during the last half-century, or are now all but extinct.'

"From such fragmentary data it is difficult to form any exact conclusions as to the affinities of this Flora, but I think it may be safely regarded as an African one, and characteristic of Southern extra-tropical Africa. The genera *Phylica, Pelargonium, Mesembryanthemum, Osteospermum,* and *Wahlenbergia* are eminently characteristic of Southern extra-tropical Africa, and I find amongst the others scarce any indication of an American parentage, except a plant referred to *Physalis.* The Ferns tell the same tale. Of twenty-six species, ten are absolutely peculiar, all the rest are African, though some are also Indian and American."

"The botany of St. Helena is thus most interesting; it resembles none other in the peculiarity of its indigenous vegetation, in the great rarity of the plants of other countries, or in the number of species that have actually disappeared within the memory of living men. In 1839 and 1843 I in vain searched for forest trees and shrubs that flourished in tens of thousands not a century before my visit, and still existed as individuals twenty years before that date. Of these I saw in some cases no vestige, in others only blasted and lifeless trunks cresting the cliffs in inaccessible places. Probably one hundred St. Helena plants have thus disappeared from the *Systema Naturae* since the first introduction of goats on the island. Every one of these was a link in the chain of created beings, which contained within itself evidence of the affinities of other species, both living and extinct, but which evidence is now irrecoverably lost. If such be the fate of organisms that lived in our day, what folly it is to found theories on the assumed perfection of a geological record which has witnessed revolutions in the vegetation of the globe, to which that of the Flora of St. Helena is as nothing" (pp. 19–20).

14

"*Ascension.*— The islet of Ascension claims a passing notice. It is much smaller than St. Helena, and 600 miles N.W. of it. St. Helena has been called a barren rock, but it is a paradise as compared with Ascension, which consists of a scorched mass of volcanic matter, in part resembling bottle glass, and in part coke and cinders. A small green peak, 800 feet above the sea, monopolises nearly all the vegetation, which consists of Purslane, a Grass, and Euphorbia, in the lower parts of the island, whilst the green peak is clothed with a carpet of Ferns, and here and there a shrub, allied to but different from any St. Helena one. There are nine Ferns, of which no less than six differ from those of St. Helena, and three of them are entirely confined to the islet" (pp. 20–1).

Kerguelen Island (Kerguelen's Land) was visited by HOOKER on the voyage of the *Erebus* and *Terror* and we have already considered his published account. Here he notes the isolated position of *Pringlea antiscorbutica* and then continues:

"It is not so with the other flowering plants; they almost without exception point to the land whence they were derived. The only other peculiar genus on the island (*Lyallia*) is decidedly an Andean form; of the remaining sixteen flowering plants, four are regarded as distinct species peculiar to Kerguelen's Land, but three of them are so nearly allied to Tierra del Fuego congeners, that they may equally rank as varieties of these, and the fourth stands in the same relation to a New Zealand plant. Of the remaining twelve, ten are Fuegian, of which four are confined to Fuegia and Kerguelen's Land, including the remarkable Umbelliferous plant, which belongs to a group that is otherwise very characteristic of the South American Andes. Five are found in all south circumpolar regions, and one alone is confined to Kerguelen's Land and Lord Auckland's Group. Three are European, and all of these are common English and Antarctic freshwater plants. They are *Callitriche verna*, *Limosella aquatica*, and *Montia fontana*."

"The affinity of the Kerguelen's Land Flora is hence extremely close to the Fuegian; so close, indeed, that it cannot be doubted that it was for the most part derived from thence. And it is all the more remarkable that this relationship should be so strong and unmistakeable, if you consider that the mother country of its Flora is not that which is nearest to it, as was the case with all the other islands we have discussed, but that which is the most distant from it; and indeed Kerguelen's Land is more distant from a continent than any other island in the Atlantic or Indian Oceans" (pp. 22–3).

HOOKER sums up his survey of the above islands or island groups as follows:

"*1*) That the Flora of no oceanic island which we have considered is an independent one; that in all cases it is quite manifestly closely allied to some one continental Flora, and that however distant it may be from the mother continent, and however it by so much approximates to another continent, it never presents more than faint traces of the vegetation of such other continent. Thus the Azores, though 1000 miles nearer to America than Madeira has not even so many American types as Madeira has. St. Helena, though 1000 miles nearer to South America than is any part of the African coast, contains scarcely any plants that are even characteristic of America; and Kerguelen's Land, though far more distant from Tierra del Fuego than it is from Africa, Australia, or New Zealand, is almost purely Fuegian in its Flora."

"2) The Floras of all these islands are of a more temperate character than those of the mother continents in the same latitude; thus, Madeira and the Canaries have a Mediterranean Flora, though they are respectively 5° and 10° south of the principal parallel of the Mediterranean region; the affinities of the St. Helena Flora are strongly South African; and the Flora of Kerguelen's Land, in lat. 48°, is what we might expect to meet with in Fuegia, were the American continent produced southward to lat. 60°."

"3) All contain many and great peculiarities, distinguishing them from the continental Floras; and these admit of the following classification:—

"a) Plants peculiar to the islands and betraying no affinity with those of the mother continent, as the Laurels, etc., of Madeira and the Canaries and Azores; the arborescent *Compositae* of St. Helena, and the Kerguelen's Land Cabbage."

"b) They contain certain genera that are very different from those of the mother continent, but are evidently allied to them; and others but slightly different. They contain species that are very different from, but allied to those of the mother continent; and others that are but slightly different from continental; and they contain varieties in the same categories."

"4) As a general rule, the species of the mother continent are proportionally the most abundant, and cover the greatest surface on the islands. The peculiar species are rarer, the peculiar genera of continental affinity are rarer still; whilst the plants having no affinity with those of the mother continent are often very common, in the temperate islands especially—at least under the conditions which the island vegetation now presents."

"5) Indigenous annual plants are extremely rare or absent; but recently introduced annuals are very abundant in those islets that have been frequented by man" (pp. 23–5).

We come now to our author's discussion of "the hypotheses that have been invented by naturalists to account for the presence of continental plants in oceanic islands, and for those various differences between insular and continental Floras that I have indicated." He frankly states that:

"These hypotheses are as yet unverified and insufficient; neither geological considerations, nor botanical affinity, nor natural selection nor all these combined, have yet helped us to a complete solution of this problem, which is at present the *bête-noire* of botanists" (p. 25).

"There are only two possible hypotheses to account for the stocking of an oceanic island with plants from a continent: either seeds were carried across the ocean by currents, or the winds, or birds, or similar agencies; or the islands once formed part of the continent, and the plants spread over intermediate land that has since disappeared."

"To a superficial observer either of these causes may appear admissible, or feasible and sufficient; but the naturalist, who takes nothing for granted, finds insuperable obstacles to the ready acceptation of either. Upon one fundamental point most of the advocates of both hypotheses are agreed, namely, that those plants which are common to the islands and continents were not independently created in both localities, but that they did pass from one to the other; and another will probably gain ready credence, *viz.*, that those peculiar insular plants which have no affinity with continental ones, are relics of a far more ancient vegetation than now prevails on the mother continents" (pp. 25–6).

HOOKER very conscientiously examines the arguments in favour of transoceanic stocking of islands and those in favour of previous land connection. He notes that DARWIN favoured the former and he outlines the reasons; that some means of transport are always in operation, as birds, winds, and ocean currents.

"Of negative evidence in favour of this view, Mr DARWIN adduces the fact that oceanic islands are poor in species, and that whole groups of continental plants are absent from them, which should not be the case had there been continental extension; that land mammals and Batrachians are absent from all oceanic islands, though winged mammals, as bats, together with birds, insects, and other transportable terrestrial creatures, are present in more or less abundance; that if we demand continental extension for some islands, it must be admitted for all, which is, according to his views of the permanence of the general outlines and dispositions of the continents and sea-beds, during the later geological epochs, quite inadmissible" (pp. 28–9).

"But though Mr DARWIN's explanations cover many of the requirements of our problem, and may eventually prove to satisfy all, there are great difficulties in the way of its full acceptance" (p. 29).

Winds, oceanic currents, etc. in the North Atlantic should favour the bringing of American and not European plants to the Azores.

"And yet we find even fewer American types in this group than in Madeira and the Canaries."

"So too, it is with St. Helena and Ascension—they have no land birds, but an African vegetation; and though nearly midway between Africa and America, they have scarcely a single American type of flowering plants: and Kerguelen's Land has a Flora of whose elements most have emigrated not from the nearest land but from the most distant."

"Another difficulty is presented by the extreme rarity of some of the plants common to several of the North Atlantic Islands; take, for example, that remarkable Canarian tree, *Bencomia caudata*, of which only two individuals have been found in the mountains of Madeira, and these a male and a female. It is almost inconceivable that individuals of both sexes should have been transported within the same lifetime from the Canaries to so great a distance; and so with the other peculiar and rare plants common to these groups; intermediate masses of land, as the Salvages (supposing these once to have been larger), and on which such plants may have abounded, afford the only conceivable means of interinsular transport; and if intermediate islands are granted (and Mr DARWIN freely grants these), why not continents?" (pp. 30–1).

HOOKER notes the mixed floras on large islands and argues:

"We should thus be forced to admit that whereas great islands which are peopled by plants through direct communication with the adjacent continents, do receive immigrants from other most distant continents, the little islets that are much nearer the continent, and over whose course the currents of migration must have swept, have been exempted from its effects. Here again, as it appears to me, the only answer is by an appeal to the very different rates in which the vegetation has changed in the islands and the continents during comparatively recent geological periods" (p. 31).

He comments upon absentees in island floras, as the gum trees and legumes of Australia that are not found in New Zealand.

"Even if we grant, with Mr. DARWIN, that the specific and sub-specific change between the Floras of oceanic islands and continents is due to the new relations into which the continental plants are brought in the narrow areas that islands present, and the ensuing sharper struggle for existence, how does it come about that the plants of the Azores, which islands are 750 miles from Europe, are less changed than those of Madeira, which is only 300? This objection seems to me to be imperfectly met by the hypothesis that the nearer island, receiving more immigrants, exhibits the sharper struggle; for this same cause should rather replenish the island with identical forms, and by cross-fertilisation tend to keep them more specifically true; as was assumed to account for the European birds of the Madeiran group being unchanged, whilst the plants of the same group have changed "(p. 32).

However, he sums up on the whole, in favour of trans-oceanic migration linked with DARWIN'S views on evolution and natural selection:

"On the other hand, to my mind, the great objection to the continental extension hypothesis is, that it may be said to account for everything, but to explain nothing; it proves too much; whilst the hypothesis of trans-oceanic migration, though it leaves a multitude of facts unexplained, offers a rational solution of many of the most puzzling phenomena that oceanic islands present—phenomena which, under the hypothesis of intermediate continents, are barren facts literally of no scientific interest—are curiosities of science, no doubt, but are not scientific curiosities."

"Thus, according to the hypothesis of trans-oceanic migration, and the theory of the derivative origin of species, we can understand why the ancient types, like ancient races of mankind, which have disappeared before the steady forward pressure of superior races on the continents, should have survived on the islands to which but few of the superior race had penetrated—we can understand how it comes about that so many continental species and genera are represented on the island by similar but not identical species and genera, and that there is such a representation of genera and species in the separate islets of the group; we can understand why we find in the Atlantic island Floras such a graduated series of forms ascending from variety to genus without those sharp lines of specific distinction that continental plants exhibit; why whole tribes are absent in the Islands; why their Floras are limited, and species are few in proportion to genera; why so many of their peculiar genera tend to grotesque or picturesque arborescent forms; and many other minor facts which it would weary you to enumerate."

"And if many of the phenomena of oceanic island Floras are thus well explained by aid of the theory of the derivative origin of species, and not at all by any other theory, it surely is a strong corroboration of that theory. Depend upon it, the slow but steady struggle for existence is taking advantage of every change of form and every change of circumstance to which plants no less than animals are exposed; and that variation and change of form are the rules in organic life, is as certain as that definite combinations and mathematical proportions are the rules in the inorganic."

"By a wise ordinance, it is ruled that amongst living beings like shall never produce its exact like; that as no two circumstances in time or place are absolutely synchronous, or equal, or similar, so shall no two beings be born alike; that a variety in the environing conditions in which the progeny of a living being may be placed shall be met by variety in the progeny itself. A wise ordinance it is, that ensures the succession of beings, not by multiplying absolutely identical forms, but by varying these, so that the right form may fill its right place in Nature's ever varying economy" (pp. 33–4).

During the second half of last century both botanists and geologists gave much attention to the general problems connected with the flora and fauna of islands. This was partly due to the interest shown by DARWIN in the subject (*see* Origin of Species, ed. 6, chapter XIII), to the publication of WALLACE's Island Life (London, 1888), and to the records of the Challenger Expedition (*see* HEMSLEY, Report on the scientific results of the expedition. Botany. Edinburgh, 1885, 3 parts). This century has seen a relative lapse in wide comparative studies of island floras though there have been many and valuable advances in our knowledge of the botany of individual island groups. It is remarkable how the subject as a whole is neglected or at least not brought up-to-date in recent text books on plant geography. The suggestion is made that some young phytogeographer should take up the comparative study of island floras especially on the basis of modern knowledge of cytogenetics and ecology not available to HOOKER and his contemporaries. The remarks that follow have to be limited to an examination of the data used by and the conclusions reached by HOOKER, in the light of more recent researches but they include ideas which can obviously be much enlarged and, with modifications, extended to island floras in all parts of the world.

The islands whose floras HOOKER considers in his Essay are: Madeira, Canary Islands, Azores, Cape Verde Islands, St. Helena, Ascension, Kerguelen. Of these we have considered recent researches on the Canary Islands and Kerguelen elsewhere in this work in connection with other publications of HOOKER. First, then, we will refer to more modern studies on the other islands and secondly deal with the general problems of island floras.

Madeira.— LOWE's Manual Flora (LOWE, 1868) was published two years after HOOKER delivered his lecture. It covers Madeira and the associated islands of Porto Santo and The Desertas in a very adequate manner. Apart from a summary of the "regions or zones of vegetation" it is taxonomic and there is no discussion of phytogeographical problems. The last sentence is true also of the later (1914) flora of MENEZES, except that this author gives a list with notes of botanists who had explored the islands and a useful bibliography. Attention may be called to GRABHAM's (1926) booklet on the plants cultivated in Madeiran gardens. Other accounts are those of BRITTEN (1904), COCKERELL (1928), DRUCE (1911), RILEY (1925), SPIX and MARTIUS (1824), VAHL (1904, 1905), and VIENNOT-BOURGIN (1938-9).

The most complete account of the vegetation and the fullest discussion of the floristic and ecological phytogeography of Madeira is that of VAHL (1905). This author notes that the lowlands of the Cape Verde Islands belong to the South Saharan steppes and of the Canaries to the North Saharan, yet the marked endemism leads to the recognition of two special floral districts. The lowlands of Madeira are not steppe but correspond to a transition area between lowland steppe, as on the Canaries, and the maquis altitudinal zone. The lower maquis region of Madeira is analogous to the maquis zone of the Canaries and the lowland and forest region of the Azores. The upper maquis zone of Madeira corresponds to the less marked belt in the Canaries between the maquis zone and the upper steppes and to the "subalpine" maquis zone of the Azores. There is no "alpine" zone in Madeira. The lower maquis zone is the most peculiar. It is characterized by maquis which differs from Mediterranean maquis by the broader leaves of the shrubs and the lack of bulbous plants. It can be designated Macronesian maquis. The shrubs of the upper maquis are narrower leaved.

Here too there are few geophytes. It is natural to group the maquis of the Azores, Madeira, and Canaries as a subdivision of the Mediterranean Region or as a separate region (the Macronesian) related to this and characterized by the almost complete absence of geophytes correlated with the long vegetation period.

The flora of Madeira is definitely Mediterranean with the addition of a Macronesian element. This latter consists of 34% of the species, the Mediterranean of 24%, and the remainder of the species occur in the Mediterranean but also have a wider range. A tropical element can be traced only in three ferns. 18 species represent the African steppe flora. The cultivated plants and the ruderal flora (of 300 to 400 species) are predominantly Mediterranean. The native flora must have migrated over the sea for the islands have been separated from the mainland since Miocene times. Birds, winds, and sea-currents provided uncertain transport with long intervals of time. The Mediterranean species came from the Iberian Peninsula and Morocco. The Macronesian element is a South European relict flora, composed partly of species originally European and partly such as have newly evolved. There is nothing in favour of a direct migration from Tropical Africa or America. Many interesting details in support of these conclusions are given by VAHL.

The Azores are much farther west in the Atlantic, and thus much nearer to America, than are Madeira and the Canaries. An up-to-date summary, with references, of the history of the botanical exploration of the islands is given by PALHINHA (1947). The flora has been worked out by WATSON (*in* GODMAN, 1870) and by TRELEASE (1897), with additions by DRUCE (1911) and TUTIN and WARBURG (1932). For a general discussion WATSON's account and the two works by GUPPY (1914, 1917) are the most important. GUPPY recognizes four zones: (*1*) the Faya zone, from the coast up to 2000 and 2500 feet, with *Myrica faya*, *Erica azorica*, and *Laurus canariensis* as the most abundant trees; (*2*) the Juniper and *Daphne* zone, 2000 to 4500 feet for the wood proper, and 4500 to 5500 feet for the scrub; (*3*) the *Calluna*, *Menziesia*, and *Thymus* zone, 5500 feet to the summit; (*4*) the upland moors, 2000 to 4000 feet. Whilst the plants of the upland moors are in the mass European species that do not occur either in Madeira or in the Canaries, most of the characteristic trees of the woods are non-European and either exist in the other two groups or are represented there by closely allied species. It is suggested that the European element in the woods was mainly derived by way of the Atlas Mountains. The woods of the Azores, as regards their component trees and shrubs, are to be compared with the "laurel belt" that forms the middle zone of vegetation on the slopes of Teneriffe. The lower African zone and the higher region of pines as displayed on that mountain are not to be found in the Azores, their absence being due to lack of the requisite climatic conditions in the former and to want of suitable soil conditions on the high levels of pine in the latter. The "marked endemism of the Canarian and Madeiran flora is but slightly displayed in that of the Azores. The revolutions in plant-life which are suggested by the presence in the other two groups of representatives of genera now exclusively American cannot be predicated for the Azores. On the contrary the Azorean plants supply us with a story of to-day for the upland moor and of yesterday for the mountain wood."

In his later paper GUPPY concludes that "Whilst with the Canaries, and to a less extent with Madeira, there were early invasions of African, American, and Asiatic plants, they made but little mark on the Azores. The Azorean flora appears not to have shared in such revolutionary changes, and its history begins with the later invasion in Upper Tertiary times from Europe and the Mediterranean region of plants that in their descendants now give character to the Laurel woods of all three Macronesian groups. The parent stocks have since been driven from their European home, and the Laurel woods of Macronesia are all that remain of a period when trees now characteristic of Asia and America formed the forests of our continent (HOOKER). The last invasion of Macronesia, which has extended down to recent times, is indicated by those plants that still exist in South Europe and North Africa. It is represented by the minority of the plants of the woods, and particularly in the Azores by the plants of the moors. The Azorean flora has shared only in the later revolutionary changes of the plant world in this region and so the means of dispersal figure prominently in any inquiry into its history." GUPPY stresses the importance of birds as transporting agents of fruits and seeds and is critical of that of wind on the basis of experiments by LLOYD PRAEGER (Proc. Roy. Irish Acad. 31: 63–80, 1911).

Cape Verde Islands:— The most complete modern account of the plant life of this archipelago is that by CHEVALIER (1935). Other references are BÉGUINOT (1917) and CHEVALIER (1938).

CHEVALIER has an interesting discussion on the Macronesian Islands as a whole and a few of his general conclusions should first be noted. He accepts the view that all the islands emerged isolated from the ocean towards the middle of the Tertiary and have (except perhaps the Canaries) never been joined to any continent. They received no mammals and a great number of the plants of the Mediterranean Region are absent from them. They have no connection with America. The flora, even including that of the Azores, is essentially Mediterranean. Later the Canaries and, still more, the Cape Verde Islands received additions from the arid African regions (Saharan and Sahelo-Sudanese zones) but their primitive flora has an exclusive Mediterranean and, in a certain measure, Atlantic North European basis. The islands have served as refuges for various plants which populated Europe in the Eocene: *Lauraceae, Myrica, Faja, Phoenix*, and *Sapotaceae*.

The indigenous flora of the Cape Verde Islands is poor, consisting of not more than 300 phanerogamic species of which 91 are endemic. More than four fifths of the latter are closely related to Mediterranean species. The families with most endemics are *Compositae* (20), *Leguminosae* (12), *Gramineae* (8), and *Cruciferae* (6). Some species have probably disappeared with the devastation of the primitive vegetation, especially through the destructive action of goats. Two species, *Cyphia stheno* and *Habenaria petromedusa* have not been found again for 150 years. CHEVALIER has no doubt that sea currents, wind, and birds account for transport across the sea to the Cape Verde Islands and gives important evidence from modern researches in support of this view. There are three elements distinguishable in the flora of the Cape Verde Islands: (*1*) The Mediterranean–Atlantic Island; (*2*) the Tropical African; (*3*) the human introductions. There is a rather ill-defined altitudinal zonation, the lowest zone being arid or sub-arid. Interesting accounts of the vegetation are given by CHEVALIER.

St. Helena:— HOOKER himself twice visited St. Helena, in 1840 and 1843 on his way out to and on the return journey from his Antarctic voyage. The best account, with very fine plates, of the flora of the island is that by MELLISS (1875). A recent account of the history of the botanical exploration of the island, with some new information regarding the survival of some of the endemics is that by TURRILL (1949) from which the following is extracted:

"The known indigenous species of vascular plants of St. Helena are 66 in number. The following tabulation gives more details.

Indigenous species of Spermatophyta	39
Indigenous species of Vascular Cryptogams	27
Endemic species of Spermatophyta	38
Endemic species of Vascular Cryptogams	12
Indigenous genera of Spermatophyta	28
Indigenous genera of Vascular Cryptogams	13
Endemic genera of Spermatophyta	5
Endemic genera of Vascular Cryptogams	0"

"The closest affinities of the flora are considered to be with the flora of Africa, and particularly of southern Africa. Of the 28 indigenous genera of Spermatophyta, 22 are represented in the flora of South

Africa. The 5 endemic genera have the following taxonomic affinities:

Nesiota (*Rhamnaceae*), near to *Phylica*, a genus of S. Africa and the islands of the Atlantic and Indian Oceans.

Commidendron, Melanodendron, and *Petrobium* (*Compositae*), placed near to South American genera.

Mellisia (*Solanaceae*) related to the Central and South American genus *Saracha*."

The St. Helena species of the fern genus *Microstaphyla* appears to be another link between the St. Helena and South American floras. The "affinities" of the arborescent *Compositae* require re-investigating by modern criteria.

Ascension Island has the poorest flora of all the islands considered by HOOKER in his lecture. The best account of the flora is that given by HEMSLEY (1885). Other references are to BROWN (1906), STAPF (1917), and WATSON (1891). HEMSLEY says that there were only two phanerogams known (in 1885) that are definitely indigenous, both endemic: *Hedyotis adscensionis* (*Sherardia fruticosa*) and *Euphorbia origanoides*. He suggests they are perhaps the remnant of a flora that is extinct, save for these two species, so far as flowering plants are concerned. He enumerates 11 ferns of which two, *Nephrodium adscensionis* and *Gymnogramme adscensionis* are endemic. The conclusion is that "the indigenous vegetation is so exceedingly meagre that it offers nothing for consideration from a phyto-geographical standpoint, except a possible relation to the flora of St. Helena." Of very great interest was the advent of the grass *Enneapogon mollis* in Ascension. This appeared suddenly in great abundance on the lower part of the island. The species was known to occur on the coast of Angola and in Great Namaqualand and has a wide range elsewhere in Tropical Africa and Madagascar and has been collected in the Punjab. It appears to be an annual and is a typical desert grass that springs up after rains. To Ascension "it was probably brought entangled in the feathers of the Sooty Tern, which nest on this Plain in millions about every eight months, and after rearing their young all depart again, either to the West Coast of Africa or elsewhere." *Enneapogon mollis* set seed in Ascension where, however, five consecutive years may pass with scarcely any heavy rain on the lower levels.

* * *

We must now briefly discuss the theoretical part of HOOKER's lecture and a number of problems connected with island floras which have assumed different aspects since 1866. The origin of the floras on the islands dealt with, and on islands generally, is HOOKER's major concern. He rightly points out that the species now found in an island or their ancestors either came over the sea or over a land connection that has now ceased. His very fair analysis of the pros and cons deserves careful study, hence the full extracts given above. He decides, on the whole, in favour of over-sea transport. Eighty years later the problem remains not radically changed and no proposed solution has been generally accepted. The geological and other objections to filling in the great oceans with lost continents (as "Atlantis") remain. The islands with which HOOKER deals in his lecture are entirely or almost entirely of volcanic origin and there is apparently no clear evidence that they once formed parts of any continental land mass— of Africa, America, "Atlantis", or "Wegeneria". The biological problems here are very different from those connected with islands on the continental shelf or in an inland sea like the Mediterranean, where geological evidence of former transgressions and regressions of the sea has to be accepted. The hypothesis of continental drift is essentially a matter for geologists and geophysicists. Amongst such are ardent protagonists and just as critical antagonists. The biologist would be wise not to place too much reliance on explanations that oscillate rather than drift. That some facts of distributions fit in with WEGENER's views is true, but others do not. It is also often forgotten that the shifting of the poles, as well as continental displacement, was an intrinsic part of WEGENER's general picture of the history of the earth's surface.

On the other hand, there is little direct and reliable evidence for long distant over-sea transport by ocean currents (except of some halophytes), winds, and birds of disseminules of Spermatophyta. There is some and the need is for more. Botanists would welcome also a re-consideration of the theoretical possibilities of lift and dispersal of small seeds, plumed seeds and fruits, etc. by winds, including cyclones, whirlwinds, and hurricanes. Much more is now known in the modern study of aeronautics about air currents in the upper atmosphere than when conclusions were first drawn regarding the rate of fall of seeds and spores in experiments in a still atmosphere. It may well be that new discoveries allow the possibility of long distant transport within reasonable periods of time of many disseminules by upward currents redressing loss in height by gravitation. Most important would be the catching of seeds in the atmosphere over oceans and it is much to be hoped that those engaged in aerobiological work will not restrict their researches entirely to spores, bacteria, viruses, and insects.

At present, HOOKER's conclusions seem the most reasonable that have been suggested—that the island groups considered in his lecture have received their plants from across the seas. It must not be forgotten that such transport is a matter of one "accident" happening about once a century or, for some islands, once in a thousand years. The "accident" is really the co-incidence of two "accidents" with often a probability that the second will happen if the first does—successful migration of a disseminule must be followed by establishment (including reproduction). The relative poverty of the flora of most oceanic islands, the peculiarities of structure and behaviour of the components, the mixture of endemics and non-endemics with often high proportion of the former are understandable, and the occurrence of elements of different (evolutionary or geological) age are explicable best on the "accident" theory.

Chapter X

SUMMARY AND CONCLUSIONS

The Hookerian publications considered above range in time from 1844 to 1904, and thus extend over a period of 60 years. It is not without significance that this practically covers the Victorian period of British history—a period of industrial and scientific progress, of peace or only strictly localized wars, and of relative wealth. Publication of such tomes as those of the Flora Antarctica could then be undertaken without undue strain on available financial resources. Transport was, of course, very much slower and less comfortable but it was much cheaper and visas and passports were only needed exceptionally if at all. If J. D. HOOKER was fortunate in his period he was still more fortunate, in his start as a botanist, in being the son of WILLIAM JACKSON HOOKER. Not only did this result in very early training in various branches of botany but to finding a home at Kew from 1843 to 1885, with continuity of residence broken only by his journeys abroad and three or four changes of houses within the village. Without his father's herbarium and library, and the constantly increasing collections that poured into Kew, HOOKER could scarcely have worked out the taxonomy of his own great collections. It was mainly on these that most, though not all, of his important phytogeographical studies were based. We have then to keep constantly in mind the double environment of HOOKER's phytogeographical researches—his own field work and the intensive studies in herbarium and library at Kew.

J. D. HOOKER was, it has already been noted, a botanist with very wide interests but was first and foremost a taxonomist. It is essential that we should consider briefly the development of botany as a whole and of taxonomy in particular during the period with which we are concerned. Without doubt the outstanding event in the biological world was the publication of CHARLES DARWIN's "Origin of Species" in 1859 (preceded by the communications of DARWIN and WALLACE to the Linnean Society of London, 1 July 1858, published in Journ. Linn. Soc. Zool. 3, No. 9:45–62, 20 August 1858). Apart from this, and limiting our outline review to botany, prominence must be given to the increasing influence of researches from laboratories in Germany. Gross morphology, anatomy, ontogeny, and physiology were greatly advanced by new methods and by stimulating theories. The researches

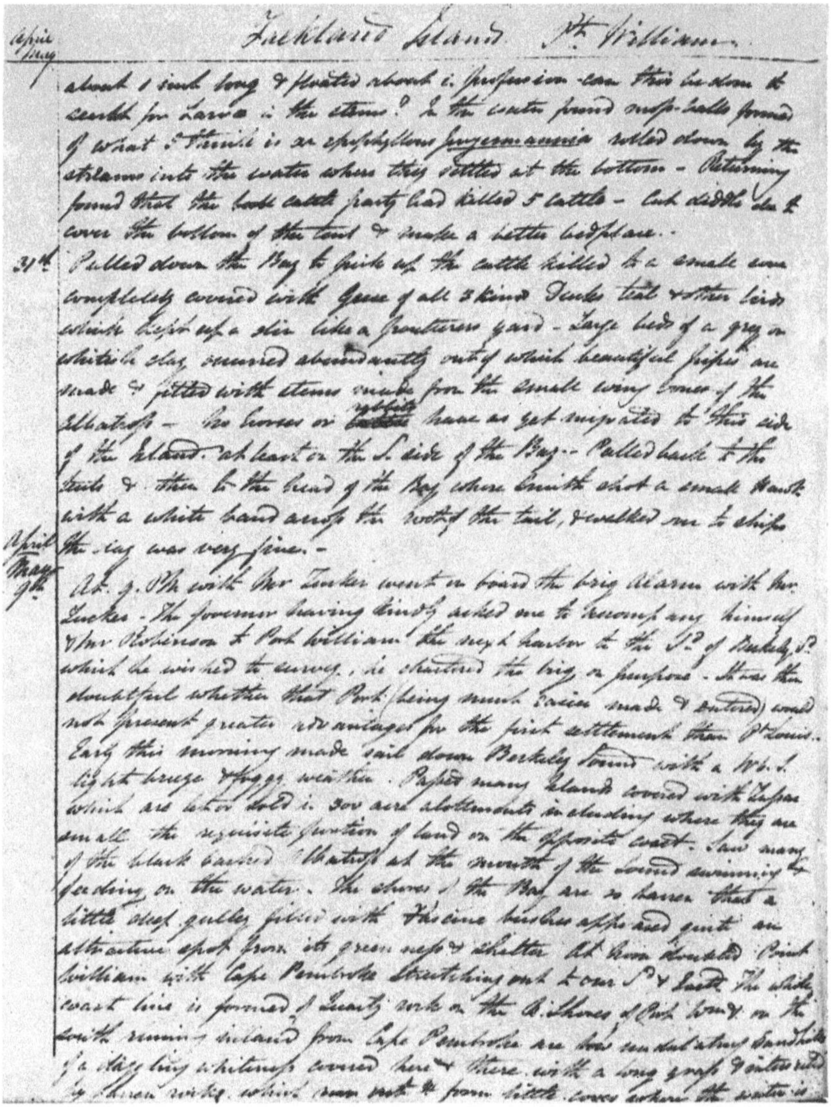
April
May
Falkland Island Pt. William

about 1 inch long & floated about in profusion — can their be done to search for Larvae in the stems. In the water found moss-balls formed of what I think is an epiphyllous Jungermannia rolled down by the streams into the water where they settled at the bottom — Returning found that the bullock catch party had killed 5 cattle — took skins to cover the bottom of the tent & make a better bedplace.

31st. Pulled down the Bay to pick up the cattle killed to a small cove completely covered with Geese of all 3 kinds Ducks teal & other birds which kept up a din like a poulterers yard — Large bed of a grey or whitish clay occurred abundantly out of which beautiful pipes are made & fitted with stems made from the small wing bones of the Albatross — No horses or rabbits have as yet migrated to this side of the Island at least on the S. side of the Bay — Pulled back to the tent & then to the head of the Bay where Smith shot a small Hawk with a white band across the root of the tail, & walked on to ships the day was very fine. —

April
May
9th. At 9. P.M. with Mr Tucker went on board the brig Alarm with Mr. Tucker — The former having kindly asked me to accompany himself & Mr Robinson to Port William the next harbor to the S. of Berkely S. which he wished to survey, he chartered the brig on purpose — It was then doubtful whether that Port (being much easier made & entered) would not present greater advantages for the first settlement than Pt Louis. Early this morning made sail down Berkely Sound with a W.S. light breeze & foggy weather. Passed many Islands covered with Tussac which are later sold in 300 acre allotments including where they are small the requisite portion of land on the opposite coast. Saw many of the black backed Albatross at the mouth of the Sound swimming & feeding on the water. The shores of the Bay are so barren that a little deep gully filled with Veronica bushes appeared quite an attractive spot from its green ness & shelter. At noon doubled Point William with Cape Pembroke stretching out to our S. & East. The whole coast line is formed of Quartz rock on the N. shores of Port Wm & on the south running inland from Cape Pembroke are low undulating Sandhills of a dazzling whiteness covered here & there with a long grass & intersected by there rocks which run out & form little coves where the water is

Plate 20. — From the Manuscript at Kew of J. D. Hooker's Antarctic Journal. A page dealing with his visit to the Falkland Islands.

Plate 21. — J. D. HOOKER IN HIS STUDY AT SUNNINGDALE, 1886, a few months after
his retirement as Director of the Royal Botanic Gardens, Kew.
From a sketch in the portrait collection at Kew.

of HOFMEISTER (published in 1851 and 1862) showed fundamental unities in the structure of all archegoniate plants and enabled botanists the more easily to accept the theory of evolution. In particular, there were tremendous advances in the understanding of the life histories of cryptogams and the foundations were laid of modern algology, mycology, lichenology, bryology, and pteridology. Theories to explain on an evolutionary basis the discovered facts of the alternation of generations extended views on metamorphoses, and interpretations of floral morphology were largely developed in Germany and Britain. In France and in England palaeobotany was greatly advanced by improvements in microscopic technique. Anatomical investigations of both phanerogams and cryptogams were undertaken in numerous laboratories and the results were published in many new journals. Cytology was born in its modern guise in this period but the fundamental importance of the nucleus was scarcely realized till the linkage of cytology and genetics after 1900. In physiology, the mechanisms of water absorption, photosynthesis, nitrogen assimilation, enzyme actions, and respiration were studied by experimental methods as were also problems of growth and plant movements. Many of the developments, whether in morphology or physiology, were stimulated if not initiated by evolutionary ideas after 1859 but apart from this very important connecting link one has the general impression of intense but rather isolated activities. "Schools" of research were localized around some master and it was only towards the end of the century that there began that synthesis which, fortunately, has so continued and increased that botany tends to become a more and more unified subject. The present writer believes that this historical fact needs emphasizing more than is usually done. There is not, and one hopes there never will be, a divorce in the study of plant life between morphology and physiology. Generally, in schools and universities, at international congresses, and in the practical application of botany in agriculture, forestry, and horticulture, plant physiology is included as an integral part of botany. During the present century the growth of ecology has tended to weld together many aspects of structure and function and it may be suggested with some confidence that the natural tendency will be for cytogenetics to assimilate, or be assimilated by physiology, or still further to combine morphology and physiology. Three names should be specially mentioned in this connection since their work falls into the latter part of the Hookerian period. E. WARMING published in 1895 a Danish work with the title "Plantesamfund". This was translated into German, and later into English with the title "Oecology of Plants" (Oxford, 1905). The Preface to A. F. W. SCHIMPER's comprehensive book on plant geography is dated "end of July 1898". The English translation has the title "Plant Geography upon a physiological basis" (Oxford, 1903). These two works in many ways laid the foundations for both ecology and plant geography in their modern forms. They remain classics that are still constantly referred to and quoted. On very different lines, was the pioneer work of DE VRIES who was largely responsible for introducing experimental methods into the study of "species problems". His "Die Mutationtheorie" (Leipzig, vol. 1 : 1901; vol. 2 : 1903) and "Species and Varieties, their origin by mutation" (London, 1905) opened up wide new possibilities. MENDEL's work, lost in obscurity till 1900, in the early part of this century came to overshadow that of DE VRIES but it may be that a greater place in the history of botany will be given to DE VRIES than has yet been accorded him.

We find in the second half of the nineteenth century that botany expanded especially by the increasing use of the microscope and by adoption of experimental methods. J. D. HOOKER was not a specialist in either of these lines of research. Let us turn then to consider advances in plant taxonomy. These were very considerable but were along orthodox lines which are continuous from RAY, LINNAEUS, ADANSON, B. and A. L. DE JUSSIEU, LAMARCK, A. P. DE CANDOLLE, and others. Within our own period special mention must be made of the great clarification in the taxonomy and description of the groups of cryptogams. No better idea of the advances made can be quickly obtained than by reading the portions dealing with the cryptogams in LINDLEY's "The Vegetable Kingdom" (London, 1846, 1847, 1853) and comparing them with the corresponding portions in any good modern text book of botany. For further details and references the reader may care to consult TURRILL, "Taxonomy and Phylogeny" in Bot. Rev. 8: 247–70, 473–532, 655–707 (1942). Within the phanerogams by far the greatest advance was the publication of the Genera Plantarum of G. BENTHAM and J. D. HOOKER (1862–83; see Journ. Linn. Soc. Bot. 20: 304–8, 1883 for the joint and separate work of the two authors). It is not the BENTHAM and HOOKER system as given in the Genera Plantarum which makes this work an indispensable classic to the student of phanerogamic taxonomy. The system is a modification of that of DE CANDOLLE. The outstanding feature of the Genera Plantarum is the excellence of the descriptions of families and genera. These descriptions taken as a whole, have never been surpassed. They have often been copied or translated or used in compilations (not always with acknowledgment). Prepared on a unified plan they are easily comparable. The descriptions in the original edition of ENGLER, Die Pflanzenfamilien are, in general, scrappy compared with those in the Genera Plantarum but the German work is more extensive since it includes the cryptogams and has the advantage of black and white text illustrations. The Englerian system for the phanerogams is different from that of BEN-THAM and HOOKER. An outline appeared in 1897 and from this and later publications it is clear that ENGLER did not consider his system to be more than partially phylogenetic at most but to consist of series of progressions including phylogenetic facts or suppositions in a patch-like manner. While the system of BENTHAM and HOOKER is a natural one in that it is based upon a consideration of all then known relevant characters, BENTHAM was not at its initiation a convert to evolutionary theory—one may even doubt if he ever was a whole-hearted or, at least, very enthusiastic convert (*see* B. DAYDON JACKSON, "GEORGE BENTHAM", pp. 184, 195, London, 1906 and J. D. HOOKER *in* Ann. Bot. 48 : 1–22, 1898). No claim must therefore been made that the BENTHAM and HOOKER system is based on phylogeny or on evolution-ary theories.

Apart from the systems of classification mentioned our period saw the publication of great floras. These followed or approximately coin-cided with the botanical exploration of large areas of the earth's surface. The size of the areas included are particularly remarkable since the general tendency in more recent times has been to produce floras of smaller areas—"regional floras" as they are sometimes rather unfor-tunately termed. Noteworthy examples of the extensive floras are MARTIUS, Flora Brasiliensis (1840–76), BENTHAM, Flora Australiensis (1863–78), J. D. HOOKER, Flora of British India (1875–97), Flora Capensis (1859–1925, Suppl. 1933), and The Flora of Tropical Africa (1868—unfinished). Monographs of families and genera were also prepared in many countries and published in various periodicals, notably in the Transactions and Journal of the Linnean Society of London, ENGLER's Botanische Jahrbücher, the Journal of Botany, and Annales des Sciences Naturelles. In addition attention should be called to the Botanical Magazine founded by WILLIAM CURTIS in 1787 and still being published as the oldest of botanical and one of the longest lived of all scientific periodicals. It was edited by J. D. HOOKER from 1865 to 1904 and during these four decades most of the articles were written by him. The Icones Plantarum, founded by W. J. HOOKER in 1837, was also edited by J. D. HOOKER from 1867 to 1890 and, again, many of the articles were from his pen during this period.

When one recalls that in addition to his researches and editorial duties, J. D. HOOKER from 1865 to 1885 was Director of the Royal Botanic Gardens, Kew, with all the administrative work and all the associated activities such a post involved, one can only say "there were giants in those days."

J. D. HOOKER, with his training and outlook as a taxonomist had

15

certain advantages as a plant geographer. He knew, what is now sometimes forgotten, that taxonomy must be the basis for the study of plant
distribution, perhaps the more obviously if such research involves
whole floras. The taxonomic units (or taxa, to use LAM's useful modern term) must be precisely defined before their ranges can be
mapped or tabulated. Again, as a taxonomist HOOKER gave due weight,
as evidence for any conclusions reached, to all taxa. There is always a
danger of missing the truth by subjective bias if the ranges of a few
species (or genera, or even families) be picked out for special consideration or their ranges emphasized as specially supporting this or that
hypothesis, unless, at the same time an analysis of the sum total flora
be provided. All evidence, positive and negative, must be presented.
J. D. HOOKER always did this fairly and lucidly to the limits of his
available data and knowledge. That modern explanations of distribution tend to include ecological and cytogenetical viewpoints which,
as we have seen, it could not have been historically possible for HOOKER
to use, in no way detracts from the importance of his fundamental
work. One can only hope that the new methods will give more complete solutions to problems which faced HOOKER, whether to confirm
or modify his tentative conclusions, and some of which still remain
acute.

Of these problems the most general is the controversy between
"extensionists" and "migrationists". The former claim that long
distant transport of disseminules (apart from those carried by man) is
not possible, or has not been proved, and that the only explanation
of wide discontinuities is that there was land connection when the
ranges were being established. The latter argue that transport by
ocean-currents, birds, and winds is sufficient to explain discontinuities
even over wide stretches of ocean. There are variations in the hypotheses put forward by both sides. For example, many extensionists
accept wide transport by ocean-currents, especially of some tropical
strand plants. Again, the theory of continental drift is not strictly
speaking an extensionist hypothesis but biologically can be classified
under that heading. DARWIN, as is well known, was a migrationist
(*see* "Origin of Species", ed. 6, chapters 12 and 13). The modern tendency has been mainly against his views in this respect, at least amongst
botanists. Nevertheless, the questions remain (*1*) do winds and birds
sometimes, even if only very occasionally, transport seeds (or other
disseminules) across the seas and oceans and other barriers; (2) if so,
do such disseminules sometimes arrive in habitats where they can become established; (3) if so, will such transport account for the known
discontinuities of range; (4) alternatively, will accepted transport by
ocean currents and by man (including prehistoric man) account for

such ranges; (5) if not, and if barrierless continuities of suitable habitats be essential for spread does connection by land bridges that have since disappeared, or continental drift, or some other hypothesis give the best explanation. J. D. HOOKER found either the extensionist or migrationist hypothesis partly possible but neither proved and there were grave objections to both. The hypothesis of continental drift was not known to him, or to DARWIN. With regard to this it is very unfortunate for biologists that geologists and geophysicists are not in agreement as to the probability, or even possibility, of former continental fusion with later breakaway and drift. Whether the new methods of examination of ocean floors by echo-sounding and by core lifting will prove or disprove WEGENER's views remains to be known but the preliminary reports of the recent *Albatross* expedition hold out some hope that they may. It has also to be noted that the continental drift hypothesis, as put forward by WEGENER and KÖPPEN is very closely linked with the hypothesis of moving poles. The evidence against this latter appears to be strong, and possibly conclusive. The present writer has given insufficient attention to this problem to be dogmatic in answering it but he would call attention to the following points: (1) Total floras should be studied. Sound results will not be obtained merely by considering examples favouring this or that hypothesis and ignoring those that are opposed to it or are neuter to it. (2) The evidence of faunas must be considered with that of floras. (3) The taxonomic units must be studied taxonomically and their ranges must not be merely compiled from books. It is very probable that taxonomic revaluation will throw further light on many phytogeographical problems. (4) Ecological and cytogenetical researches with a phytogeographical background must be extended. (5) The attention of geologists, ornithologists, meteorologists, archaeologists, historians, and others must be directed to biological problems and their help solicited. (6) The possibilities of extension of ranges by early migrations of man must not be forgotten. (7) The problems of wide discontinuities within what are now continental areas have not received sufficient attention compared with those of transoceanic discontinuities.

Taking now the areas studied by J. D. HOOKER, in the sequence in which they are dealt with above, we must outline his main contributions to their phytogeography. His study of the distribution of Arctic plants was particularly important in that it showed the essential unity of the problems of the Arctic flora in Europe, Asia, and America and yet pointed out the secondary differences in the northern parts of these continents. For Syria and Palestine, HOOKER summarized very clearly what was then known regarding the flora and its phytogeography. He had insufficient data to appreciate the detailed problems of plant distribution in the Nearer East, in particular those connected with the desert and semi-desert floras. J. D. HOOKER's work on the Indian flora and vegetation, in so far as we are concerned with it here, is divisible into two parts: (a) his own field explorations and (b) his summaries of the phytogeography of the subcontinent and associated areas. With regard to his field work, it is of outstanding importance in consisting of pioneer researches in the hitherto botanically unknown eastern Nepal and Sikkim. The originality and wide interest of his observations are exemplified especially from the quotations given above from his letters and the "Himalayan Journals." The accuracy of his observations has been proved by later travellers, both botanists and non-botanists. With regard to his wide general summaries it can only be said that they have not yet been replaced as a whole by anything better. That they now need revision is certain; that they will receive it adequately in the near future is doubtful; that they will ever be entirely superseded is improbable. The present position with regard to these wide general summaries illustrates well the modern dilemma. It would be extremely valuable if they could be brought up-to-date, for sound generalizations are essential for the full progress of phytogeography. On the other hand the subject is so vast and so much detailed research obviously remains to be done that those best qualified to untertake the task (and they are few) hesitate to do so.

In Morocco, HOOKER was partnered by JOHN BALL and GEORGE MAW and the former was largely responsible for the detailed studies resulting from the expedition. Again, the field investigations were of a pioneer nature in areas previously unexplored botanically. Fortunately, what is now French North Africa has been, and is being, well worked by French botanists and the taxonomy and ranges of species and their ecology is becoming adequately known and is being incorporated in adequate phytogeographical accounts. In Tropical African plant geography HOOKER drew attention to several problems which are still of great interest. These include the floristic links between West and East Africa, the "temperate" flora of tropical mountain peaks and its origin, and the geographically wide affinities of many African genera.

HOOKER's personal knowledge of the North American flora was restricted to the one tour with ASA GRAY. Their joint conclusions, however, were important in that they were published in such a form as to give a general picture of the phytogeographical composition and history of the flora and thus to serve as a stimulus to and a basis for further regional researches. There is internal evidence that HOOKER was particularly responsible for comparisons of the North American flora with what he had termed the "Scandinavian." The wealth of researches, by old and new methods, on the flora and vegetation of North America since the days of ASA GRAY and J. D. HOOKER could not be summarized here. The result has been that the continent is now better known phytogeographically than any other except the much smaller Europe (excluding Russia).

The papers by HOOKER on the flora of the Galapagos Islands drew attention to the floristic peculiarities of the group and raised acutely the question of migration over stretches of ocean versus previous land connections. He, cautiously, accepts ocean currents, winds, birds, and man as transporting agents. His analysis of the flora is remarkably clear, except for his peculiar use of the phrase "West Indian" to include Panama and the adjacent lowlands of the continent. On the whole, recent researches have supported HOOKER's views as to the composition and origin of the flora of the Galapagos Archipelago.

HOOKER's botanical work on lands of the southern hemisphere is pre-eminent for the taxonomic researches involved and for the phytogeographical conclusions he draws from their analysis. The distances covered by his own "Antarctic" travels were very great and he made very large collections of dried plants. His journal and field notes are at Kew but have never been published as such though much of the information contained in them appears in the works referred to in our chapter VIII. In the introductory essays to his floras of New Zealand and Tasmania HOOKER dealt at some length with the phytogeography of these islands and his conclusions formed the foundation for all future studies on the constitution and origins of their floras, just as his taxonomic studies are basic to their systematic investigation. The same remarks hold also for the various smaller southern islands and island groups he visited or whose floras he studied in detail at Kew. Many of these last were previously unknown or scarcely known botanically. The great extent of land and sea embraced in the volumes of the "Flora Antarctica" raised many questions. The connections between the floras of New Zealand, South America, and, to a less degree, South Africa, their past histories, their origins, their migrations, and similar problems naturally intrigued HOOKER's alert and ever questioning mind. His analyses of the then known facts, many of which were first discovered by himself, were followed by the syntheses of his "introductory essays" and other papers. While his conclusions may now need many modifications none of the major ones is entirely "out of court" and some have been widely or even completely accepted.

Attention must here be called to the interesting, and to us nowadays in some ways peculiar, discussion in the Flora Novae-Zelandiae on species. It is evident from this that HOOKER, in common with some other thoughtful biologists, was not content with the view that species were immutable. He was groping for a solution of the species problems which confronted him both in his taxonomic and his phytogeographical researches. In the Flora Tasmaniae, HOOKER accepts the theory of evolution as based on natural selection by CHARLES DARWIN. His comments on "The general phenomena of variation in the vegetable kingdom" are of extraordinary interest to the student of the history of botany and of the theory of evolution, especially when compared with the corresponding paragraphs (written about seven years earlier) in the Flora Novae-Zelandiae. Many of the phrases we familiarly associate with Darwinism are absent, though the underlying ideas are present. It is evident, too, from considering both the above essays, that HOOKER had thought matters out for himself and did not merely follow in DARWIN's footsteps. This is clear from the examples he quotes. These are not only botanical but often what one might almost call personal. It is, however, clear that HOOKER believed that Darwinism was a most important key to many of the problems that had beset him, since it allowed for the evolution not only of species but also of floras, the latter not merely by migration but also by *in situ* development and by development along migrational routes. We like to think that we know more about variation than did either DARWIN or HOOKER but it is only fair to draw attention to some of HOOKER's statements at this early date in the history of the Darwinian theories. Thus he notes that "the variable cultivated species present us with the most important phenomena for investigating the laws of mutability and permanence". Well, MENDEL used garden peas and thereby established modern genetics. "Variation is effected by graduated changes" is a statement we should now say is too sweeping since some mutations are "large" and some forms of polyploidy result in the establishment of what may become species at one step only. Yet it is somewhat significant that recent authors tend to emphasize small mutations as most important in evolution and the multifactorial or cumulative hypothesis, in part under the newest name of polygenic theory, appears to be finding increasing favour. HOOKER notes that some species may have had their origin in hybridization but, rightly, emphasizes the distinction between the results of hybridization and what would now be called variation by mutation. Even more interesting, for our main thesis in this book, is the evidence for evolution derivable from a study of distribution, as HOOKER expounds it. That species and other taxonomic units have defined ranges favours the hypothesis of similar kinds of plants having had one parent (or one pair of parents). Yet HOOKER recognizes the limitations of range studies and was quite aware of competition, ecesis, and plant succession—

though his published views long antedated these terms as they are now used by botanists. It is worth noting here that not only have new terms been introduced for ideas to be found in HOOKER's writings but that there has been a refinement of terms used by him. For example, readers will have noticed that he uses the term "parasites" to include what are now, rightly, distinguished as epiphytes. "Natural Orders" are now usually referred to as "Families".

In his miscellaneous papers on the geographical distribution of plants, HOOKER deals especially with the evidence this provides for the theory of evolution and with such wide or general problems as the origin of floras in the northern hemisphere, the peculiarities of the floras of oceanic islands, and the possibilities of transoceanic migration. HOOKER's dilemmas in these matters are still present with us. The conclusive evidence has not yet come to hand and the modern phytogeographer would appeal to his colleagues in other branches of science to provide sound and reliable data—to geologists for evidence of land extensions and connections where now there is sea; to palaeontologists for direct evidence of the evolution of floras; to ornithologists for the facts of birds on migration sometimes or never carrying disseminules; and to students of meteorology and aerodynamics for knowledge of the possibilities of transport of seeds etc. through the atmosphere. It may well be that neither the hypothesis of previous land connections nor that of transoceanic (and transcontinental) migration by itself gives the truth, the whole truth, and nothing but the truth of how the ranges of plants have come to be as now we find them. The hypotheses are not necessarily antagonistic and may be complementary.

J. D. HOOKER found great interest in the large problems of plant distribution. It is greatly to be hoped that students of plant life will retain and extend this interest. The unavoidable trend towards greater and greater specialization must be counterbalanced by recalling, at least from time to time, such wider issues as appear in phytogeography. The newer methods evolved for the solution of special problems can often be extended to aid in solving more general ones. The student concerned in detailed research will the better keep a balance between over-specialization and dispersal of energies over too wide a field by considering the phytogeographical researches and conclusions of JOSEPH DALTON HOOKER as outlined in this book.

REFERENCES

ABEYESUNDERE, L. A. J. and ROSAYRO, R. A. DE (1939): Draft of first descriptive check-list for Ceylon (Imp. For. Inst., Oxford, Check-Lists No. 4).

AGHARKAR, S. (1920): Die Verbreitungsmittel der Xerophyten... des nordwestlichen Indiens (Engl. Bot. Jahrb. Bd. 56, Beibl. 124: 1–42).

AIYAR, T. V. V. (1932): The Sholas of the Palghat Division—a study in the ecology and silviculture of the tropical rain-forests of the Western Ghats, Part 1 (Ind. For. 58: 417–32).

ALLAN, H. H. (1936): Indigene versus alien in the New Zealand plant world (Ecology 17: 187–93).

ALLAN, H. H. (1937): A consideration of the "Biological Spectra" of New Zealand (Journ. Ecol. 25: 116–52).

ALLAN, H. H. (1937): The origin and distribution of the naturalized plants of New Zealand (Proc. Linn. Soc. 150th Session, pp. 25–46).

ALLAN, H. H. (1940): A handbook of the Naturalized Flora of New Zealand (Dept. Sci. Indust. Research, N. Zeal., Bull. No. 83).

ALSTON, A. H. G. (1938): The Kandy Flora (Colombo).

ANDERSON, A. W. (1931): Some plant affinities of New Zealand and South America (Gard. Chron. 3rd ser., 93: 108–9, 137–8).

ANDERSON, T. (1863): On the flora of Behar and the mountain Parashnath, with a list of the species collected by Messrs. HOOKER, EDGEWORTH, THOMSON and ANDERSON (Journ. Asiat. Soc. Bengal 32: 189–218).

ANDREÁNSZKY, B. G. (1932): Adatok Eszak-Afrika. Flóréja Ismeretéhez (with full summary in German under the title "Beiträge zur Pflanzengeographie Nordafrikas") (Index Horti Bot. Univ. Budapest., pp. 61–147).

ANDREÁNSZKY, J. (1939): Der Baumwuchs und seine klimatischen Grenzen in Nordafrika (Engl. Bot. Jahrb. 70: 153–88).

ATKINSON, D. J. (1948): Forests and forestry in Burma (Journ. Roy. Soc. Arts 96: 478–91).

BAKER, J. R. (1938): Rain-forest in Ceylon (Kew Bull., pp. 9–16).

BAMBER, C. J. (1916): Plants of the Punjab (Lahore).

BANERJI, J. (1948): The Tamur Valley Expedition I (Indian Forester 74: 96–101).

BANNERMAN, D. A. (1922): The Canary Islands (London).

BARRINGTON, A. H. M. (1931): Forest soil and vegetation in the Hlaing Forest Circle, Burma (Burma Forest Bull. No. 25 [Ecology Series, No. 1]: 1–42).

BATTISCOMBE, E. (1926): A descriptive catalogue of some of the common trees and woody plants of Kenya Colony (London).

BAUR, G. (1891): On the origin of the Galapagos Islands (Amer. Nat. 25: 217–29, 307–26).

BAUR, G. (1897): New observations on the origin of the Galapagos Islands, with remarks on the geological age of the Pacific Ocean (Amer. Nat. 31: 661–80, 864–96; incomplete).

BEEBE, W. (1924): Galápagos World's End (London and New York).

BEETLE, A. A. (1943): Phytogeography of Patagonia (Bot. Rev. 9: 667–79).

BÉGUINOT, A. (1917): Contributo alla flora delle isole del Capo verde e notizie sulla sua affinita ed origine (Ann. del Mus. civico di storia naturale di Genova, ser. 3a, 8 (48): 8–73).

BERRY, E. W. (1927): Links with Asia before the mountains brought aridity to the Western United States (Scientific Monthly 25: 321–8).

BIRGER, S. (1906–7): Die Vegetation bei Port Stanley auf den Falklands-Inseln (Engl. Bot. Jahrb. 39: 275–305).

BISWAS, K. (1932): Glimpses of the vegetation of South Burma (Journ. Bombay Nat. Hist. Soc. 36: 285–7).

BISWAS, K. (1933): The distribution of wild conifers in the Indian Empire (Journ. Ind. Bot. Soc. 12: 24–47).

BISWAS, K. (1934): The vegetation of the neighbouring areas of the Raniganj and Gharia coalfields (Trans. Mining and Geol. Inst. India, 29: 61–3).

BISWAS, K. (1943): Systematic and taxonomic studies on the flora of India and Burma (Proc. 30th Ind. Sci. Congr. Part 2, pp. 101–52).

BLATTER, E., McCANN C., and SABNIS, T. S. (1929): The flora of the Indus Delta (Madras).

BOLLE, C. (1893): Botanische Rückblicke auf die Inseln Lanzarote und Fuertaventura (Engl. Bot. Jahrb. 16 : 224–61).

BOR, N. L. (1938a): The vegetation of the Nilgiris (Ind. Forester 64: 600–9).

BOR, N. L. (1938b): A sketch of the vegetation of the Aka Hills, Assam, A synecological study (Ind. For. Records 1 (N.S.), No. 4: 103–221).

BOR, N. L. (1938c): A list of the grasses of Assam (Ind. For. Records 1, No. 3 : 47–102).

BOR, N. L. (1942a): The relict vegetation of the Shillong Plateau—Assam (Ind. For. Records 3, No. 6: 152–95).

BOR, N. L. (1942b): Some remarks upon the geology and the flora of the Naga and Khasi Hills (150th Anniv. Vol. Roy. Bot. Gard. Calcutta, pp. 129–35).

BÖRGESEN, F. (1924): Contributions to the knowledge of the vegetation of the Canary Islands (Mém. Acad. Roy. Sci. et Lettres de Danemark, Copenhagen).

BOULOUMOY, L. (1930): Flore du Liban et de la Syrie (Paris, 2 vols: 1 with text, 1 with plates).

BOURDILLON, T. F. (1908): The Forest Trees of Travancore (Trivandrum).

BOYKO, H. (1945): On forest types of the semi-arid areas at lower latitudes (Palestine Journ. Bot. R. Ser., 5: 1–21).

BOYKO, H. (1947): A laurel forest in Palestine (Palestine Journ. Bot. R. Ser., 6: 1–13).

BOYSON, V. F. (1924): The Falkland Islands (Oxford).

BRANDIS, D. (1872): On the geographical distribution of forests in India (Journ. Bot. 10: 283–5).

BRANDIS, D. (1884?): Der Wald des äusseren Nordwestlichen Himalaya (Verhandl. natur. Ver. preuss. Rheinl. u. Westph. 42: 153–80).

BRANDIS, D. (1906): Indian Trees (London; ed. 2: 1911, London).

BRAUN-BLANQUET, J. et MAIRE, R. (1921): Etudes sur la végétation et la flore marocaines (Bull. Soc. Bot. France, 68: 1–224; also published in Mém. Soc. Sc. Nat. du Maroc, No. 8, part 1, 1924).

BRITTEN, J. (1904): List of Madeira plants (Journ. Bot. 42: 1–8, 39–46, 175–82, 197–200).

BROCKMANN-JEROSCH, H. (1928): Die südpolare Baumgrenze (Festschrift HANS SCHINZ, pp. 705–18, Zürich).

BROWN, R. N. (1906a): Contributions towards the botany of Ascension (Trans. and Proc. Bot. Soc. Edinb. 23: 199–201; also published in Report of the scientific results of the voyage of S. Y. Scotia, Edinburgh, 1912).

BROWN, R. N. R. (1906b): Antarctic Botany: its present state and future problems (Scott. Geogr. Mag. 22: 473–84).

BROWN, R. N. R. (1912): The problems of Antarctic plant life (Sci. Res. Scott. Nat. Antarctic Exped. 3: 3–20).

BROWN, R. N. R. (1923): Plant Life in the Antarctic (Discovery 4: 149–53).

BROWN, R. N. R. (1928): Antarctic Plant Life (Nature 122: 144; Abstract).

BURCHARD, O. (1929): Beiträge zur Ökologie und Biologie der Kanarenpflanzen (Bibl. Bot. Heft 98).

BURKILL, I. H. (1906): Notes from a journey to Nepal (Rec. Bot. Surv. India 4, No. 3).

BURKILL, I. H. (1907): Alpine notes from Sikkim (Kew Bull., pp. 92–4).

BURKILL, I. H. (1908): Some autumn observations in the Sikkim Himalaya (Journ. and Proc. Asiat. Soc. Bengal (N.S.) 4: 180–95).

BURKILL, I. H. (1916): A note on the Terai forests between the Gandak and the Teesta (Journ. and Proc. Asiat. Soc. Bengal (N.S.) 12: 267–72).

BURKILL, I. H. (1924): The botany of the Abor Expedition (Rec. Bot. Surv. India 10: 1–420).

BURKILL, I. H. and HOLTTUM, R. E. (1923): A botanical reconnaissance upon the main range of the Peninsula at Fraser Hill (The Gardens' Bull. Straits Settlement 3: 19–110).

BÜSGEN, M. (1910): Der Kameruner Küstenwald (Zeitschr. f. Forst- u. Jagdw. 42: 264–83).

CAIN, S. A. (1944): Foundations of Plant Geography (New York and London).

CERECEDA, J. D. (1916): La zone espagnole du Maroc (Anns. de Géogr. 25: 366–73).

CHAMPION, H. G. (1936): A preliminary survey of the forest types of India and Burma (Ind. For. Records 1 (N.S.), No. 1).

CHANEY, R. W. (1947): Tertiary centres and migration routes (Ecol. Mongr. 17: 139–48).

CHAPMAN, V. J. (1947): The application of aerial photography to ecology as exemplified by the natural vegetation of Ceylon (Ind. Forester 73: 287–314).

CHATTERJEE, D. (1940): Studies on the endemic flora of India and Burma (Journ. Roy. Asiat. Soc. Bengal, 5, 1939, No. 1: 19–67).

CHEESEMAN, T. F. (1888): On the flora of the Kermadec Islands (Trans. Proc. New Zeal. Inst. 20: 151–81).

CHEESEMAN, T. F. (1925): Manual of the New Zealand Flora, ed. 2 (Wellington, N.Z.).

CHENGAPA, B. S. (1944): The Andaman forests and their regeneration (Ind. Forester 70: 297–304, 339–51, 380–5, 421–30).

CHEVALIER, A. (1935): Les îles du Cap Vert. Flore de l'Archipel (Rev. Bot. Appl. 15: 733–1090).

CHEVALIER, A. (1938): L'extension du Sahara aux îles du Cap Vert. (Mém. Soc. Biogr. 6: 323–4).

CHILTON, C. (1909): The Subantarctic Islands of New Zealand (Wellington, N.Z., 2 vols.; Botany by CHEESEMAN, COCKAYNE, LAING, and others).

CHRIST, H. (1885): Vegetation und Flora der Canarischen Inseln (Engl. Bot. Jahrb. 6: 458–526).

CHRIST, H. (1886): Eine Frühlingsfahrt nach den Canarischen Inseln (Basel, Genf, und Lyon).

CHUBB, L. J. (1933): The origin of the Galapagos Islands (Bernice P. Bishop Museum Bull. 110, Honolulu).

CLARKE, C. B. (1876): Botanic notes from Darjeeling to Tonglo (Journ. Linn. Soc. Bot. 15: 116–59).

CLARKE, C. B. (1885): Botanic notes from Darjeeling to Tonglo and Sundukphoo (Journ. Linn. Soc. Bot. 21: 384–91).

CLARKE, C. B. (1886): Botanical observations made in a journey to the Naga Hills (between Assam and Muneypore), in a letter addressed to Sir J. D. HOOKER, K.C.S.I., F.R.S. by C. B. CLARKE, Esq. F.R.S., F.L.S. (Journ. Linn. Soc. Bot. 22: 128–36).

COCKAYNE, L. (1928): The vegetation of New Zealand (ENGLER und DRUDE, Die Vegetation der Erde, 14, ed. 2).

COCKAYNE, L. (1929): The vegetation and flora of Rainbow Mountain (Ann. Rep. Scenery-Preservation, Appendix C).

COCKAYNE, L. (1930): The flora and vegetation of New Zealand (separate from New Zealand Official Year-book).

COCKAYNE, L. and CHALDER, J. W. (1932): The present vegetation of Arthur's Pass (New Zealand) (Journ. Ecol. 20: 270–83).

COCKAYNE, L., SIMPSON, G., and THOMSON, J. S. (1932): Some New Zealand indigenous-weeds and indigenous-induced modified and mixed plant-communities (Journ. Linn. Soc. Bot. 49: 13–44).

COCKAYNE, L., SIMPSON, G., and THOMSON, J. S. (1932): The vegetation of South Island New-Zealand (KARSTEN und SCHENCK, Vegetationsbilder 22, Heft 5/6).

COCKAYNE, L. and SLEDGE, W. A. (1932): A study of the changes following the removal of subalpine forest in the vicinity of Arthur's Pass, Southern Alps, New Zealand (Journ. Linn. Soc. Bot. 49: 115–31).

COCKAYNE, L. and TEICHELMANN, E. (1930): The glacial scenic reserves in Westland (Ann. Rep. Scenery-Preservation, 11 pp.).

COCKAYNE, L. and TURNER, E. P. (1928): The Trees of New Zealand (Wellington).

COCKERELL, T. D. A. (1928): Aspects of the Madeira flora (Bot. Gaz. 85: 66–73).

COLLETT, H. (1902): Flora Simlensis (Calcutta and Simla).

Contribution a l'étude du peuplement des Isles Atlantiques (Soc. Biogéogr. Mem. 8, Paris, 1946).

COOKE, T. (1903–8): Flora of the Bombay Presidency (2 vols., London).

COOKSON, I. C. (1946): Pollen-analysis of lignite from the Kerguelen Archipelago (Nature 157: 658).

COOKSON, I. C. (1947): Plant microfossils from the lignites of the Kerguelen Archipelago (Rep. Brit., Austrl. and N. Zeal. Antarctic Res. Exped. 1929–31, 2, Adelaide).

COPLESTON, W. E. (1925): The Bombay Forests (Bombay).

CORNER, E. J. H. (1940): Wayside trees of Malaya (Singapore).

COTTON, A. D. (1930): A visit to Kilimanjaro (Kew Bull., pp. 97–121).

COVENTRY, B. O. (1929): Denudation of the Punjab Hills (Ind. For. Records, Silv. Series, 14: 1–30).

COWAN, J. J. M. (1928): The flora of the Chakaria sundarbans (Rec. Bot. Surv. India 11, No. 2).

COWAN, J. M. (1929): The forests of Kalimpong (Rec. Bot. Surv. India, 12, No. 1).

COWAN, J. M. (1942): The rhododendrons of India (150th Anniv. Vol. Roy. Bot. Gard. Calcutta, pp. 105–8).

CRANWELL, L. M. (1938): Fossil pollens (New Zeal. Journ. Sci. and Technol. 19: 628–45).

CROCKER, R. L. and WOOD, J. G. (1947): Some historical influences on the development of the South Australian vegetation communities and their bearing on concepts and classification in ecology (Trans. Roy. Soc. S. Austrl. 71: 91–136).

CUBITT, G. E. S. (1920, 1924): Forestry in the Malay Peninsula (Kuala Lumpur).

CUMBERLAND, K. B. (1941): A century's change: natural to cultural vegetation in New Zealand (Geogr. Rev. 31: 529–54).

DALZIEL, J. M. (1928): [Meeting of Brit. Ecol. Soc. at Kew] (Journ. Ecol. 16: 182).

DALZIEL, J. M. (1930): Cameroon Mountain (Scot. Geogr. Magn. 46: 257–74).

DAS, A. (1942): Floristics of Assam—a preliminary sketch (150*th* Anniv. Vol. Roy. Bot. Gard. Calcutta, pp. 137–47).

DAUBENMIRE, R. F. (1943): Vegetation zonation in the Rocky Mountains (Bot. Review 9: 325–93).

DAVIES, W. (1939): The grasslands of the Falkland Islands (Stanley, Falkland Is.).

DAVIES, W. (1940): The grasslands of the Argentine and Patagonia (Herb. Publ. Ser., Bull. 30, Aberystwyth).

DAVIS, C. (1941): Preliminary survey of the vegetation near New Harbour, southwest Tasmania (Papers and Proc. Roy. Soc. Tasmania, 1940, pp. 1–9).

DICE, L. R. (1943): The biotic provinces of North America (Ann Arbor, Univ. of Michigan Press).

DIELS, L. (1906): Die Pflanzenwelt von West-Australien südlich des Wendelkreises (ENGLER und DRUDE, Die Vegetation der Erde 7, Leipzig).

DONAT, A. (1931): Über Pflanzenverbreitung und Vereisung in Patagonien (Ber. deutsch. bot. Ges. 49: 403–13).

DONAT, A. (1932): Zur regionalen Gliederung der Vegetation Patagoniens (Ber. deutsch. bot. Ges. 50: 429–36).

DONAT, A. (1934): Zur Begrenzung des Magellanischen Florengebietes (Ber. deutsch. bot. Ges. 52: 131–42).

DONAT, A. (1935): Problemas fitogeográficos relativos a la Región Magallanica (Revista Argent. Agron. 2: 86–95).

DRUCE, G. C. (1911*a*): Madeira and the Azores (Chemist and Druggist, pp. 175–6).

DRUCE, G. C. (1911*b*): Plants of the Azores (Journ. Bot. 49: 23–8).

DRUDE, O. (1890): Handbuch der Pflanzengeographie (Stuttgart; French edition: Manuel de Géographie, Paris, 1897).

DUDGEON, W. (1920): A contribution to the ecology of the Upper Gangetic Plain (Journ. Ind. Bot. Soc. 1: 296–324).

DUDGEON, W. and KENOYER, L. A. (1925): The ecology of Tehri Garhwal: a contribution to the ecology of the Western Himalaya (Journ. Ind. Bot. Soc. 4: 233–85).

DU RIETZ, G. E. (1940): Problems of Bipolar Plant Distribution (Acta Phytogeogr. Suecica 13: 215–82).

DUSÉN, P. (1898): Über die Vegetation der feuerländischen Inselgruppe (Engl. Bot. Jahrb. 24: 179–96).

DUSÉN, P. (1903): The vegetation of Western Patagonia (Reports of the Princeton Univ. Expeds. to Patagonia, 1896–1899, 8: 1–33).

DUSÉN, P. (1905): Die Pflanzenvereine der Magellanswälder (Wissensch. Ergeb. d. schwed. Exped. Magellansländern 1895–7, 3).

DUSÉN, P. und NEGER, F. W. (1908): Chilenisch-patagonische Charakterpflanzen (KARSTEN u. SCHENCK, Vegetationsbild. 6, H. 8).

DUTHIE, J. F. (1893–4): Report on a botanical tour in Kashmir (Records Bot. Surv. India, 1, No. 1, No. 3).

DUTHIE, J. F. (1898): The botany of the Chitral Relief Expedition, 1895 (Rec. Bot. Surv. Ind. 1: 139–81).

DUTHIE, J. F. (1903–29): Flora of the Upper Gangetic Plain (Calcutta).

DYER, W. T., THISELTON (1912): On the supposed Tertiary Antarctic Continent (Journ. Acad. Nat. Sci. Philadelphia, 2nd Ser., 15: 237–9).

EDWARDS, D. C. (1935): The grasslands of Kenya. I. Areas of high moisture and low temperature (Empire Journ. Expt. Agric. 3: 153–9).

EDWARDS, D. C. (1940): A vegetation map of Kenya with particular reference to grassland types (Journ. Ecol. 28: 377–85).

EIG, A. (1926): A contribution to the knowledge of the flora of Palestine (Inst. Agric. and Nat. Hist. Bull. 4).

EIG, A. (1927a): On the vegetation of Palestine (Inst. Agric. and Nat. Hist. Bull. 7).

EIG, A. (1927b): A second contribution to the knowledge of the flora of Palestine (Inst. Agric. Nat. Hist. Bull. 6).

EIG, A. (1931-2): Les éléments et les groupes phytogéographiques auxiliaires dans la flore palestinienne (FEDDE, Repert. sp. nov. Beih. 63: 1-201, et l.c. 63, 2).

EIG, A. (1939): The vegetation of the light soils belt of the Coastal Plain of Palestine (Palestine Journ. Bot. Jer. Ser., 1 : 255-308).

EMBERGER, L. (1927): La végétation des montagnes du Maroc central (Compt. Rend. Acad. Sci., Paris, 185: 1152-54).

EMBERGER, L. (1932): Recherches botaniques et phytogéographiques dans le Grand-Atlas oriental (Mém. Soc. Sc. Nat. du Maroc, No. 33, 1-30).

EMBERGER, L. (1934): La végétation et la flore du Maroc (La Science au Maroc, Cassablanca).

EMBERGER, L. (1936): Remarques critiques sur les étages de végétation dans les montagnes marocaines (Ber. schweiz. bot. Ges. 46: 614-31).

EMBERGER, L. (1938): Les arbres du Maroc et comment les reconnaître (Paris).

EMBERGER, L. (1938): Contribution à la connaissance des Cèdres et en particulier du Déodar et du Cèdre de l'Atlas (Rev. Bot. Appl. 18: 77-92).

EMBERGER, L., FONT-QUER, P., et MAIRE, R. (1928): La végétation de l'Atlas rifain occidental (Compt. Rend. séanc. Soc. Biogéogr., No. 42: 70-5).

EMBERGER, L. et MAIRE, R. (1927): Spicilegium Rifanum (Mém. Soc. Sc. Nat. du Maroc, No. 17: 1-59).

ENGLER, A. (1879, 1882): Versuch einer Entwicklungsgeschichte der extra-tropischen Florengebiete der nördlichen Hemisphäre (Leipzig).

ENGLER, A. (1892): Über die Hochgebirgsflora des tropischen Afrika (Abhandl. Preuss. Akad. Wiss. Berlin 1891, 461 pp.).

ENGLER, A. (1895): Die Pflanzenwelt Ost-Afrikas und der Nachbargebiete, Theil A. Grundzüge der Pflanzenverbreitung in Deutsch-Ost-Afrika und den Nachbargebieten (Berlin).

ENGLER, A. (1904a): Plants of the Northern Temperate Zone in their transition to the high mountains of Tropical Africa (Anns. Bot. 18: 523-40).

ENGLER, A. (1904b): Über das Verhalten einiger polymorpher Pflanzentypen der nördlich gemässigten Zone bei ihrem Übergang in die afrikanischen Hochgebirge (Festschrift zu P. ASCHERSON's siebzigstem Geburtstage, Berlin, pp. 552-68).

ENGLER, A. (1910): In ENGLER und DRUDE, Die Vegetation der Erde 9, Die Pflanzenwelt Afrikas 1: 681-768 (for Cameroons), 758-767 (for Cameroons Mountain); also (1925) 5: 189-94 (for N.W. Cameroons).

ENGLER, A. (1919): Die Vegetationsverhältnisse des Kongoa-Gebirges und der Bambuto-Berge in Kamerun (Engl. Bot. Jahrb. 55, Beibl. 122: 24-32).

EXELL, A. W. (1935): The botany of the islands in the Gulf of Guinea (Proc. Linn. Soc. 147th Session: 120-6).

EXELL, A. W. (1944): Catalogue of the vascular plants of S. Tomé (London).

EXELL, A. W. (1947): Discussion on the percentage relationship calculated by Dr. WILLIAMS (Proc. Linn. Soc. 158th Session: 108-10).

FABER, F. C. VON, (1908): Vegetationsbilder aus Kamerun (Beih. Bot. Centralbl. 23: 26-42).

FISCHER, C. E. C. (1918): Preliminary note on the flora of the Anaimalais (Journ. and Proc. Asiat. Soc. Bengal (N.S.) 14: 379-88).

FISCHER, C. E. C. (1921): A survey of the flora of the Anaimalai Hills in the

Coimbatore District, Madras Presidency (Rec. Bot. Surv. India, 9, No. 1: 1–218).

FISCHER, C. E. C. (1938): The flora of the Lushai Hills (Rec. Bot. Surv. Ind. 12, No. 2: 75–161).

FLINT, E. A. (1938): A preliminary study of the phytoplankton of Lake Sarah ,(New Zealand) (Journ. Ecol. 26: 253–8).

FONT-QUER, P. (1929): El pino rodeno en la zona española de Marruecos (Mem. R. Soc. Españ. Hist. Nat. 15: 203–6).

FOURY, F. (1934–5): La question forestière au Cameroun (Rev. Bot. Appl. (Actes et Compt. Rend. Assoc. Colon.–Sci.) 14: 65–76, 103–8, 123–32, 151–6, 175–80, 199–206, 223–9; 15: 12–20, 37–42.)

FOURY, F. (1935): La question forestière au Cameroun (Assoc. Colon.–Sci. et Comité Nat. des Bois Coloniaux, Paris).

FOXWORTHY, F. W. (1910): Distribution and utilisation of the mangrove swamps of Malaya (Ann. Jard. Bot. Buitenzorg 2e sér. Supplément 3: 319–44)).

FRANKEL, O. H. (1940): Studies in *Hebe*. II (Journ. Gen. 40: 171–84).

FRANKEL, O. H. and HAIR, J. B. (1937): Studies on the cytology, genetics and taxonomy of New Zealand *Hebe* and *Veronica* L. (New Zeal. Journ. Sci. and Technol. 18: 669–87).

FRESHFIELD, D. W. (1905): The Sikkim Himalaya (Scot. Geogr. Mag. 21: 173–82).

FYSON, P. F. (1915–20): The Flora of the Nilgiri and Pulney Hill-tops (Madras, 3 vols.).

FYSON, P. F. (1932): The Flora of the South Indian Hill Stations (Madras, 3 vols.).

GAGE, A. T. (1904): The vegetation of the district of Minbu in Upper Burma (Rec. Bot. Surv. India 3, No. 1).

GAMBLE, J. S. (1875): The Darjeeling forests (Ind. For. 1: 73–99).

GAMBLE, J. S. and FISCHER, C. E. C. (1918–35): Flora of the Presidency of Madras (3 vols., London).

GAMMIE, G. A. (1894a): Report on a botanical tour in Sikkim (Rec. Bot. Surv. India 1, No. 2).

GAMMIE, J. (1894b): Vegetation (article on) (The Gazetteer of Sikkim, Calcutta).

GAMMIE, G. A. (1895): Report on a botanical tour in the Lakhimpur District Assam (Rec. Bot. Surv. Ind. 1: 61–88).

GAMMIE, G. A. (1898): A botanical tour in Chamba and Kangra (Rec. Bot. Surv. India 1: 183–214).

GANZENMÜLLER, K. (1881): Über Klima, Pflanzen- und Tierwelt in dem Central-zug des nordwestlichen Himalaya (Zeitschr. Ges. f. Erdk. 16: 385–420).

GIBBS, L. S. (1920): Notes on the phytogeography and flora of the mountain summit plateaux of Tasmania (Journ. Ecol. 8: 1–17; 89–117).

GODMAN, F. DU CANE, (1870): Natural History of the Azores or Western Islands (London).

GOOD, R. (1933): A geographical survey of the flora of Temperate South America (Anns. Bot. 47: 691–725).

GOOD, R. (1947): The geography of the flowering plants (London).

GOODSPEED, T. H. (1945): Notes on the vegetation and plant resources of Chile (VERDOORN, Plants and Plant Science in Latin America, pp. 145–9, Waltham, Mass. U.S.A.).

GORRIE, E. M. (1933): The Sutlej deodar, its ecology and timber production (Ind. For. Records 17: 1–140).

GRABHAM, M. C. (1926): The garden interests of Madeira (London).

GRAEBNER, P. (1929): Lehrbuch der allgemeinen Pflanzengeographie (Leipzig).

GRANDIDIER, G. (1934): Afrique équatoriale française et Cameroun (Atlas des Colonies Françaises, Paris).

GRIFFITH, A. L. (1946): The vegetation of the Thar Desert of Sind (Ind. For. 72: 307–9).

GUPPY, H. B. (1914): Notes on the native plants of the Azores as illustrated on the slopes of the mountain of Pico (Kew Bull.: 305–21).

GUPPY, H. B. (1917): Plants, seeds, and currents in the West Indies and Azores (London).

GUPPY, H. B. (1921): The testimony of the endemic species of the Canary Islands in favour of the age and area theory of Dr. WILLIS (Anns. Bot. 35: 513–21).

GUPTA, J. N. SEN (1939): Dipterocarpus (Gurjan) forests in India and their regeneration (Ind. For. Records (N.S.), Silv. Series 3, No. 4: 61–164).

HAINES, H. H. (1910): A Forest Flora of Chota Nagpur (Calcutta).

HAINES, H. H. (1921–5): The botany of Bihar and Orissa (London, 6 parts).

HAIR, J. B. (1942): The chromosome complements of some New Zealand plants (Trans. Roy. Soc. New Zeal. 71: 271–6).

HANDEL-MAZZETTI, H. (1927): Das nord-ost-westyünnanesische Hochgebirgsgebiet (KARSTEN und SCHENCK, Vegetationsbild. 17, H. 7–8).

HARSHBERGER, J. W. (1911): Phytogeographic Survey of North America (ENGLER und DRUDE, Die Vegetation der Erde 13, Leipzig and New York).

HAUMAN, L. (1928): Etude phytogéographique de la Patagonie (Bull. Soc. Roy. Bot. Belgique, 58: 105–79).

HAUMAN, L., BURKART, L., PARODI, L. R., y CABRERA, A. L. (1947): La Vegetación de la Argentina (Geografía de la República Argentina 8, 349 pp.).

HAY, T. (1934): Plants of Nepal (Journ. Roy. Hort. Soc. 59: 459–62).

HAYEK, A. (1926): Allgemeine Pflanzengeographie (Berlin).

HÉDIN, L. (1930): Etude sur la forêt et les bois du Cameroun sous mandat français (Paris).

HEDLEY, C. (1912): The palaeographical relations of Antarctica (Proc. Linn. Soc. 124th Session: 80–92 and Ann. Report Smithsonian Inst. 1912: 443–53).

HEMSLEY, W. B. (1885): In Report of the Scientific Results of the Voyage of H. M. S. CHALLENGER 1873–6 (Botany 1: 31–4, 211–4; also pp. 187–91, 245–54, 259–62).

HENDERSON, M. R. (1930): Notes on the flora of Pulau Tioman and neighbouring islands (The Gardens' Bull. Straits Settlement 5: 80–93).

HENDERSON, M. R. (1939): The flora of the limestone hills of the Malay Peninsula (Journ. Roy. Asiatic Soc. Malayan Branch 17: 13–87).

HESS, E. (1925): Forstliches aus dem Mittleren Atlas (Veröff. Geobot. Inst. Rübel in Zürich, 3: 778–93).

HILL, A. W. (1929): Antarctica and problems in geographical distribution (Proc. Internat. Congr. Plant Sci. 2: 1474–80).

HIRMER, M. (1924–6): Beiträge zur Kenntnis der Geholz-Formationen auf Teneriffa (KARSTEN und SCHENCK, Vegetationsbilder 16, H. 8).

HOLM, T. (1922): Contributions to the morphology, synonymy, and geographical distribution of Arctic plants (Rep. Can. Arctic Exped. 1913–8, 5, Part B).

HOLMES, C. H. (1945): The natural regeneration of Ceylon forests (Tropical Agriculturist 101: 84–90; 136–52).

HOLTTUM, R. E. (1924): The vegetation of Gunong Belumut in Johore (The Gardens' Bull. Straits Settlement, 3: 145–257).

HOWARD, S. H. (1928): Ecological distribution of Indian forests (Third Brit. Empire For. Conference 234–43).

HUBBLE, G. D. (1946): A soil survey of part of Waterhouse Estate, County of

Dorset, north-east coast, Tasmania (B. Agr. Sci. Bull. No. 204, Melbourne).

HULTÉN, E. (1937): Outline of the history of Arctic and Boreal biota during the Quaternary Period (Stockholm).

HURÉ, B. (1945–6): La cedraie du moyen Atlas morocain (Rev. des Eaux et Forêts 83: 706–18; 84: 1–10, 79–92).

HUTCHINSON, J. and DALZIEL, J. M. (1927–36): Flora of West Tropical Africa (London).

IRMSCHER, E. (1922, 1929): Pflanzenverbreitung und Entwicklung der Kontinente (Hamburg; separates from Mitt. Instit. f. allg. Bot. in Hamburg).

JACCARD, P. (1926): Les dunes de Mogador et leur fixation (Journ. Forest. Suisse 77: 196–202).

JAHANDIEZ, E. (1923): Contributions à l'étude de la flore du Maroc (Mém. Soc. Sc. Nat. du Maroc, 3, No. 1: 1–123).

JAHANDIEZ, E. et MAIRE, R. (1931–4): Catalogue des plantes du Maroc (3 vols., Alger).

JENTSCH, F. (1911): Der Urwald Kameruns (Beih. Tropenpflanzer 12: 1–199).

JENTSCH, F. und BÜSGEN, M. (1909): Forstwirtschaftliche und forstbotanische Expedition nach Kamerun und Togo (Beih. Tropenpflanzer 10: 185–310).

JOUBERT, A. (1932): La forêt Marocaine (Rev. des Eaux et Forêts 70: 851–8).

JOUBERT, A. (1933): Formations forestières Marocaines (Rev. des Eaux et Forêts 71: 96–107, 673–87).

KANJILAL, U. N., KANJILAL, P. C., DAS, A., DE, R. N., and BOR, N. (1934–40): Flora of Assam (5 vols., Calcutta).

KASHYAP, S. R. (1925): The vegetation of Western Himalayas and Western Tibet in relation to their climate (Journ. Ind. Bot. Soc. 4: 327–34).

KASHYAP, S. R. and JOSHI, A. C. (1936): Lahore District Flora (Lahore).

KENOYER, L. A. (1923): Waldformationen des westlichen Himalaya (KARSTEN und SCHENCK, Vegetationsbilder 15, H. 1).

KING, G. (1878): Sketch of the flora of Rajputana (Calcutta).

KJELLMAN, F. R. (1882–3): Fanerogamflora pa Novaja Semlja och Wajgatsch. Växt-geografisk studie. Ur Vega Expeditionens vetensk. iakttagelser. Bd. I (Stockholm, pp. 321–52). Die Phanerogamenflora von Nowaja-Semlja und Waigatsch. Eine pflanzengeographische Studie (Die wissensch. Ergebnisse d. Vega Expedition. Hrsg. v. Nordenskiöld. Deutsche Ausgabe, Lief. 3/4, p. 156 ff., Leipzig: Brockhaus).—(Bot. Centrlbl. 15: 139–42, 1883).

KJELLMAN, F. R. (1883): The Algae of the Arctic Sea (K. Sven. Vet. Akad. Handb. 20, No. 5: 1–350).

KNOCHE, H. (1923): Vagandi Mos I.

KROEBER, A. L. (1916): Floral relations among the Galapagos Islands, (Univ. Calif. Publ. in Botany, 6: 199–220).

LACAITA, C. C. (1916): Plants collected in Sikkim, including the Kalimpong District, April 8th to May 9th, 1913 (Journ. Linn. Soc. Bot., 43: 457–92).

LACK, D. (1947): DARWIN's Finches (Cambridge).

LAING, R. M. and OLIVER, W. R. B. (1929): Vegetation of the Upper Bealey River basin, with a list of the species (Trans. Proc. New Zeal. Inst. 59: 715–30).

LESTER-GARLAND, L. V. (1927): The flora of Simla (Journ. Bot. 65: 97–102).

LEWIS, F. (1920): Notes on a visit to Kunadiyaparawita, Ceylon (Journ. Linn. Soc. Bot. 45: 143–53).

LEWIS, F. (1926): The altitudinal distribution of the Ceylon endemic flora (Ann. Roy. Bot. Gard. Perad. 10: 1–130. *See* full abstract *in* Journ. Ecol. 16, Suppl. 17–18, No. 53).

LINDINGER, L. (1926): Beiträge zur Kenntnis von Vegetation und Flora der kanarischen Inseln (Hamburg).

LIVINGSTON, B. E. and SHREVE, F. (1921): The distribution of vegetation in the United States, as related to climatic conditions, Carnegie Institution of Washington.

LOGAN, M. C. (1934): Plant succession on the Oreti River sand dunes (Trans. and Proc. Roy. Soc. New Zeal. 64: 122–39).

LOWE, R. T. (1868): A manual flora of Madeira (London).

MACLOSKIE, G. (1904–6): Flora Patagonica (1903–6 Reports of the Princeton Univ. Expeds. to Patagonia, 1896–9, 8: 139–920).

MACLOSKIE, G. (1906): Character and origin of the Patagonian Flora (Reports Princeton Univ. Expeds. to Patagonia, 1896–8; 945–60).

MACLOSKIE, G. and DUSÉN, P. (1914): Revision of Flora Patagonica (Reports Princeton Univ. Expeds. to Patagonia, 1896–9, 8, Suppl.: 1–307).

MAHABALE, T. S. and KHARADI, R. G. (1946): On some ecological features of the vegetation at Mt. Abu (Proc. Nat. Acad. Sci. Ind. 16, Pt. 1: 13–23).

MAIRE, R. (1921): Coup d'œil sur la végétation du Maroc (in PERROT, Sur la productions végétales du Maroc, Paris).

MAIRE, R. (1939): Les arganiers des Beni-Snassen (Bot. Not.: 477–84).

MAITLAND, T. D. (1932): The grassland vegetation of the Cameroons Mountain (Kew Bull.: 417–25).

MARDNER, W. (1902): Die Phanerogamen-Vegetation der Kerguelen in ihren Beziehungen zu Klima und Standort (Inaugural-Dissertation, Mainz).

MARTIN, D. (1939): The vegetation of Mt. Wellington, Tasmania. The plant communities and a census of the plants (Papers and Proc. Roy. Soc. Tasmania: 100–13).

MARTIN, W. (1946): Geographic range and internal distribution of the mosses indigenous to New Zealand (Trans. Proc. Roy. Soc. N. Zeal. 76: 162–84).

MARTONNE, E. DE (1927): Biogéographie (Traité de Géographie physique, 3, éd. 4, Paris).

MATTICK, F. (1933–5): Vegetationsbilder von Teneriffa (KARSTEN und SCHENCK, Vegetationsbilder 24, H. 7).

McINDOE, K. C. (1932): An ecological study of the vegetation of the Cromwell District, with special reference to root habit (Trans. Proc. New Zeal. Inst. 62: 230–66).

MEAD, J. P. (1912): The mangrove forests of the west coast of the Federated Malay States (Kuala Lumpur).

MEEBOLD, A. (1909): Eine botanische Reise durch Kaschmir (Engl. Bot. Jahrb. 43, Beibl. 99: 63–90).

MELLISS, J. C. (1875): St. Helena: a physical, historical, and topographical description of the island, including its geology, fauna, flora, and meteorology (London).

MENEZES, C. A. DE (1914): Flora do Archipelago da Madeira (Funchal).

MERRILL, E. D. (1941): The Upper Burma plants collected by Captain F. KINGDON WARD on the Vernay-Cutting Expedition, 1938–9 (Brittonia 4: 20–188).

MIGEOD, P. W. H. (1928): Plant collecting in British Cameroons (Journ. Bot. 66: 356–9).

MILBRAED, J. (1922): Wissenschaftliche Ergebnisse der zweiten deutschen Zentral-Afrika-Expedition 1910–1911, Band 2, Botanik (Leipzig).

MILBRAED, J. (1930): Sample plot surveys in the Cameroons rain-forest (Empire Forestry Journ. 9: 242–66).

MILBRAED, J. (1932): Zur Kenntnis der Vegetationsverhältnisse Nord-Kameruns (Engl. Bot. Jahrb. 65: 1–52).

16

MISRA, R. (1944): The vegetation of the Rajghat Ravines (Journ. Ind. Bot. Soc. 23: 113–21).

MONNET (1923): Les forêts en Syrie et au Liban (La Géographie, 40: 453–8).

MOONEY, H. F. (1938): A synecological study of the forests of Western Singhbhum with special reference to their geology (Ind. For. Records N.S. 2: 259–356).

MOONEY, H. F. (1941): A short account of the geology and flora of the hill zamindaries in Kalahandi State (Ind. For. Records, N.S. 3: 131–43).

MOONEY, H. F. (1942): A sketch of the flora of the Bailadilla Range in the Bastar State (Ind. For. Records, N.S. 3, No. 7: 197–253).

MOREHEAD, F. T. (1944): The forests of Burma (Burma Pamphlets No. 5, London)

MOUTERDE, P. (1947): La végétation arborescente des pays du Levant (Publ. techn. et scient. de l'Ecole franç. d'Ingén. de Beyrouth, no. 13).

NANNFELD, J. A. (1940): On the polymorphy of *Poa arctica* R. Br. with special reference to its Scandinavian forms (Symb. Bot. Upsal. 4, 4: 1–85).

NATHORST, A. G. (1891): Kritische Bemerkungen über die Geschichte der Vegetation Grönlands (Engl. Bot. Jahrb. 14: 183–220).

NEGER, F. W. (1901): Pflanzengeographisches aus den südlichen Anden und Patagonien (Engl. Bot. Jahrb. 28: 231–58).

OLIVER, W. R. B. (1910): The vegetation of the Kermadec Islands (Trans. Proc. New Zeal. Inst. 42: 118–75).

OLIVER, W. R. B. (1911): List of lichens and fungi collected in the Kermadec Islands (Trans. Proc. New Zeal. Inst. 44: 86–7).

OLIVER, W. R. B. (1928): The flora of the Waipaoa Series (Later Pliocene) of New Zealand (Trans. and Proc. New Zeal. Inst. 59: 287–302).

OLIVER, W. R. B. (1930): New Zealand epiphytes (Journ. Ecol. 18: 1–50).

OLIVER, W. R. B. (1936): The Tertiary flora of the Kaikorai Valley, Otago, New Zealand (Trans. and Proc. Roy. Soc. New Zeal. 66: 284–304).

OOSTING, H. J. (1948): The Study of Plant Communities (San Francisco, California).

OPPENHEIMER, H. R. (1938): An account of the vegetation of the Huleh swamp (Palestine Journ. Bot. R. Ser., 2 : 34–9).

OPPENHEIMER, H. R. (1940): A contribution to the desert flora south and south-west of the Dead Sea (Palestine Journ. Bot. R. Ser., 3: 144–53).

OPPENHEIMER, H. R. (1949): Sand, swamp and weed vegetation at the estuary of the Rubin River (Palestine) (Vegetatio (Acta Geobot.) 1: 155–74).

OSMASTON, A. E. (1922): Notes on the forest communities of the Garhwal Himalaya (Journ. Ecol. 10: 129–67).

OSMASTON, A. E. (1927): A forest Flora for Kumaon (Allahabad).

OSTENFELD, C. H. (1921): The flora of Greenland (1: 277–90).

OSTENFELD, C. H. (1925): Flowering plants and ferns from North-Western Greenland collected during the Jubilee Expedition 1920–2 and some remarks on the phytogeography of North Greenland (Meddel. om Grönland 68: 1–42).

OSTENFELD, C. H. (1926): The flora of Greenland and its origin (Det Kgl. Danske Vid. Sels. Biol. Medd. 6, 3).

PALACKÝ, J. (1904): Über Vegetationsgrenzen in Palästina und Syrien (Mag. Bot. Lap. 3: 196–205).

PALHINHA, R. T. (1947): Explorações botânicas nos Açores (Bol. Soc. Brot. 2 sér. 21: 37–52).

PARKER, R. N. (1918): A Forest Flora for the Punjab with Hazara and Delhi (Lahore; ed. 2, 1924, Lahore).

PARKER, R. N. (1931): List of plants collected in West Nepal (For. Bull., Botany Series, No. 76, Calcutta).

PARKER, R. N. (1942): The ecological status of the Himalayan fir forests (150*th* Anniv. Vol. Roy. Bot. Gard. Calcutta: 125–8).

PARKIN, J and PEARSON, H. H. W. (1903): The botany of the Ceylon patanas (Journ. Linn. Soc. Bot. 35: 430–63).

PARKINSON, C. E. (1923): A forest flora of the Andaman Islands (Simla).

PEARSON, H. H. W. (1899): The botany of the Ceylon patanas (Journ. Linn. Soc. Bot. 34: 300–65).

PENSELER, W. H. A. (1930): Fossil leaves from the Waikato District (Trans. and Proc. New Zeal. Inst. 61: 452–77).

PIROTTA, R. e CORTESI, F. (1912): Relazione sulle piante raccolte nel Karakoram dalla spedizione di S.A.R. il Duca degli Abruzzi (Bologna).

PITARD, C. J. (1913): Exploration scientifique du Maroc. Botanique (Paris).

PITARD, J. et PROUST, L. (1908?): Les îles Canaries (Paris).

POLUNIN, N. (1940): Botany of the Eastern Arctic. 1 (Bull. 92, Nat. Mus. Canada).

POST, G. E. (1888): The botanical geography of Syria and Palestine (Victoria Inst. Proc.).

POST, G. E. (1932–3): Flora of Syria, Palestine and Sinai, ed. 2 (J. E. DINSMORE, American Press, Beirut).

PRAEGER, R. L. (1932): An account of the *Sempervivum* group (London).

PRAIN, D. (1903*a*): Flora of the Sundribuns (Rec. Bot. Surv. India, 2, No. 4: 231–370).

PRAIN, D. (1903*b*): Bengal Plants (Calcutta, 2 vols.).

PRAIN, D. (1905): The vegetation of the districts of Hughli-Howrah and the 24-Pergunnahs (Rec. Bot. Surv. India 3, No. 2 : 143–339).

PRING, N. G. (1947): The afforestation of the dry and desert areas of north-west India (Ind. Forester, 73: 170–5).

PURI, G. S. (1947): Fossil plants and the Himalayan uplift (M. O. P. IYENGAR Commem. Vol. Journ. Ind. Bot. Soc., 1946: 167–84).

RAMASWAMI, M. S. (1914): A botanical tour in the Tinnevelly Hills (Rec. Bot. Surv. India, 6: 105–71).

RAUNKIAER, C. (1934): The Life Forms of Plants and Statistical Plant Geography (Oxford).

RAUP, H. M. (1941): Botanical Problems in Boreal America (Bot. Review 7: 147–208, 209–48).

REICHE, K. (1907): Grundzüge der Pflanzenverbreitung in Chile (ENGLER und DRUDE, Die Vegetation der Erde, 8, Leipzig).

REIFENBERG, A. and WHITTLES, C. L. (1938): The soils of Palestine (London).

RIDLEY, H. N. (1900): The flora of Singapore (Journ. Roy. Asiat. Soc. 33: 27–196).

RIDLEY, H. N. (1901): The flora of Mount Ophir (Journ. Roy. Asiat. Soc. Straits Branch, No. 35: 1–28).

RIDLEY, H. N. (1906): An expedition to Christmas Island (Roy. Asiat. Soc. Straits Branch, No. 45: 137–271).

RIDLEY, H. N. (1909): The flora of the Telôm and Batang Padang Valleys (Journ. Federated Malay States Museum 4, No. 1: 1–98).

RIDLEY, H. N. (1910): A scientific expedition to Temengoh, Upper Perak (Journ. Roy. Asiatic Soc. Straits Branch, No. 57: 5–122).

RIDLEY, H. N. (1922–5): The flora of the Malay Peninsula (London, 5 vols.).

RIDLEY, H. N. (1942): Distribution areas of the Indian floras (150*th* Anniv. Vol. R. Bot. Gard. Calcutta: 49–52).

RIKLI, M. (1933): Das Ausklingen der Pteridophytenflora in der Polaris (Ber. d. Schweiz. Bot. Ges. 42: 339–56).

RIKLI, M. (1933–35): Marokko (KARSTEN und SCHENCK, Vegetationsbilder 23, H. 4/5).

RIKLI, M. (1943, 1946, 1948): Das Pflanzenkleid der Mittelmeerländer (Bern, 3 vols.).

RILEY, L. A. M. (1925): Notes on Madeira plants (Kew Bull.: 26–33).

ROBINSON, B. L. (1903): Flora of the Galapagos Islands (Proc. Amer. Acad. Arts and Sci. 38: 77–269).

RODWAY, L. (1914): Botanic evidence in favour of land connection between Fuegia and Tasmania during the present floristic epoch (Papers and Proc. Roy. Soc. Tasmania: 32–4).

RODWAY, L. (1923): The endemic phanerogams of Tasmania (Proc. of the Pan-Pacific Sci. Congr. Australia: 283–6).

ROSAYRO, R. A. DE, (1942): The soils and ecology of the wet evergreen forests of Ceylon. I and II (Tropical Agriculturist 98 (April–June): 4–14, (July–Sept.): 13–35. See full abstract in Journ. Ecol. 33: 109–10, 1945).

ROSAYRO, R. A. DE, (1945–6): The montane grasslands (Patanas) of Ceylon. An ecological study with reference to afforestation. I (The Tropical Agriculturist 101: 206–13; 102: 4–16, 81–94, 139–48).

ROTHKUGEL, M. (1916): Los Bosques Patagónicos (Buenos Aires).

RÜBEL, E. (1930): Pflanzengesellschaften der Erde (Bern–Berlin).

RÜBEL, E. und LUDI, W. (1939): Ergebnisse der internationalen pflanzengeographischen Exkursion durch Marokko und Westalgerien 1936 (Veröff. Geobot. Inst. Rübel in Zürich 14 H; for full abstract see Journ. Ecol. 27: 539–46, 1939).

RÜE, E. A. DE LA, (1932): Etude géologique et géographique de l'archipel de Kerguelen (Rev. Géogr. Phys. et de Géol. Dyn. 1: 1–231; also published separately, Paris, 1932).

RUTTLEDGE, H. (1933): Everest (London, 1934).

SAMUELSSON, G. (1943): Die Verbreitung der *Alchemilla*-Arten aus der vulgaris-Gruppe in Nordeuropa (Acta Phytogeogr. Suec. 16: 1–159).

SAXTON, W. T. (1922): Mixed formations in time: a new concept in oecology (Journ. Ind. Bot. Soc. 3: 30–3).

SAXTON, W. T. (1924): Phases of vegetation under monsoon conditions (Journ. Ecol. 12: 1–38).

SAXTON, W. T. and SEDGWICK, L. J. (1918): Plants of Northern Gujarat (Records Bot. Surv. India 6, No. 7: 203–323).

SCHENCK, H. (1905): Vergleichende Darstellung der Pflanzengeographie der subantarktischen Inseln insbesondere über Flora und Vegetation von Kerguelen (Wiss. Erg. deut. Tiefsee-Exped. auf den Dampfer "Valdivia" 1898–1899, Jena).

SCHENCK, H. (1906): Kerguelen (E. VON DRYGALSKI, Deutsche Südpolar-Expedition 1901–1903, VIII Band, Botanik, pp. 102 *seq.*, Berlin).

SCHIMPER, A. F. W. (1903): Plant Geography (English Translation, Oxford).

SCHRÖTER, C. (1909): Eine Exkursion nach den Canarischen Inseln (Zürich).

SCOTT, J. (1874): Notes on the tree ferns of British Sikkim (Trans. Linn. Soc. 30: 1–44).

SEWARD, A. C. (1931): Plant Life through the Ages (Cambridge).

SEWARD, A. C. (assisted by CONWAY, V.) (1934): A phytogeographical problem: fossil plants from the Kerguelen Archipelago (Anns. Bot. 48: 715–41).

SIMMONS, H. G. (1913): A survey of the phytogeography of the Arctic American Archipelago (Lunds Univ. Årsskr. N.F. 9, Nr. 19).

SIMPSON, G. and THOMSON, J. S. (1938): The Dunedin Sub-district of the South Otago Botanical District (Trans. and Proc. Roy. Soc. New Zeal. 67: 430–42).

SKOTTSBERG, C. (1915): Notes on the relations between the floras of Subantarctic America and New Zealand (The Plant World 18: 129–42).

SKOTTSBERG, C. (1931): Zur Pflanzengeographie Patagonicus (Ber. deutsch bot. Ges. 49: 481–93).

SKOTTSBERG, C. (1936): Antarctic plants in Polynesia (Essays in Geobotany in honor of WILLIAM ALBERT SETCHELL, Berkeley, California).

SKOTTSBERG, C. (1940): Nagra av den antarkiska kontinentens biologiska historia (D.K.N.V.S. Forhandl. 12: 45–55).

SKOTTSBERG, C. (1942): The Falkland Islands (Chron. Bot. 7: 23–6).

SMITH, W. W. (1913): The alpine and sub-alpine vegetation of South-East Sikkim (Rec. Bot. Surv. India, 4, No. 7; see review in Journ. Ecol. 2: 268–70, 1914).

SMITH, W. W. and CAVE, G. H. (1911): The vegetation of the Zemu and Llonakh valleys of Sikkim (Rec. Bot. Surv. India 4, No. 5).

SMYTHE, F. S. (1938): The Valley of Flowers (London).

SMYTHIES, E. A. (1921): Note on the miscellaneous forests of the Kumaon Bhabar (For. Bull. No. 45, Calcutta).

SPIX, J. B. and VON MARTIUS, C. F. P. (1824): Travels in Brazil (notes on flora of Madeira and Canaries in Chapter IV and pp. 126–9).

SPRAGUE, T. A. and HUTCHINSON, J. (1913): A botanical expedition to the Canary Islands (Kew Bull.: 287–99).

SRIVASTAVA, G. D. (1944): The biological spectrum of the Allahabad flora (Journ. Ind. Bot. Soc. 23: 1–7).

STAMP, L. DUDLEY (1923): The ecology of the riverine tract of Burma (Journ. Ecol. 11: 129–59).

STAMP, L. DUDLEY (1924): Notes on the vegetation of Burma (Geogr. Journ. 64: 231–7).

STAMP, L. DUDLEY (1925a): The aerial survey of the Irrawaddy Delta forests (Burma) (Journ. Ecol. 13: 262–76).

STAMP, L. DUDLEY (1925b): The vegetation of Burma (Calcutta).

STAPF, O. (1917): Enneapogon mollis in Ascension Island (Kew Bull.: 217–9; see also l.c., p. 342).

STEBBING, E. P. (1922): The forests of India. 1 (London).

STEFFEN, H. (1924): Versuch einer Gliederung der arktischen Flora in geographische bzw. genetische Florenelemente (Bot. Archiv. 6: 7–49).

STEFFEN, H. (1937): Gedanken zur Entwicklungsgeschichte der arktischen Flora (Beih. Bot. Centrlbl. 56 B: 409–47 and 57: 367–430).

STEFFEN, H. (1939): Die Grundzüge der Entwicklung der arktischen Flora (Chron. Bot. 5: 176–7).

STEPHENS, C. G. and CANE, R. F. (1939): The soils and general ecology of the north-east coastal regions of Tasmania (Papers and Proc. Roy. Soc. Tasmania: 201–5).

STEWART, A. (1911): A botanical survey of the Galapagos Islands (Proc. Calif. Acad. Sci. 4th series, 1: 7–228).

STEWART, A. (1916): Some observations concerning the botanical conditions of the Galapagos Islands (Trans. Wisconsin Acad. Sci. Arts and Letters 18, Part I: 272–340).

SUTTON, C. S. (1923): Cradle Mountain (Tasmania) and its flora (Victorian Naturalist 40: 131–7).

SUTTON, C. S. (1928): A sketch of the vegetation of the Cradle Mountain, Tasmania, and a census of its plants (Papers and Proc. Roy. Soc. Tasmania: 132–59).

SVENSON, H. K. (1935): Plants of the Astor Expedition, 1930 (Galapagos and Cocos Islands) (Amer. Journ. Bot. 22: 208–68).

Svenson, H. K. (1945): A brief review of the Galapagos flora (Plants and Plant Science in Latin America (*ed.* F. Verdoorn), pp. 149–50, Waltham, Mass.).

Svenson, H. K. (1946): Vegetation of the coast of Ecuador and Peru and its relation to the Galapagos Islands (Amer. Journ. Bot. 33: 394–426; 427–98).

Swarth, H. S. (1934): The bird fauna of the Galapagos Islands in relation to species formation (Biol. Rev. 9: 213–34).

Tansley, A. G. and Fritsch, F. E. (1905): The flora of the Ceylon littoral (New Phyt. 4: 1–17, 27–55).

Thomson, J. S. (1935): Some aspects of the vegetation and flora of South Island (Journ. New Zeal. Inst. Hortic. 4, No. 4: 1–18).

Thomson, J. S. and Simpson, G. (1939): Some characteristic South Island mountain plants (Quart. Bull. Alpine Gard. Soc. 7: 297–315).

Tolmachev, A. I. (1930): On the origin of the flora of Vaigacha and Nova Zembla (Trav. Mus. Bot. Acad. Sci. URSS. 22: 181–205; Russian).

Trelease, W. (1897): Botanical observations on the Azores (Missouri Bot. Gard. Eighth Ann. Rep., pp. 77–220).

Trimen, H. (1893–1900): A hand-book to the flora of Ceylon (London, 5 vols. (parts). Suppl. (part 6) by A. H. G. Alston, London, 1931).

Troll, C. (1939): Das Pflanzenkleid des Nanga Parbat: Begleitworte zur Vegetationskarte der Nanga-Parbat-Gruppe (Nordwest-Himalaya) (Wiss. Veröffentl. deut. Mus. f. Landark. No. 7 (N.S.), Leipzig).

Troup, R. S. (1911): A note on some statistical and other information regarding the teak forest of Burma (Ind. For. Rec. 3: 1–73).

Troup, R. S. (1921): The silviculture of Indian trees (Oxford).

Troup, R. S. (1926): Problems of forest ecology in India (Tansley and Chipp, Aims and methods in the study of vegetation, London, pp. 283–313).

Turrill, W. B. (1919): Botanical results of Swedish South American and Antarctic expeditions (Kew Bull.: 268–79).

Turrill, W. B. (1920): Botanical exploration in Chile and Argentina (Kew Bull., 57–66, 223–4).

Turrill, W. B. (1949): On the flora of St. Helena (Kew Bull. 1948: 358–62).

Tutin, T. G. and Warburg, E. F. (1932): Contributions from the University Herbarium, Cambridge—notes on the flora of the Azores (Journ. Bot. 70: 7–13, 38–46).

Unwin, A. H. (1920): West African forests and forestry (London, pp. 415–46).

Vahl, M. (1904): Madeiras Vegetation (Geografisk monografi, København and Kristiania).

Vahl, M. (1905): Über die Vegetation Madeiras (Engl. Bot. Jahrb. 36: 253–349).

Vallentin, E. F. (1921): Illustrations of the Flowering Plants and Ferns of the Falkland Islands. With descriptions by Mrs. E. M. Cotton (London).

Viennot-Bourgin, G. (1939): Contribution à la connaissance de la mycoflore de l'Archipel de Madère (Ann. de l'Ecole Nat. d'Agric. de Grignon, ser. 3, 1: 69–169, 6 pl., 21 figs.).

Waddell, L. A. (1899): Among the Himalayas (Westminster).

Wallace, A. R. (1880): Island Life (London: Chapter XII: The Azores).

Ward, F. Kingdon, (1921): In farthest Burma (London).

Ward, F. Kingdon (1923): The mystery rivers of Tibet (London).

Ward, F. Kingdon (1924a): The flora of the Upper Irrawaddy (Journ. Roy. Hort. Soc. 49: 148–56).

Ward, F. Kingdon (1924b): The romance of plant hunting (London).

Ward, F. Kingdon (1924c): From China to Hkamti Long (London).

WARD, F. KINGDON (1927): The Sino-Himalayan flora (Proc. Linn. Soc. 139*th* Session: 67–74).

WARD, F. KINGDON (1930*a*): The forests of the North-East frontier of India (Empire For. Journ. 9: 11–30).

WARD, F. KINGDON (1930*b*): The distribution of Primulas from the Himalaya to China, with descriptions of some new species (Anns. Bot. 44, 111–25).

WARD, F. KINGDON (1930*c*): Plant hunting on the edge of the world (London).

WARD, F. KINGDON (1931): Botanical Exploration: Mishmi Hills, Assam (Proc. Linn. Soc. 142*nd* session: 60–4).

WARD, F. KINGDON (1932*a*): Botanical exploration on the Burma-Tibet frontier (Proc. Linn. Soc. 144*th* Session: 140–3).

WARD, F. KINGDON (1932*b*): Explorations on the Burma-Tibet Frontier (Geogr. Journ. 80: 465–83).

WARD, F. KINGDON (1933): Plant collecting at the source of the Irrawaddy (Journ. Roy Hort. Soc. 58: 103–14).

WARD, F. KINGDON (1941): The Vernay-Cutting Expedition, November, 1938, to April, 1939: report on the vegetation and flora of the Hpimaw and Htawgaw Hills, Northern Burma (Brittonia 4: 1–19).

WARD, F. KINGDON (1942): An outline of the vegetation and flora of Tibet (150*th* Anniv. Vol. R. Bot. Gard. Calcutta: 99–103).

WARD, F. KINGDON (1944–5): A sketch of the botany and geography of North Burma (Journ. Bombay Nat. Hist. Soc. 44: 550–74; 45: 16–30, 133–48).

WARD, F. KINGDON (1946*a*): Botanical explorations in North Burma (Journ. Roy. Hort. Soc. 71: 318–25).

WARD, F. KINGDON (1946*b*): The Talok Pass (Burma) (Gard. Chron. 3*rd* ser. 119: 302).

WARD, F. KINGDON (1946*c*): Additional notes on the botany of North Burma (Journ. Bombay Nat. Hist. Soc. 46: 381–90).

WARD, F. KINGDON (1946*d*): The Mekong River (Gard. Chron. 3*rd* ser. 120: 42–3).

WARMING, E. (1888): Om Grönlands Vegetation (Meddel. om Grönland, 12: 1–223).

WARMING, E. (1909): Oecology of Plants (English Translation, Oxford).

WARMING, E. und GRAEBNER, P. (1930–3): Lehrbuch der ökologischen Pflanzengeographie (Berlin).

WATSON, J. G. (1928): Mangrove forests of the Malay Peninsula (Malayan Forest Records, No. 6).

WATSON, S. (1891): Notes upon a collection of plants from the island of Ascension (Contrib. Amer. Bot. 18: 161–2; and Proc. Amer. Acad. Arts and Sci. 26).

WATT, G. (1881): Notes on the vegetation of Chumba State and British Lahoul (Journ. Linn. Soc., Bot. 18: 368–82).

WEAVER, J. E. and CLEMENTS, F. E. (1929, *ed.* 2, 1938): Plant Ecology (New York and London).

WEGENER, A. (1924): The Origin of Continents and Oceans (English Translation, London).

WERTH, E. (1911): Die Vegetation der subantarktischen Inseln Kerguelen, Possession- und Heard Inseln (E. VON DRYGALSKI, Deutsche Südpolar-Expedition 1901–1903, VIII. Band, Botanik, 2: 223–371, Berlin).

WILLIAMS, C. B. (1947): The logarithmic series and the comparison of island floras (Proc. Linn. Soc. 158*th* session: 104–8).

WILLIS, J. C. (1915): The endemic flora of Ceylon, with reference to geographical distribution and evolution in general (Phil. Trans. Roy. Soc. Lond. B. 206: 307–42).

WILLIS, J. C. (1916): The evolution of species in Ceylon with reference to the dying out of species (Anns. Bot. 30: 1–23).

WOODROW, G. M. (1894): Notes on a journey from Haveri to Kumta (Rec. Bot. Surv. India 1: 49–57).

WRIGHT, H. (1905): Foliar periodicity of endemic and indigenous trees in Ceylon (Anns. Roy. Bot. Gard. Peradeniya, 2: 415–515).

WRIGHT, H. L. (1931): The forests of Kashmir (Empire For. Journ. 10: 182–9).

WULFF, E. V. (1943): An introduction to Historical Plant Geography (English Translation, Waltham, Mass., U.S.A.).

WULFF, E. V. (1944): Historical flora of the countries of the globe (Moscow-Leningrad; Russian).

ZOHARY, M. (1937): Die Verbreitungsökologischen Verhältnisse der Pflanzen Palästinas (Beih. Bot. Centrlbl. 56: 1–155).

ZOHARY, M. (1940a): Forest and forest remnants of *Pistacia atlantica* Desf. in Palestine and Syria (Palestine Journ. Bot. R. Ser., 3: 158–61).

ZOHARY, M. (1940b): Geobotanical analysis of the Syrian Desert (Palestine Journ. Bot. Jer. Ser., 2: 46–96).

ZOHARY, M. (1942): The vegetational aspect of Palestine soils (Palestine Journ. Bot. Jer. Ser., 2: 200–46).

ZOHARY, M. (1944): Vegetational transects through the desert of Sinai (Palestine Journ. Bot. Jer. Ser., 3: 57–78).

ZOHARY, M. (1945): Outline of the vegetation in Wadi Araba (Journ. Ecol. 32: 204–13).

ZOHARY, M. (1947): A geobotanical soil map of Western Palestine (Palestine Journ. Bot. Jer. Ser., 4: 24–35).

ZOHARY, M. and ORSHANSKY, G. (1947): The vegetation of the Huleh Plain (Palestine Journ. Bot. Jer. Ser., 4: 90–104).

ZOTOV, V. D. (1938): Some correlations between vegetation and climate in New Zealand (New Zeal. Journ. Sci. Technol. 19: 474–87).

ZOTOV, V. D. (1939): Survey of the tussock-grasslands of the South Island, New Zealand (New Zeal. Journ. Sci. Technol. 20: 212a; Bull. 73, Dep. Sci. Industr. Res.).

ZOTOV, V. D., ELDER, N. L., BEDDIE, A. D., SAINSBURY, G. O. K., and HODGSON, E. A. (1938): An outline of the vegetation and flora of the Tararua Mountains (Trans. Proc. Roy. Soc. New Zeal. 68: 259–324).

INDEX

17